BTEC
Level 2

edexcel
advancing learning, changing lives

APPLIED SCIENCE LEVEL 2

BTEC First

Christine Brain | Peter Gale | David Goodfellow | Sue Hocking
Christine Kitchin | Roy Llewellyn | Julie Matthews
Ismail Musa | Patricia Rhodes

A Pearson Company

Published by Pearson Education Limited, a company incorporated in England and Wales, having its registered office at Edinburgh Gate, Harlow, Essex, CM20 2JE. Registered company number: 872828

www.pearsonschoolsandfecolleges.co.uk

Edexcel is a registered trademark of Edexcel Limited

Text © 2010
First published 2010

13 12 11 10
10 9 8 7 6 5 4 3

CITY OF WOLVERHAMPTON COLLEGE
LEARNING RESOURCES

Soo BRA   PR

2 4 JAN 2013   8833621

£ 18·50   WOLO53574

British Library Cataloguing in Publication Data
A catalogue record for this book is available from the British Library

ISBN 978 1 846906 09 1

Copyright notice
All rights reserved. No part of this publication may be reproduced in any form or by any means (including photocopying or storing it in any medium by electronic means and whether or not transiently or incidentally to some other use of this publication) without the written permission of the copyright owner, except in accordance with the provisions of the Copyright, Designs and Patents Act 1988 or under the terms of a licence issued by the Copyright Licensing Agency, Saffron House, 6–10 Kirby Street, London EC1N 8TS (www.cla.co.uk). Applications for the copyright owner's written permission should be addressed to the publisher.

Edited by Ashwell Enterprises Ltd, Lucy Tritton, Lindsey Williams, Tim Jackson and Martin Brooks
Designed by Wooden Ark
Typeset by HL Studios
Original illustrations © Pearson and Sean @ kja-artists.com
Cover design by Visual Philosophy, created by eMC Design
Cover photo/illustration © Tek Images/Science Photo Library
Picture research by Susi Paz
Back cover photos © Surfer: Epic Stock/Shutterstock, Welder: Marten Czamanske/Shutterstock
Printed in Spain by Grafos, Barcelona

Every effort has been made to contact copyright holders of material reproduced in this book. Any omissions will be rectified in subsequent printings if notice is given to the publishers.

Acknowledgements
We would like to thank Frances Annets, BTEC External Verifier, and Andy Skepper, Edexcel reviewer and Advanced practitioner, Leeds County College, for their invaluable help in the development of this course.

Hotlinks
There are links to relevant websites in this book. In order to ensure that the links are up to date, the links work, and that the sites are not inadvertently linked to sites that could be considered offensive, we have made the links available on the following website www.pearsonschoolsandfecolleges.co.uk/hotlinks. When you access the site, search for the BTEC Applied Science student book using key terms, the express code 6091V or the ISBN (9781846906091). Go to the unit you are interested in and click on the name of the hotlink you want to visit.

Disclaimer
This material has been published on behalf of Edexcel and offers high-quality support for the delivery of Edexcel qualifications. This does not mean that the material is essential to achieve any Edexcel qualification, nor does it mean that it is the only suitable material available to support any Edexcel qualification. Edexcel material will not be used verbatim in setting any Edexcel examination or assessment. Any resource lists produced by Edexcel shall include this and other appropriate resources.

Copies of official specifications for all Edexcel qualifications may be found on the Edexcel website: www.edexcel.com

Contents

Credits

The authors and publisher would like to thank the following individuals and organisations for permission to reproduce photographs:

p.1 Kristin Duvall/Getty Images, **p.3** Pearson Education Ltd/Studio 8/Clark Wiseman, **p.4** left (**l**) Photodisc/John A. Rizzo, middle (**m**) Stockbyte/Getty Images, right (**r**) Comstock Images, **p.5** VeryBigAlex/Shutterstock, **p.7** David Goodfellow, **p.8 l** Charles D. Winters/Science Photo Library, **r** Photodisc/Steve Cole, **p.11** Reuters/CORBIS, **p.13** Photodisc/Photolink/Tomi, **p.14** Pearson Education Ltd/Trevor Clifford, **p.15** Andrew Lambert Photography/Science Photo Library, **p.16 l** Marten Czamanske/Shutterstock, **r** hotodisc/StockTrek, **p.17** top (**t**) Blue Wave, bottom (**b**) Olaru Radian-Alexandru/Shutterstock, **p.18 l, m** and **r** Pearson Education Ltd/Trevor Clifford, **p.19** Andrew Lambert Photography/Science Photo Library, **p.20** Greg McCracken/Shutterstock, **p.22** Martyn F. Chillmaid/Science Photo Library, **p.24** Photodisc/Photolink, **p.26** National Geophysical Data Center/Department of Natural Resources, State of Washington, **p.28** Lee Prince/Shutterstock, **p.29** Stocktrek Images, Inc./Alamy, **p.30 t** Jordan Tan/Shutterstock, **b** Yarygin/Shutterstock, **p.32** Photodisc, **p.33** Joanne Beech, **p.35** NASA/NOAA/SPL/Getty Images, **p.37** Ismail Musa, **p.38** Rosenfeld Images Ltd/Science Photo Library, **p.39 t** Nicholas Monu/iStockphoto, **m** Emanuel/Shutterstock, **b** Christian Lagerek/Shutterstock, **p.40 t** age fotostock/Photolibrary, **b** Natali Glado/Shutterstock, **p.42** Fancy/Photolibrary, **p.44** Ted Kinsman/Science Photo Library, **p.45** Kevin Wright, **p.46** Maciej Korzekwa/iStockphoto, **p.47** Richard Hobson/iStockphoto, **p.48** Epic Stock/Shutterstock, **p.49** Lisa F. Young/Shutterstock, **p. 50** Photodisc/Photolink, **p. 51 t** JupiterImages/Goodshoot/Alamy, **m** Gary James Calder/Shutterstock, **b** Chad McDermott/Shutterstock, **p.52** Health Protection Agency/Science Photo Library, **p.54** Color Day Production/Stone/Getty Images, **p.56** dirimage/Alamy, **p.57** sarka/Shutterstock, **p.60** Leif Norman/iStockphoto, **p.62** Photodisc/StockTrek, **p.64 t** Photodisc/StockTrek, **b** Mark Garlick/Science Photo Library, **p.67** Steve Bly/Getty Images, **p.69** Julie Matthews, **p.77** Michael W. Tweedie/Science Photo Library, **p.80** James Robinson/Photolibrary, **p.82** G P Bowater/Alamy, **p.83** Brian Woodward, **p.84** Jules Selmes/Dorling Kindersley, **p.86** Eddie Lawrence/Dorling Kindersley, **p.88** Eye Of Science/Science Photo Library, **p.90 t** James King-Holmes/Science Photo Library, **b** Science Photo Library, **p.92** Peter Weber/Shutterstock, **p.97** Victor Habbick Visions/Science Photo Library, **p.99** Pearson Education Ltd/Gareth Boden, **p.102** Maria Suleymenova/Shutterstock, **p.103** Franck Boston/Shutterstock, **p.104** Peter Gould, **p.106** Pearson Education Ltd/Trevor Clifford, **p.108** Peter Gould, **p.110** Charles D. Winters/Science Photo Library, **p.111** Alena Root/Shutterstock, **p.112 t** Stephen Mulcahey/Shutterstock, **b** Christ Priest & Mark Clarke/Science Photo Library, **p.114** Victor Habbick Visions/Science Photo Library, **p.115** Colin Benison, **p.117** Martin Bennett/Alamy, **p.119** Ismail Musa, **p.120** TommL/iStockphoto, **p.121** John Prescott/iStockphoto, **p.122** vario images GmbH & Co. KG/Alamy, **p.124** Daboost/Shutterstock, **p.126 t** David Woods/Corbis, **b** Trajche Roshkoski/Shutterstock, **p.127** Howard Taylor/Alamy, **p.128 t** Patti McConville/Alamy, **b** Andreas Steinhart/iStockphoto, **p.131** Jim Varney/Science Photo Library, **p.132** Tonylady/Shutterstock, **p.133** Photos.com, **p.134** Seth Solesbee/Shutterstock, **p.135** peresanz/Shutterstock, **p.137** Patrick Landmann/Science Photo Library, **p.139** Ismail Musa, **p.140** Alexander Raths/Shutterstock, **p.142** Digital Vision/Getty Images, **p.144** Lisa F. Young/Shutterstock, **p.146** Photodisc/Photolink, **p.148** Juergen Berger/Science Photo Library, **p.150** Sergei Telegin/Shutterstock, **p.152** Imane/Science Photo Library, **p.153 l** . Kolesnik/Shutterstock, **p.154** Michael Taylor/Shutterstock, **p.155** Imperial College Healthcare NHS Trust, **p.156** Alex Bartel/Science Photo Library, **p.157** Phototake Science/PhotoLibrary, **p.159** Hazel Appleton, Centre for Infections/Health Protection Agency/Science Photo Library, **p.160** Simon Fraser/RVI, Nevcastle-Upon-Tyne/Science Photo Library, **p.161** CNRI/Science Photo Library, **p.163** ason/Shutterstock, **p.165** Pearson Education Ltd/Gareth Boden, **p.167** Laurence Gough/Shutterstock, **p.168** Jaimie Duplass/Shutterstock, **p.169** Meena Kaur, **p.171** Alexis Rosenfeld/Science Photo Library, **p.173** Pearson Education Ltd/Jules Selmes, **p.174 t** NHS, **m** Kevin L Chesson/Shutterstock, **b** Pearson Education Ltd/Ben Nicholson, **p.176 t** Patrick Landmann/Science Photo Library, **b** Maximilian Stock Ltd/Science Photo Library, **p.177** (from top to bottom) Maximilian Stock Ltd/Science Photo Library, Opla/Shutterstock, Monkey Business Images/Shutterstock, Mark Yuill/Shutterstock, Monkey Business Images/Shutterstock, **p.178** Health Protection Agency/Science Photo Library, **p.181 t** James Steidl/Shutterstock, **b** Joanne Young, **p.183** Patrick Landmann/Science Photo Library, **p.185** Ismail Musa, **p.186** Paula Solloway/Alamy, **p.187** Maciej Oleksy/Shutterstock, **p.189** Blend Images/Alamy, **p.190** UpperCut Images/Alamy, **p.191** Photos.com, **p.193** Sebastian Kaulitzki/Shutterstock, **p.195** Julie Matthews, **p.197** Laguna Design/OSF/Photolibrary, **p.205 t** Suponev Vladimir Mihajlovich/Shutterstock, **b** Photodisc/Kevin Peterson, **p.207** John Bavosi/Science Photo Library, **p.209** Pascal Goetgheluck/Science Photo Library, **p.211** Julie Matthews, **p.212** Pakhnyushcha/Shutterstock, **p.216** Charles D. Winters/Science Photo Library, **p.217** Martin Shields/Science Photo Library, **p.218** Martyn F. Chillmaid/Science Photo Library, **p.220** Sandy Maya Matzen/Shutterstock, **p.221** David Noton Photography/Alamy, **p.222** Morozova Tatyana (Manamana)/Shutterstock, **p.223** Pearson Education Ltd/Lord and Leverett, **p.225** Photodisc/Photolink/Rim Light, **p.227** Pearson Education Ltd/Rob Judges, **p.230** Santje/Shutterstock, **p.232** Ng Wei Keong/Shutterstock, **p.235** Mark William Richardson/Shutterstock, **p.239 t** Artville/Dennis Nolan, **b** Eduardo Jose Bernardino/iStockphoto, **p.241** Philippe Psaila/Science Photo Library, **p.243** Pearson Education Ltd/Gareth Boden, **p.244 t** Jack Dagley Photography/Shutterstock, **b** Peter Menzel/Science Photo Library, **p.245** Michael Donne/Science Photo Library, **p.246 l** Louise Murray/Science Photo Library, **r** Mauro Fermariello/Science Photo Library, **p.249 t** Mauro Fermariello/Science Photo Library, **b** Paul Rapson/Science Photo Library , **p.250 l** Philippe Psaila/Science Photo Library, **r** Pasieka/Science Photo Library, **p.252** scoutingstock/Shutterstock, **p.253** John Archer/iStockphoto, **p.257** Photodisc, **p.259** Photodisc, **p.260 t** Photos.com **b** Chris Priest/Science Photo Library, **p.261** Living Art Enterprises, LLC/Science Photo Library, **p.262** Steve Gschmeissner/Science Photo Library, **p.263** Saturn Stills/Science Photo Library, **p.264** Paula Solloway/Alamy, **p.265 t** Jaimie Duplass/Shutterstock, **b** Ann Baldwin/Shutterstock, **p.266 t** Brasiliao/Shutterstock, **b** Custom Medical Stock Photo/Science Photo Library, **p.268 t** S.P. Rayner/iStockphoto, **b** Dr P. Marazzi/Science Photo Library, **p.269** Bodenham, LTH NHS Trust/Science Photo Library, **p.270** Photos.com, **p.271** Pearson Education Ltd/Jules Selmes, **p.273** Nikada/iStockphoto, **p.275** Pearson Education Ltd/Jules Selmes, **p.276 t** Photodisc/StockTrek, **b** Sebastian Kaulitzki/Shutterstock, **p.278** Hazel Appleton, Centre For Infections/Health Protection Agency/Science Photo Library, **p.280** Cultura/Alamy, **p.281** Arvind Balaraman/Shutterstock, **p.283** Pearson Education Ltd/Jules Selmes, **p.284 t** Lee Prince/Shutterstock, **b** Dr. Gladden Willis/Getty Images, **p.285** Stockdisc/Getty Images, **p.286** NASA/Science Photo Library, **p.291** NASA/Science Photo Library, **p293** Pearson Education/Jules Selmes, **p.294** Sean Locke/iStockphoto, **p.295** Brownie Harris/Corbis, **p.296** Tek Image/Science Photo Library, **p.297** Bridget Bowen, **p.299** Dr. Jurgen Scriba/Science Photo Library, **p.301** Pearson Education Ltd/Gareth Boden, **p.303** filmfoto/Shutterstock, **p.304 t** jocicalek/Shutterstock, **b** GustoImages/ Science Photo Library, **p.306** Andrew Lambert Photography/Science Photo Library, **p.309 t** Dr. P. Marazzi/Science Photo Library, **b** Adam Hughes, **p.311** Photolibrary.com, **p.313** Pearson Education Ltd/Gareth Boden, **p.314** Ptolemy, **p.318** Christian Darkin/Science Photo Library, **p.319** John McCormick, **p.320** NASA, **p.323 t** Eckhard Slawik/Science Photo Library, **m** George East/Science Photo Library, **b** Photodisc/StockTrek, **p.325** John Kroetch/Shutterstock, **p.327** Pearson Education Ltd/MindStudio, **p.328** Valentin Casarsa/iStockphoto, **p.330** Stocktrek Images, Inc./Alamy, **p.334** yuyangc/Shutterstock, **p.336** Igoraul/Shutterstock, **p.337** Jan Stromme/Getty Images, **p.338** Patricia Barber/Custom Medical Stock Photo/Science Photo Library, **p.340** Yenka, **p. 341** Pearson Education Ltd/Jules Selmes, **p.343** L. Willatt, East Anglian Regional Genetics Service/Science Photo Library, **p.345** Julie Matthews, **p.346 l** Pamela Moore/iStockphoto, **r** teekaygee/Shutterstock, **p.347** Hubert Raguet/Look At Sciences/Science Photo Library, **p.349** Photos.com, **p.351** iStockphoto, **p.353** Pearson Education Ltd/Jules Selmes, **p.354** Jose Gil/Shutterstock, **p.356** Fancy/Veer/Corbis, **p.357** Small Town Studio/Shutterstock, **p.358** Matthew Polak/Sygma/Corbis, **p.359** Pearson Education Ltd/MindStudio, **p.361** iStockphoto, **p.363** Pearson Education Ltd/Studio 8/Clark Wiseman, **p.365** Robert Dant/iStockphoto, **p.366** Britain on View/Photolibrary, **p.368** Pearson Education Ltd/Arnos Design, **p.370** Pearson Education Ltd/Rob Judges, **p.371** Pearson Education Ltd/Studio 8/Clark Wiseman.

About the authors

Christine Brain is an experienced teacher and senior examiner, and is an Associate Lecturer for the Open University, a Chartered Psychologist, Chartered Scientist and trainee counsellor. She has contributed to the BTEC First Applied Science course and BTEC National Applied Science, bringing more psychology into both courses. Christine has written numerous resources, including textbooks, study guides and teacher guides.

Peter Gale studied applied physics and electronics before training as a physics teacher. Peter is an Assistant Head and an experienced teacher of science and design and technology at a large secondary school in the West Midlands. He has been involved with developing a range of initiatives for the Local Authority, including vocational subjects, secondary literacy and science. He has taught science at all levels and is an experienced author.

David Goodfellow is a freelance writer and examiner, having previously taught chemistry at all levels for over twenty years. He led the development of the AS Science qualification for 2007 and is an experienced author.

Sue Hocking has been an examiner for almost 30 years. She has delivered BTEC science and health studies courses in FE colleges, as well as GCSE and A level biology courses in secondary schools and a sixth form college. Her specialist fields are biology, biochemistry and health promotion. Sue is a series editor for Pearson and has written many books and teacher support resources.

Christine Kitchin is a senior external verifier, trainer, standards writer and reviewer and she was involved in the development of the 2010 specification for BTEC Applied Science. Christine is a freelance educational consultant and was previously Head of Science in a comprehensive school.

Roy Llewellyn is a trainer and external verifier for BTEC Applied Science. He is a senior science teacher, GCE physics examiner and head of vocational studies at a secondary school in South Wales. Roy has helped develop a number of publications and was involved in the development of the BTEC Level 2 and Level 3 specifications in Applied Science for 2006 and 2010.

Julie Matthews is an external verifier and was involved in the development of the 2010 BTEC specification and accompanying study guide. Julie is a Lead/Advanced Learning Coach (ALC) and has taught applied science for the past 4 years including BTEC first certificates, diplomas and national courses. Before this Julie delivered science to all ages in a secondary school.

Ismail Musa is an external verifier, an examiner for A level Physics and was involved in the development of the 2010 specification for BTEC Applied Science. Ismail has been teaching vocational applied science courses for over ten years. As well as teaching and examining he has been working as a Subject Learning Coach for Science (SLC), coaching students and staff and organising and running teaching and learning sessions.

Patricia Rhodes has been a senior verifier for Applied Science for 10 years. She has worked in education for over 35 years and has extensive experience of designing, delivering and assessing vocational programmes. Patricia delivers training and support sessions to help schools and colleges introduce and deliver successful BTEC programmes. She is a Fellow of two professional bodies including the Chartered Institute of Educational Assessors.

About your BTEC Level 2 First Applied Science

Choosing to study for a BTEC Level 2 First Applied Science qualification is a great decision to make for lots of reasons. More and more employers are looking for well-qualified people to work within the fields of science, technology, engineering and maths. The applied sciences offer a wide variety of careers, such as forensic scientist, drug researcher, medical physics technician, science technician and many more. Your BTEC will sharpen your skills for employment or further study.

Your BTEC Level 2 First Applied Science is a **vocational** or **work-related** qualification. This doesn't mean that it will give you *all* the skills you need to do a job, but it does mean that you'll have the opportunity to gain specific knowledge, understanding and skills that are relevant to your chosen subject or area of work.

What will you be doing?

The qualification is structured into **mandatory units** (ones you must do) and **optional units** (ones you can choose to do). This book contains all 22 units, so you can be sure that you are covered whichever qualification you are working towards.

- BTEC Level 2 First **Certificate** in Applied Science: 3 mandatory (M) units that provide a combined total of 15 credits

- BTEC Level 2 First **Extended Certificate** in Applied Science: 3 mandatory units and optional (O) units that provide a combined total of 30 credits

- BTEC Level 2 First **Diploma** in Applied Science: 3 mandatory units and optional units that provide a combined total of 60 credits

Unit number	Credit value	Unit name	Cert	Ex. Cert	Diploma
1	5	Chemistry and our Earth	M	M	M
2	5	Energy and our Universe	M	M	M
3	5	Biology and our environment	M	M	M
4	5	Applications of chemical substances		O	O
5	5	Applications of physical science		O	O
6	5	Health applications of life science		O	O
7	5	Practical scientific project		O	O
8	5	Science and the world of work		O	O
9	5	Working in a science-based organisation		O	O
10	10	The living body		O	O
11	10	Monitoring the environment		O	O
12	10	Growing plants for food		O	O
13	10	Investigating a crime scene		O	O
14	10	Science in medicine		O	O
15	5	Using mathematical tools in science		O	O
16	5	Designing and making useful devices in science		O	O
17	10	Chemical analysis and detection		O	
18	10	Exploring our Universe		O	O
19	10	Electronics in action		O	O
20	10	Biotechnology procedures and applications		O	O
21	5	Science in the world		O	O
22	10	Investigating human behaviour		O	O

How to use this book

This book is designed to help you through your BTEC Level 2 First Applied Science course. It is divided into 22 units to reflect the units in the specification. This book contains many features that will help you use your skills and knowledge in work-related situations and assist you in getting the most from your course.

Introduction

These introductions give you a snapshot of what to expect from each unit – and what you should be aiming for by the time you finish it!

Assessment and grading criteria

This table explains what you must do in order to achieve each of the assessment criteria for each unit. Each unit contains a number of assessment activities to help you with the assessment criterion, shown by the grade button **P1**.

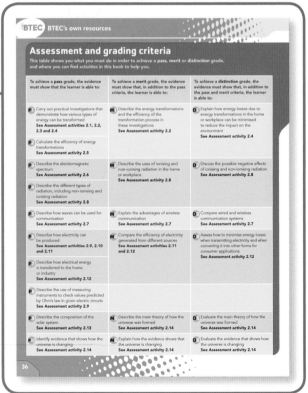

Assessment

Your tutor will set **assignments** throughout your course for you to complete. These may take the form of projects where you research, plan, prepare, make and evaluate a piece of work or activity, case studies and presentations. The important thing is that you collect evidence of your skills and knowledge to date.

Stuck for ideas? Daunted by your first assignment? These students have all been through it before…

Unit 2 Energy and our Universe

How you will be assessed

Your assessment could be in the form of:

- an observation sheet e.g. from your practical investigations into energy
- a report or presentation e.g. on the properties and applications of waves and radiation
- a leaflet e.g. on the production and distribution of electricity
- a presentation e.g. on the Solar System and the changing Universe

Tariq, 18 years old

I enjoyed this unit and I particularly liked the section on the Solar System as looking at the night sky fascinates me. It is amazing how we can see objects that are millions of miles away from us. Our class took a trip to the National Space Centre, in Leicester, which was fantastic and it brought this unit together.

I found the section on using light in communication very useful as it showed me that there are lots of technologies, some better than others. We experimented with laser light, which I found really interesting.

Energy issues are always on the news and the section on energy allowed me to be part of this debate. I feel that completing this unit has improved my practical skills and made me more aware of the world we live in.

Catalyst

Have you got the energy?

Imagine our lives without energy. How could we work without eating or drinking? How could a bus move from one bus stop to another without the fuel its engine needs? How could your mp3 player work without the electrical energy it requires to power it up?

In small groups discuss some other things you have used recently that require energy. In your groups work out what type of energy has been used.

37

Activities

There are different types of activities for you to do: **assessment activities** are suggestions for tasks that you might do to help build towards your assignment and will help you develop your knowledge, skills and understanding. Each one has **grading tips** that clearly explain what you need to do in order to achieve a pass, merit or distinction grade.

BTEC Assessment activity 2.10

1. Explain the difference between a rechargeable and a non-rechargeable battery. Give five examples of uses of each. **P6**
2. Draw a labelled diagram of a primary cell. **P6**
3. Discuss with a partner the advantages and disadvantages of rechargeable and non-rechargeable batteries. **P6**

Grading tip

To meet part of the grading criterion for **P6**, make sure that you include a diagram for the primary cell. To get all of the **P6** criterion you need to also describe another way of generating electricity.

 Activity A

Imagine heating up some baked beans in a metal saucepan. You stir the beans with a metal spoon. Using the idea of conduction, explain why the spoon gets hot.

There are also suggestions for activities for you to show your knowledge and understanding.

Worked examples

Worked examples provide a clear idea of what is required for calculations.

 Worked example

The frequency of microwaves used by a microwave oven is 2000 MHz. What is the period of the microwaves?

First remember to change the frequency prefix to a standard number. 2000 MHz is 2 000 000 000 Hz.

$$period = \frac{1}{frequency} = \frac{1}{(2\,000\,000\,000)}$$

$period = 0.5 \times 10^{-9}$ seconds (half of a billionth of a second)

Personal, learning and thinking skills

Throughout your BTEC Level 2 First Applied Science course there are lots of opportunities to develop your personal, learning and thinking skills. Look out for these as you progress.

PLTS

Team workers and Self-managers

Producing the leaflet in your groups will help you develop team-working and self-management skills

Functional skills

It's important that you have good English, maths and ICT skills – you never know when you'll need them, and employers will be looking for evidence that you've got these skills too.

Functional skills

English

In discussing nuclear energy and fossil fuels as a way of producing electricity, you will develop both speaking and listening skills, as you present your arguments and listen to the views of others.

Key terms

Technical words and phrases are easy to spot, and definitions are included. There is also in the glossary at the back of the book with additional terms.

Key term

Orbit – the path of an object moving through space, such as the path of the Earth as it goes round the Sun.

Did you know?

Where you see these boxes, look out for interesting snippets of information related to the science topics.

Did you know?

The honey bee can see ultraviolet light. Snakes such as the viper can see infrared.

Safety and hazards

This useful box alerts you to possible hazards and safety issues when using science in the real world.

Safety and hazards

Electrical current is dangerous as it could cause the heart to stop working. You can also get burns from where the current enters and leaves the body. Before working with electrical equipment make sure you ask for a safety briefing from your supervisor.

Science focus

Highlights specific scientific concepts and ideas to focus on.

Science focus

A battery produces electricity by the chemical reactions that take place inside it. The chemical inside a battery is called an **electrolyte**. Batteries can be rechargeable or non-rechargeable.

Case studies

Case studies show you examples of specific scientific topics applied in the workplace

Case study: Energy-efficient flight

Jenny is a trainee engineer working for an aerospace company. She is working with other engineers to design a more efficient engine for the planes. They want the engine to transform as much energy as possible into useful forms and to reduce the amount of energy that is wasted.

Which types of energy given out by the engine are wasted?

WorkSpace

Case studies provide snapshots of real workplace issues, and show how the skills and knowledge you develop during your course can help you in your career. Where you see the STEM ambassadors logo, the person featured is a working scientist who is part of the Science, Technology, Engineering and Mathematics Network. See the **STEMNET website** for more information.

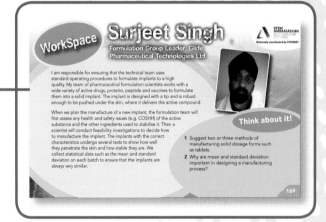

WorkSpace

Surjeet Singh
Formulation Group Leader, Glide Pharmaceutical Technologies Ltd

I am responsible for ensuring that the technical team uses standard operating procedures to formulate implants to a high quality. My team of pharmaceutical formulation scientists works with a wide variety of active drugs, proteins, peptide and vaccines to formulate them into a solid implant. The implant is designed with a tip and is robust enough to be pushed under the skin, where it delivers the active compound.

When we plan the manufacture of a new implant, the formulation team will first assess any health and safety issues (e.g. COSHH) of the active substance and the other ingredients used to stabilise it. Then a scientist will conduct feasibility investigations to decide how to manufacture the implant. The implants with the correct characteristics undergo several tests to show how well they penetrate the skin and how stable they are. We collect statistical data such as the mean and standard deviation on each batch to ensure that the implants are always very similar.

Think about it!

1 Suggest two or three methods of manufacturing solid dosage forms such as tablets.

2 Why are mean and standard deviation important in designing a manufacturing process?

169

Just checking

When you see this sort of activity, take stock! These quick questions are there to check your knowledge. You can use them to see how much progress you've made or as a revision tool.

Edexcel's assignment tips

At the end of each unit you'll find hints and tips to help you get the best mark you can, such as the best websites to go to, checklists to help you remember processes and useful facts and figures.

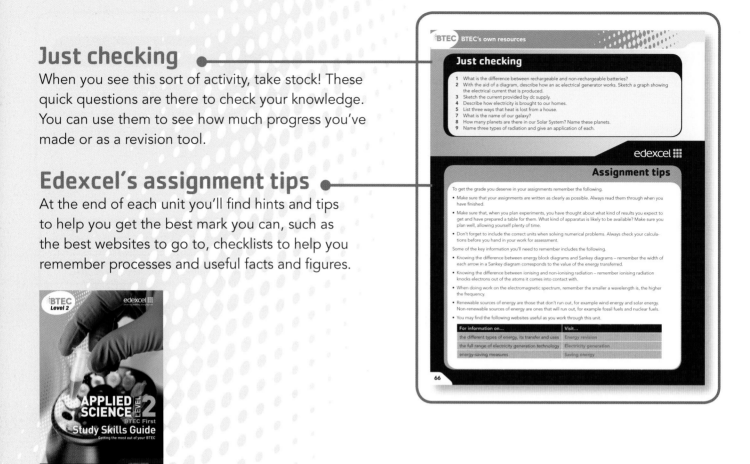

Have you read your **BTEC Level 2 First Study Skills Guide**? It's full of advice on study skills, putting your assignments together and making the most of being a BTEC Applied Science student.

Your book is just part of the exciting resources from Edexcel to help you succeed in your BTEC course. Visit www.edexcel.com/BTEC or www.pearsonfe.co.uk/BTEC 2010 for more details.

Hotlinks

Where you see a **hotlink reference** in this book go to the website www.pearsonschoolsandfecolleges.co.uk and search for the BTEC Level 2 First Applied Science student book using key terms, the express code 6091V or the ISBN. Go to the unit you are interested in and click on the name of the hotlink you want to visit.

Credit value: 5

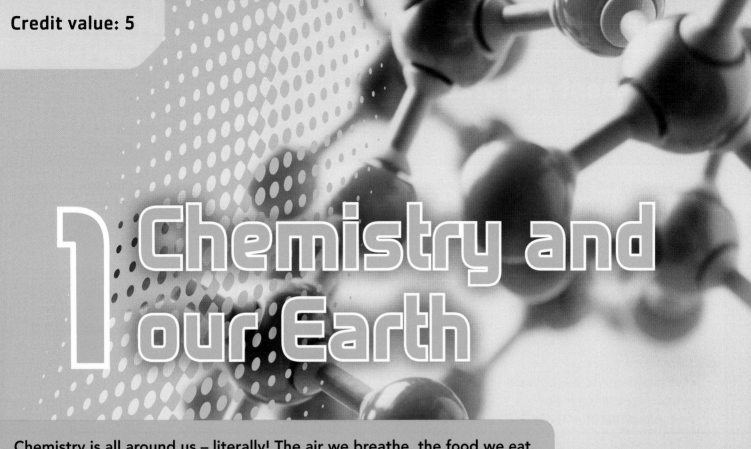

1 Chemistry and our Earth

Chemistry is all around us – literally! The air we breathe, the food we eat, the ground we stand on are all made up of chemicals. All of these things are made up from just over 100 different types of particles, called atoms, arranged and bonded together in different ways. Chemistry explains how this amazing fact can be true.

If chemists understand the way that these particles fit together then they can explain why the substances made from these particles look and behave so differently. But not only that – it means that chemists can work out how to design new substances and find new ways of making these from the substances that we find naturally in the earth and the sea.

In this unit you will explore some of the most important ideas in chemistry and find out how chemistry is used in industry to produce important substances. You will be able to carry out some practical investigations to discover for yourself how industrial processes can be made quicker and cleaner. From your own research you will discover more about the chemistry of environmental problems and discuss ways of helping to solve these problems.

Learning outcomes

After completing this unit you should:

1 be able to investigate different types of chemical substances related to their physical properties

2 be able to investigate the properties of elements relating to their atomic structure

3 be able to investigate the factors involved in the rate of chemical reactions

4 know the factors that are affecting the Earth and its environment.

Assessment and grading criteria

This table shows you what you must do in order to achieve a **pass**, **merit** or **distinction** grade, and where you can find activities in this book to help you.

To achieve a **pass** grade, the evidence must show that the learner is able to:	To achieve a **merit** grade, the evidence must show that, in addition to the pass criteria, the learner is able to:	To achieve a **distinction** grade, the evidence must show that, in addition to the pass and merit criteria, the learner is able to:
P1 Identify different types of chemical substances **See Assessment activity 1.1**	**M1** Describe the differences between types of chemical substances **See Assessment activity 1.1**	**D1** Explain how the structure of different chemicals affects their properties **See Assessment activity 1.3**
P2 Carry out a practical investigation into the physical properties of chemicals **See Assessment activity 1.2**	**M2** Explain how the physical properties of chemicals make them suitable for their uses **See Assessment activities 1.2 and 1.3**	
P3 Describe atomic structures of elements 1–20 found in the periodic table **See Assessment activities 1.4 and 1.5**	**M3** Describe the trends within the atomic structure of groups 1 and 7 in the periodic table **See Assessment activity 1.5**	**D2** Explain the trends in the chemical behaviour of the elements of groups 1 and 7 in relation to their electronic structure **See Assessment activity 1.8**
P4 Carry out an investigation into the chemical properties of elements in groups 1 and 7 **See Assessment activity 1.6**	**M4** Explain why the elements of groups 1 and 7 are mostly used in the form of compounds **See Assessment activity 1.7**	
P5 Carry out an investigation to establish how factors affect the rates of chemical reactions **See Assessment activities 1.9 and 1.10**	**M5** Explain how different factors affect the rate of industrial reactions **See Assessment activities 1.9, 1.10 and 1.11**	**D3** Analyse how different factors affect the yield of industrial reactions **See Assessment activity 1.11**
P6 Identify the human activities that are affecting the Earth and its environment **See Assessment activity 1.13**	**M6** Describe how the choices humans make have an effect on the Earth and its environment **See Assessment activities 1.13 and 1.14**	**D4** Explain possible solutions to the effect humans have on the Earth and its environment. **See Assessment activities 1.14 and 1.15**
P7 Identify natural factors that have changed the surface and atmosphere of the Earth. **See Assessment activity 1.12**	**M7** Describe the ways that natural factors have changed the surface and atmosphere of the Earth over millions of years. **See Assessment activity 1.12**	

How you will be assessed

Your assessment could be in the form of:

- a poster e.g. explaining the differences between elements, mixtures and compounds
- a table e.g. showing the electronic structure of the first 20 elements
- data from experiments on chemical reactions
- an article e.g. on the impact on the environment of human and natural factors.

Jasmine, 16 years old

I was really amazed at how many different areas we covered in this unit. I've always enjoyed practical work, but it was good to see how the results of our experiments could be useful in the chemical industry. We were shown some new ways of investigating rates of reaction, like weighing a beaker to see how much gas had been given off. I got lots of practice in drawing graphs of the results and found how these could be used to compare the rates of reactions in different experiments.

The section we did about changes to the Earth was really interesting – finding out about how volcanoes erupt and why earthquakes and tsunamis happen was particularly good.

We had a really good debate in class about whether we ought to be doing more about global warming. I was really surprised about how difficult it is to decide how to tackle it. Because I spent a lot of time researching this I was much more confident when I had to speak about it in front of the rest of the group.

Catalyst

Chemistry is all around you

Take a look around you. You are surrounded by different chemical substances – in the air, in the things used to construct and furnish your classroom and in your clothes and possessions.

In a small group, make a list of as many names as you can of chemical substances which are around you as you sit in your classroom or laboratory.

1.1 Elements, compounds, mixtures and molecules

In this section: P1 M1

Key terms

Element – a substance which cannot be broken down into anything simpler. It contains just one type of atom.

Compound – a substance made up of two or more different elements that are chemically bonded together.

Mixture – a substance made up of two or more simpler substances that are not chemically bonded. Mixtures can contain both elements and compounds.

Water, air and iron are all familiar substances, but they are very different.

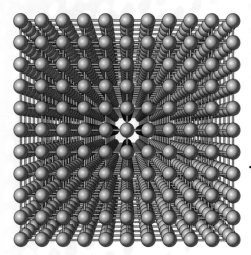

The structure of an element.

Elements

Iron is an example of an **element**. It is represented by the symbol Fe. You can identify elements by remembering:

- elements contain just one type of atom
- elements cannot be split up into simpler substances
- elements all have one-word names (and a symbol) and are found in the periodic table.

Unit 1: You can find the periodic table on page 10.

Other examples of elements are potassium, oxygen and bromine.

Activity A

Use the periodic table to find the symbol for the elements potassium, oxygen and bromine.

Compounds

Water is a **compound** of the elements hydrogen and oxygen. The hydrogen and oxygen are bonded together and *a lot* of energy is needed if you want to separate them. You can identify compounds by remembering that:

- compounds contain two or more different types of atom bonded together
- compounds can be split up into simpler substances only by chemical reactions.

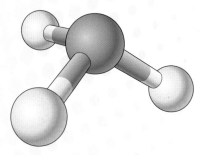

The structure of a compound.

The name or the formula of a compound often tells you what elements it contains. Some examples of compounds are sodium chloride (NaCl), ammonia (NH_3) and sodium hydroxide (NaOH).

Activity B

Look at this list of substances and decide whether you think they are elements or compounds:

Sulfur; calcium oxide; nitric acid (HNO_3); platinum.

Mixtures

The air we breathe is a **mixture** of nitrogen, oxygen, argon and small amounts of other gases.

You can identify mixtures by remembering that:

- mixtures contain several different substances that are not bonded together

- mixtures can be quite easily split up into simpler substances using physical processes like filtration, evaporation or distillation.

The names of mixtures often sound quite familiar, but do not give very much information about what they contain. Examples of mixtures are brine (salt water) and crude oil.

Many common substances are **solutions**. These are special cases of mixtures. They contain a substance called a solute. A solute dissolves and mixes with a solvent such as water or alcohol. So brine (salt water) is a mixture of the two substances: sodium chloride and water.

Did you know?

It is quite easy, given the right conditions, to separate the different gases in air. Several big companies supply pure oxygen, nitrogen and other gases which have been made in this way. Nitrogen is often used to fill packets of crisps. It stops the bags from getting crushed when they are packed together but unlike air it will not react with the fat in the crisps.

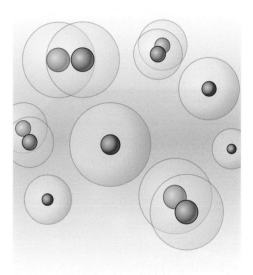

Particle model of a mixture.

BTEC Assessment activity 1.1 P1 M1

1 In groups of three, choose some common chemical substances which you know the names of. Some good examples might be sodium, crude oil, hydrochloric acid, sea water and rust. Use these or any other substances you know. Research whether they are elements, mixtures or compounds using the Internet and books. P1

2 Make a PowerPoint presentation for the rest of the class based on what you have found out. Include a description of the difference between elements, compounds and mixtures. P1 M1

PLTS

Creative thinkers

This assignment gives you an opportunity to develop your skills as creative thinkers. You can make your presentation more interesting for the audience if you choose some unusual or important chemicals.

Grading tip

For P1 you need to be able to identify elements, compounds and mixtures – you could use the periodic table to check whether any of your substances are elements.

When working towards M1 you might find it difficult to describe the difference between a mixture, compound and element. You could try making an identification key to help you – you may have used a key to help you identify different types of biological organisms. For example, one question you could include in your key might be 'Is the name of the substance in the periodic table?'

1.2 Substances and their uses: physical properties

In this section: **P2** **M2**

Key term

Physical property – something we can observe about a substance that doesn't involve a chemical change happening; for example: melting point, boiling point, electrical conductivity.

Street lights are often filled with sodium vapour.

Activity A

Pentane is a substance which has a melting point of –130°C and a boiling point of 36°C.

Chemical engineers working in the petrol industry sometimes add pentane to petrol because it boils at a low temperature. This can make it easier to start a car. Discuss whether pentane would be a good substance to add to petrol which is being used in Australia in the summer (temperatures up to 40°C) or Russia in the winter (temperatures down to –50°C).

The three forms of matter

You have probably seen solid sodium used in the laboratory – it is a soft grey solid which is very reactive. Liquid sodium is used to help cool nuclear reactors and sodium gas (vapour) is used in street lights.

Solids, liquids and gases look and behave very differently. This is because of the way in which the particles are arranged.

(a)   **(b)**   **(c)**

The arrangement of sodium atoms in **(a)** solid sodium, **(b)** liquid sodium and **(c)** sodium vapour.

Solid	Liquid	Gas
Particles close together	Particles close together	Particles a long way apart
Particles arranged regularly	Particles arranged randomly	Particles arranged randomly
Particles not moving, just vibrating	Particles moving past each other	Particles moving in all directions and colliding with each other

Melting and boiling points

We can look up or measure the melting point and boiling point of substances. This will help us decide whether a substance is likely to be a solid, liquid or gas.

Melting point and boiling point are examples of **physical properties**.

Room temperature is normally about 20°C. If a substance has a melting point above 20°C it will be a solid at room temperature. If a substance has a melting point below 20°C it will be a liquid. If its boiling point is also below 20°C it will be a gas.

Chemists working in quality control often measure the melting point of samples to find out how pure they are. If a substance is not pure then it will melt at a lower temperature than a pure substance.

 ## Case study: Checking the purity

Sally is a technician working at a company that extracts the compound vanillin from vanilla pods to use as a flavouring for ice-cream. They can't sell the vanillin to the ice-cream manufacturers if it isn't pure enough. The melting point of pure vanillin is 82.0°C. The melting point of the product must be within 1% of this figure before Sally can authorise its sale. She measures the melting point as 81.6°C.

Is this pure enough?

Mary uses measuring point equipment to monitor the melting point of the product.

Other physical properties

Thermal conductivity

You may have noticed that if you touch the metal part of the handlebars of a bike on a winter morning they feel very cold but the plastic handles feel much warmer. They aren't actually warmer but they have a very low **thermal conductivity**. This means that the heat from your hands isn't conducted away by the plastic. Metals, though, have high thermal conductivities.

The liquid sodium used in nuclear reactors needs to be able to conduct heat away from the reactor very quickly.

Solubility

When you add sugar to hot tea it dissolves to form a solution. You can dissolve a lot of sugar in a small amount of hot water so sugar has a high solubility.

Viscosity

Liquids like water flow well when you pour them. Other liquids, like olive oil and syrup, seem much thicker and don't flow as well. Scientists call these liquids viscous and they measure the viscosity – how viscous they are.

 ## Did you know?

It is really important to know the viscosity of oils which are used to lubricate moving parts in car engines. If the oil is too viscous then the moving parts will not move so easily.

 ## BTEC Assessment activity 1.2

Motor oil is a mixture of several different liquids used to lubricate the moving parts of engines.

1 Find out some of the physical properties of motor oil (you may be able to investigate some of these practically). **P2**

2 Produce a poster which explains these properties and why they are important in the use of motor oil. **M2**

Grading tip

You can use websites or the labels on canisters of lubricating oil to help you find information for **P2**. To achieve **M2** you should certainly think about boiling point and viscosity, but you may like to find out whether the oil is soluble in water.

1.3 Choosing a substance: more physical properties

In this section: D1 M2

Key terms

Polymer – a long molecule formed from many small molecules bonded together.

Composite material – a material made from two or more different substances with different physical properties.

Electrical conductivity

Some substances can conduct an electric current when a voltage is applied to them. You can look up the **electrical conductivity** of a substance, or investigate it practically. Metals, especially copper and aluminium, have very high electrical conductivity – they are called conductors. If a substance does not conduct electricity very well it is called an insulator.

The materials used in overhead power cables are chosen because of their physical properties.

Two-component polyurethane foam is a composite material with just the right properties for use in fridge walls.

Unit 2: You will find out more about electrical conductivity on pages 54–55.

Unit 4: You will find out why the structure of metals allows them to conduct electric currents on page 101.

Composite materials – the best of both worlds

Chemists have developed a wide range of materials in the last fifty years or so. One important group of materials are **polymers**.

Unit 4: You can find out more about polymers on page 110.

Polymers are light (low density) but are often quite flexible. This is useful if you are manufacturing plastic bags, but sometimes engineers need a different range of properties. One answer is to use **composite materials**. These materials contain two or more different substances. They are not chemically bonded, so we can call them a mixture.

Polyurethane foam is a composite. Bubbles of gas are blown into the polymer as it forms. This gas makes the foam light and also makes it a very good thermal insulator – it does not conduct heat. It is used in the walls of fridges and freezers.

Gases are often good thermal insulators because the particles are a long way apart. This makes it harder for heat energy to be transferred between the particles. Some solids are conductors of heat, because the particles are close enough to pass on energy as they vibrate together.

Electrical conductivity – the right stuff for the job

Scientists and engineers need to know data for the physical properties of substances when they are choosing suitable materials to use for particular tasks.

The materials used in overhead power lines or in power cables in the walls of new buildings need to have specific properties:

- high electrical conductivity

- strength

- low density (lightness)

- low cost.

In buildings the power cables must be surrounded by an insulating cover.

Engineers look at tables of physical properties to help them choose suitable materials.

Activity A

Formula 1 racing cars are built using a composite material. Use the Internet to find out what substances are combined in this material and why they are chosen.

Did you know?

Humans have been using composite materials for thousands of years. In many parts of the world, houses are traditionally built from bricks made from mud mixed with straw. Straw adds strength and the dried mud holds the straw together.

Activity B

Power cables are hung from pylons by strings of ceramic discs. Suggest an important physical property that these ceramic discs must have.

BTEC Assessment activity 1.3 D1 M2

1 Power cables used to be made of copper. A mixture of aluminium and iron is used in most power cables now.

Use the data in the table to discuss with a partner why power cables are made from a mixture of iron and aluminium.

2 Imagine you are writing a report to help an engineering company decide which materials to use to manufacture new cables to replace old power lines. Explain why you would recommend a mixture of aluminium and iron. **M2**

3 Power cables need to be good thermal conductors to avoid 'hot spots' building up in the cables. Add a short note to your report to explain why the metals in the cables will have a suitable thermal conductivity. **D1**

Table: The physical properties of some metals

Metal	Electrical conductivity (%, Cu = 100)	Strength (%, Al = 100)	Density (grams per cm³)	Cost (2009) per tonne
Copper	100	300	8.9	£5300
Aluminium	61	100	2.7	£900
Iron (steel)	15	1000	7.9	£500

Grading tip

To achieve **M2** first explain what physical properties the materials need to have for the job, in this case what properties are needed for power cables. Then use the information you are given to show that the chosen materials, in this case a mixture of aluminium and steel, have these properties.

For **D1** find a diagram showing the structure of a solid, like a metal. You should be able to find some help in this spread to explain why the structure of solids helps a material to be a good conductor of heat (thermal conductor).

1.4 The periodic table

Key terms

Proton – a sub-atomic particle with a mass of 1 unit and a charge of +1.

Neutron – a sub-atomic particle with mass of 1 unit and no charge.

Electron – a sub-atomic particle with very little mass and a charge of –1.

Isotopes – atoms with the same number of protons but different numbers of neutrons.

1	2	3	4	5	6	7	8
1 **H** Hydrogen 1							4 **He** Helium 2
7 **Li** Lithium 3	9 **Be** Beryllium 4	11 **B** Boron 5	12 **C** Carbon 6	14 **N** Nitrogen 7	16 **O** Oxygen 8	19 **F** Fluorine 9	20 **Ne** Neon 10
23 **Na** Sodium 11	24 **Mg** Magnesium 12	27 **Al** Aluminium 13	28 **Si** Silicon 14	31 **P** Phosphorus 15	32 **S** Sulfur 16	35.5 **Cl** Chlorine 17	40 **Ar** Argon 18
39 **K** Potassium 19	40 **Ca** Calcium 20	70 **Ga** Galium 31	73 **Ge** Germanium 32	75 **As** Arsenic 33	79 **Se** Selenium 34	80 **Br** Bromine 35	84 **Kr** Krypton 36
85 **Rb** Rubidium 37	88 **Sr** Strontium 38	115 **In** Indium 49	119 **Sn** Tin 50	122 **Sb** Antimony 51	128 **Te** Tellurium 52	127 **I** Iodine 53	131 **Xe** Xenon 54
133 **Cs** Caesium 55	137 **Ba** Barium 56	204 **Tl** Thalium 81	207 **Pb** Lead 82	209 **Bi** Bismuth 83	209 **Po** Polonium 84	210 **At** Astatine 85	222 **Rn** Radon 86
223 **Fr** Francium 87	226 **Ra** Radium 88						

The columns are called *groups*. Elements in the same group have similar properties.

Page 378: A full periodic table is shown at the back of this book.

The rows are called *periods*. Elements in the same period do not have similar properties, but the properties often change in a regular pattern as you go across.

All of the elements that exist are arranged in the periodic table. This is a shortened version.

The periodic table

You may have seen the element potassium reacting with water. It fizzes and catches fire and can even explode if a big lump is used. But what would happen if you used another element, like sodium? Would it react in the same way? Perhaps it would react faster or slower?

All scientists like to look for patterns in the properties of substances. Patterns mean that you can often predict how something is going to behave before you use it in an experiment.

The best way chemists have found to show patterns in the properties of chemical elements is to arrange the elements into a table called the periodic table.

A Russian scientist named Dmitri Mendeleev published the first periodic table in 1869. Since then there have been additions as new chemicals have been discovered, but the modern periodic table still has the same structure; the elements are arranged in order of their atomic number. The lightest atoms have the smallest atomic number.

Activity A

1 Name an element in **(a)** group 1 and **(b)** group 7. Give the chemical symbol for each of these elements.

2 Name an element in **(a)** period 2 and **(b)** period 4. Give the chemical symbol for each of these elements.

Activity B

Work out the number of protons, neutrons and electrons in these atoms:

$^{12}_{6}C$ $^{14}_{6}C$ $^{35}_{17}Cl$ $^{2}_{1}H$

What's in an atom?

Atoms are incredibly small – much too small to see even with microscopes. Scientists now know that atoms are made up of even smaller particles: **electrons**, **protons** and **neutrons**. Protons and neutrons are also known as nucleons.

Electrons have a negative charge, protons have a positive charge and neutrons have no charge. Protons and neutrons each have a mass of one atomic mass unit. Electrons are so light that the mass doesn't really count.

Did you know?

All carbon atoms have the same chemical properties, but the nuclei of some carbon atoms are slightly different. Carbon-14 ($^{14}_{6}C$) has a heavier nucleus than normal carbon-12 ($^{12}_{6}C$) that we see in the periodic table. The nucleon number of carbon-14 is 14 instead of 12. This tells you that carbon-14 has two extra neutrons. $^{14}_{6}C$ and $^{12}_{6}C$ are two different **isotopes** of carbon. The nucleus of $^{14}_{6}C$ is unstable and breaks down over time – this means it is radioactive and how much is present can be easily measured.

(b) The atomic mass, or **nucleon**, number tells you the mass of the nucleus. We already know there are 3 protons so there must be 4 neutrons (7 – 3 = 4).

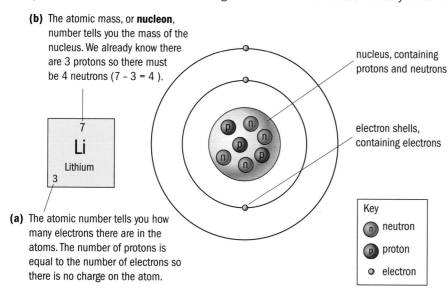

nucleus, containing protons and neutrons

electron shells, containing electrons

7
Li
Lithium
3

(a) The atomic number tells you how many electrons there are in the atoms. The number of protons is equal to the number of electrons so there is no charge on the atom.

Key
- **n** neutron
- **p** proton
- ○ electron

An atom of lithium contains a nucleus surrounded by shells of **electrons**.

BTEC Assessment activity 1.4   P3

Draw up a table for the first 20 elements in the periodic table to show the number of protons and electrons in the atoms. You can also add information about the normal number of neutrons in an atom of each element. Choose three of the elements and describe their electronic structure in words. **P3**

Grading tip

Remember that you can work out the number of neutrons in an atom by taking away the number of protons from the mass number. The periodic table on the previous page usually shows the mass number of the most common isotope of each element. But there are some exceptions like chlorine, which has two common isotopes – its mass number is the average mass number of the two isotopes.

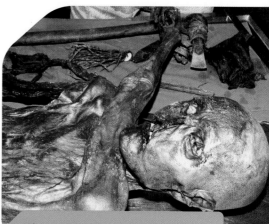

Chemists have worked out the age of the Iceman, Ötzi, from the amount of carbon-14 isotope left in his body. This is because the amount of $^{14}_{6}C$ falls in a regular way as soon as a living organism dies.

1.5 Electron arrangements

In this section:

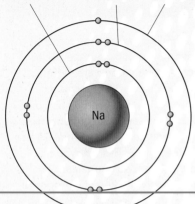

two electrons can fit into the first shell

eight electrons can fit into the second shell

the final electron goes into the third shell

The electron shells of the sodium atom.

Electron shells

Electrons surround the nucleus in shells. There is a limit to how many electrons can fit into a shell before it becomes full.

The atomic number tells you how many electrons there are in the atom. Electrons fill up the shells. When one shell is full, the electrons must go into the next shell.

Atomic number	Element	First shell	Second shell	Third shell	Fourth shell
1	hydrogen	1			
2	helium	2			
3	lithium	2	1		
4	beryllium	2	2		
5	boron	2	3		
6	carbon	2	4		
7	nitrogen	2	5		
8	oxygen	2	6		
9	fluorine	2	7		
10	neon	2	8		
11	sodium	2	8	1	
18	argon	2	8	8	
19	potassium	2	8	8	1
20	calcium	2	8	8	2

The electron arrangement of some of the first 20 elements in the periodic table.

Did you know?

The only atom which doesn't contain any neutrons is the hydrogen atom. Neutrons are used to 'glue' the protons together and stop them flying apart. There is only one proton in a hydrogen atom so it does not need any glue!

Activity A

(i) The electron arrangement of a sodium atom can be written as 2,8,1. Use the information in the table to write the electron arrangements of:

(a) oxygen

(b) potassium.

(ii) Work out the electron arrangement of:

(a) aluminium (atomic number 13)

(b) chlorine (atomic number 17).

Case study: The doping of silicon

James works in a company manufacturing semiconductors. Silicon is used for semiconductors because it is in the middle of the periodic table – it is halfway between being a metal and a non-metal.

To change the properties of silicon, James finds out the effect of adding small amounts of phosphorus to it. He knows that this should increase the conductivity because the arrangement of electrons in phosphorus is very similar to silicon – but there is one more electron in the outer shell. This electron helps to carry the current through the silicon and improves its conductivity.

Work out the arrangement of electrons in silicon (atomic number 14) and phosphorus (atomic number 15).

Electron arrangements and the periodic table

You may have noticed that there is a connection between the electron arrangement of an atom and where it appears in the periodic table:

- the number of electrons in the outer shell is the same as the group number

- the number of shells in the atom is the same as the period number.

For example, phosphorus has an electron arrangement 2,8,5 and is in group 5, period 3.

You may notice that some very big atoms, like those at the bottom of groups 1 and 7, can have more than eight electrons in some of their shells.

Activity B

(i) An element has electron arrangement 2,8,18,6.

(a) What group is it in?

(b) What period is it in?

(c) Use the full periodic table on page 378 to find out what the element is.

(ii) Rubidium is in group 1 and period 5.

(a) How many electrons are there in its outer shell?

(b) How many shells of electrons does it have?

(c) Find out from a periodic table website what its full electron arrangement is.

 BTEC ## Assessment activity 1.5

1 In groups of four make a large poster or a set of cards to show the atomic structure of the first 20 elements in the periodic table. For each element you should show the number of protons and neutrons in the nucleus and the arrangement of electrons in shells. **P3**

2 Look at your poster or cards and describe how the atomic structure of the elements changes as you go across the periodic table. **P3**

3 Work out the number of neutrons, protons and electrons, as well as the number of electrons in each shell, for

(a) the first three elements in group 1 (H, Li and Na)

(b) the first three elements in group 7 (F, Cl, Br).

4 Describe how the atomic structure changes as you go down each group. **M3**

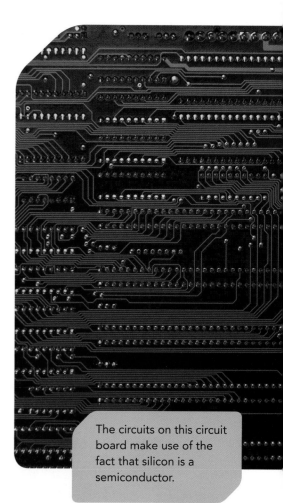

The circuits on this circuit board make use of the fact that silicon is a semiconductor.

 ## PLTS

Team workers

This assignment will give you a chance to develop your skills as team workers. You will need to agree how to divide up the elements and make sure that every member of the group is confident with the tasks which they have to carry out before starting work.

Grading tip

Don't forget that some elements have several isotopes. You should find at least two examples of these and include both isotopes in your poster or card set. **P3 M3**

1.6 Chemical reactions

In this section: P4

Key terms

Chemical formula – a way of showing which elements a compound contains and the ratio of atoms of each element.

Exothermic – a reaction which gives out heat energy is exothermic.

You will find out more about energy changes in chemical reactions in Unit 4.

Iron oxide reacts violently with aluminium in the thermite reaction.

 Did you know?

The molten iron produced by the thermite reaction can be used to repair broken railway track. Engineers carry out the reaction above the break in the rails, and then pour the molten iron into the gap. When it solidifies, it welds the broken track together.

Telling the story of a chemical reaction

Word equations are a quick and easy way to describe what happens in a chemical reaction. They tell you, in words, what the reactants and products of a chemical reaction are.

For example the word equation for the reaction of iron oxide with aluminium is:

$$\underbrace{\text{iron oxide + aluminium}}_{\text{reactants}} \rightarrow \underbrace{\text{iron + aluminium oxide}}_{\text{products}}$$

The **reactants** react with each other and are changed into the **products**. A lot of energy is given out in the form of heat so this is a very **exothermic** reaction. This is why the reaction is known as the thermite reaction. The heat of the reaction melts the iron and this is a way of making molten iron.

 Activity A

In the iron-making industry, iron is obtained from the reaction of iron oxide with carbon. Carbon dioxide is given off as a by-product. Write a word equation for this.

 Case study: Making bromine

Bryn is a chemical engineer working for a company which manufactures bromine from sea water. Bromine is used in the manufacture of pesticides and photographic film.

He controls the addition of chlorine gas to the sea water. The chlorine reacts with sodium bromide which is dissolved in sea water. Bromine is formed, as well as sodium chloride, which can be returned safely to the sea.

Write a word equation to describe this reaction.

Chemical formulas

The periodic table lists the symbols for all of the chemical elements. When these elements join together to make compounds, you can also combine the symbols to make a **chemical formula**.

water

carbon dioxide

chlorine

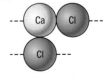
part of the structure of calcium chloride

The structures of water (H_2O), carbon dioxide (CO_2), chlorine (Cl_2) and calcium chloride ($CaCl_2$). The formulas show the ratio of atoms of each element present.

Hydrogen, oxygen and chlorine are elements. The formulas of the elements are written as H_2, O_2 and Cl_2. This shows you that the atoms in these elements always go around in pairs.

Activity B

Sulfur dioxide contains two oxygen atoms for every sulfur atom. Write down its formula.

Functional skills

Mathematics

You may need to remind yourself of the work you have done on ratios in maths. You need to understand that a ratio of 4:2 or 200:100 is the same as a ratio of 2:1. Chemical formulas are often written using the simplest possible ratio.

Chemical equations

Word equations tell you which substances react and what the products are. A balanced chemical equation gives you more information: it tells you how many atoms of each element are involved in the reaction.

Some equations you may come across in your practical work are shown in this table.

Reaction	Balanced equation
Group I metals reacting with water e.g. sodium + water → sodium hydroxide + hydrogen	$2Na + 2H_2O \rightarrow 2NaOH + H_2$ **(a)**
A reactive halogen displacing a less reactive halogen e.g. chlorine + potassium bromide → bromine + potassium chloride	$Cl_2 + 2KBr \rightarrow Br_2 + 2KCl$ **(b)**

BTEC **Assessment activity 1.6**   P4

Imagine you are a chemical engineer looking at ways of extracting useful substances from sea water. When you add chlorine to sea water you notice that the solution becomes orange-brown.

1 Write a short report for the manager of a chemical company to explain what has happened in your experiment.

   Include word and chemical equations in your report for this reaction.

2 Explain to the manager what useful substance you have made during your experiment and whether you think this would be a good product for the company to make. P4

Grading tip

When you do your own reactions, in order to achieve P4, you will need to describe what you see. Make sure you use phrases in your report such as 'the colour of the solution changed from colourless to …'

When chlorine is added to potassium bromide, the solution goes orange-yellow as bromine is formed.

1.7 Chemical properties

In this section: **M**4

Lithium oxide reacts with carbon dioxide in the air. It is used to purify the air in spacecraft and submarines.

Activity A

Name an element that is:

(i) unreactive

(ii) reactive.

Welding is often carried out using argon to shield the metal from the air.

Activity B

(i) The arrangement of electrons in a sodium atom is 2,8,1. Write down the arrangement of electrons in a sodium ion which is formed when sodium reacts with other substances.

(ii) A potassium ion has electron arrangement 2,8,8. Write down the electron arrangement in a potassium atom.

Chemical properties

If you observe a chemical reaction taking place, such as a fuel burning or a metal fizzing when you put it in water, you are observing a chemical property. One chemical property of a substance is how reactive it is.

Chemists can use the periodic table to help them predict chemical properties. For example, elements in group 8, 'the noble gases', are all unreactive. Elements in group 1 and 7 are much more reactive.

If an atom has a full outer shell (normally eight electrons) it means it will be stable and unreactive. All elements in group 8 have full outer shells so they will be very unreactive.

Case study: Arc welding

Sam works as a welder in a construction company. Arc welding means to join pieces of metal together by melting them with an electric current and then letting them cool. If hot metal is in contact with air it can react, which may be dangerous or affect the weld joint. Sam prevents this by surrounding the hot metals with argon gas.

Argon is an unreactive gas. Unlike the oxygen in air it will not react with metals even at very high temperatures. This makes it ideal for welding.

Can you suggest any other elements which could be used as alternatives to argon?

Properties of group 1 (the alkali metals)

Group 1 elements have one electron in their outer shell (see pages 12–13). This electron can easily be lost to produce a positive ion, leaving a full shell. This means that group 1 elements are very reactive. There are not many uses of the elements themselves because they will quickly change into other compounds when they react with water or air.

The compounds of group 1 can be very useful, for example:

- sodium hydroxide is used as a strong cleaning liquid because it reacts with grease and fat

- potassium nitrate is used as a fertiliser because potassium (and nitrogen) is an important plant nutrient.

Properties of group 7 (the halogens)

Group 7 elements have seven electrons in their outer shell. It is quite easy for the elements to gain an electron, so that there is a full outer shell, and produce a negative ion. This means that group 7 elements are quite reactive. For example they will react with some of the substances in living organisms and so they are toxic substances.

They can be used as antiseptics and in water treatment to kill bacteria.

The compounds of group 7 are not toxic and not as reactive as the elements. Salt (sodium chloride) is an example of a compound of a group 7 element. It's not reactive and it's not toxic (except in very large amounts).

Photoreactivity

Silver chloride is a white solid. When it is exposed to light it goes darker, because it breaks down into silver and chlorine. The grains of silver give the solid a grey colour. Something similar happens for silver bromide.

This is the way in which photographic film works. A piece of plastic is covered in a film of silver bromide. It produces a negative image – the darker areas in a negative show you where the silver bromide has been exposed to most light.

Digital sensors have now mostly replaced film, but you can observe a similar effect yourself if you wear photoreactive sunglasses. These get darker when you go into the sun.

Tablets containing chlorine are added to the water of swimming pools.

Did you know?

Because group 7 elements can react to make salt and other similar substances, they are often called halogens (a Greek word meaning salt-maker). Compounds of halogens are called halides.

The silver chloride in the lenses of photoreactive sunglasses darkens when it is exposed to sunlight.

BTEC Assessment activity 1.7 **M4**

1 Use the Internet or textbooks to find out about some compounds of group 7 elements and how they are used. **M4**

2 Make a short PowerPoint presentation to the rest of the class to explain why the compounds you researched are so useful and why group 7 elements themselves do not have many uses. **M4**

3 You might look for the way in which the compounds are used in X-ray photography, agriculture and refrigerators. **M4**

Grading tip

Make sure that you explain how the properties of the compounds are different to the elements to achieve **M4**.

Functional skills

English

Preparing and giving your presentation will help you to develop your skills in communicating information. Always practise your presentation before you give it to the class.

1.8 Trends in chemical properties

 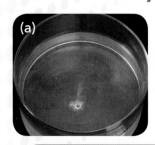

Key term

Displacement reaction – a chemical reaction in which a less reactive element is replaced in a compound by a more reactive one.

The reactivity of group 1 elements

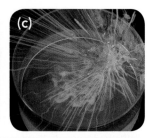

Group 1 elements all react with water but, as you go down the group, the elements become more reactive: **(a)** lithium, **(b)** sodium and **(c)** potassium.

Elements in the same group have similar properties. These properties change gradually as you go down the group; for example group 1 elements become more reactive. You can explain this if you look at the arrangement of electrons in the atoms.

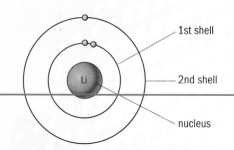

1st shell

2nd shell

nucleus

When a group 1 atom reacts, for example, with water, it loses its outer electron. As you go down group 1, the outer electron gets further away from the nucleus.

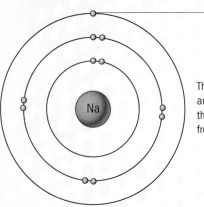

The more shells there are, the further away the outer electron is from the nucleus

Activity A

(i) In what way is the arrangement of electrons the same for all the elements in group 1?

(ii) How does the arrangement change as you go down the group?

The nucleus attracts electrons because it is positive and the electrons are negative. So if the electron is further away the attraction is weaker and it is easier for the electron to be lost. This means that the elements become more reactive as you go down the group.

The reactivity of group 7 elements

For the group 1 elements you can easily see the trend in reactivity simply by adding them to water. It isn't as easy to see the trend in reactivity of group 7 elements – they do not react with water in the same way. You need to carry out displacement reactions to investigate the trend.

An example of a **displacement reaction** is the reaction of chlorine with potassium bromide:

chlorine + potassium bromide → bromine + potassium chloride

$$Cl_2 \quad + \quad 2KBr \quad \rightarrow \quad Br_2 \quad + \quad 2KCl$$

The arrangement of electrons in group 1 elements.

Chlorine is more reactive than bromine so it displaces the bromine from the compound.

When group 7 elements react they gain an electron. Group 1 elements lose electrons when they react. It is easier to gain electrons if an atom is small because the electron is closer to the positive nucleus and is more attracted to it. This means that the atoms at the top of group 7 are more reactive.

You can tell whether a halogen has been displaced by looking at the colour of the solution after the reaction.

	KCl (colourless)	KBr (colourless)	KI (colourless)
Cl_2 (pale green)	–	Orange Br_2 formed	Purple I_2 formed
Br_2 (orange)	X (stays orange)	–	Purple I_2 formed
I_2 (purple)	X (stays purple)	X (stays purple)	–

The more reactive halogens at the top of the group displace the less reactive halogens at the bottom of the group from solutions of potassium halides.

When the halogens dissolve in hexane, the colour of each halogen is very different. Chlorine is pale green, bromine is orange and iodine is purple.

Safety and hazards

Caesium is the most reactive of all the group 1 metals – it explodes when it touches water. Even lithium and sodium will explode if large lumps are added to water. Great care must be taken when adding group 1 metals to water.

Did you know?

Iodine and bromine both look quite similar when they are dissolved in water. If you dissolve them in the solvent hexane then it is easy to see the difference.

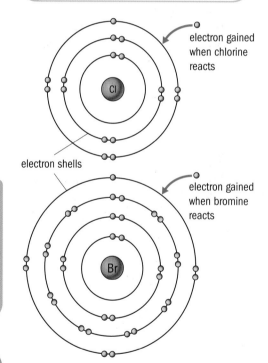

electron gained when chlorine reacts

electron shells

electron gained when bromine reacts

The arrangement of electrons in some group 7 atoms.

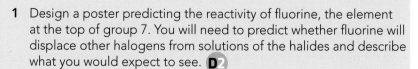
Assessment activity 1.8 D2

1 Design a poster predicting the reactivity of fluorine, the element at the top of group 7. You will need to predict whether fluorine will displace other halogens from solutions of the halides and describe what you would expect to see. **D2**

2 Use the electron arrangements of fluorine and the other halogens to explain the reactivity of fluorine. **D2**

Grading tip

To meet **D2** remember to include some diagrams of electron arrangements to explain any trend that you describe.

1.9 Rates of chemical reactions

In this section:

Key term

Rate of reaction – the rate at which reactants are converted into products. If a lot of product is produced in a short time we say that a reaction has a fast rate.

Activity A

(i) Two reactions produce the same amount of product. Reaction 1 takes 20 s, reaction 2 takes 40 s. Which one is reacting at the faster rate?

(ii) Reaction 3 produces 20 cm^3 of product in 100 s, reaction 4 produces 60 cm^3 in 200 s. Which one is reacting at the faster rate?

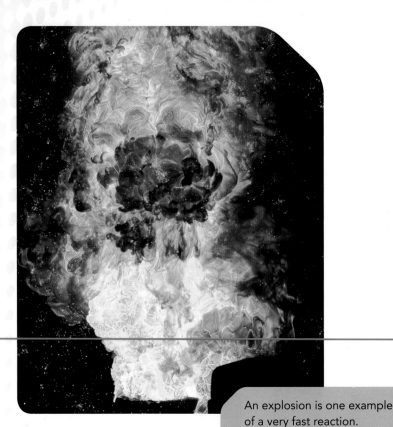

An explosion is one example of a very fast reaction.

Fast and slow reactions

When gunpowder explodes, a lot of gas is produced in a fraction of a second. We say that the reaction is happening at a very fast rate. Other reactions are very slow, like the rusting of iron. It can take months or years before you notice any rust being produced on the iron. We talk about reactions having a fast or a slow **rate of reaction**.

Speeding up and slowing down reactions

One important challenge for chemists is to control the rates of reactions. In the chemical industry it is important to make as much product as possible in the shortest time. This must also be done safely, which sometimes means slowing the reaction down.

There are several things a chemist can do to speed up a chemical reaction:

Activity B

Write down three things that you could do to slow a reaction down.

- heat up the reaction (increase the temperature)

- if there are solutions in the reaction, make them more concentrated (increase the concentration)

- if there are solids in the reactions, grind them up into smaller pieces (increase the surface area of the solid)

Unit 1: You will learn more about catalysts on page 22.

- add a chemical called a catalyst.

Particle theory and reaction rates

If a reaction is going to happen between two substances, then the particles of the substances must collide. They have to collide with quite a lot of force (or energy) so that the particles can break up and rejoin to form new products. This is often called the particle collision theory.

You can use collision theory to help you understand some of the reasons why reactions speed up (or slow down) when the conditions are changed.

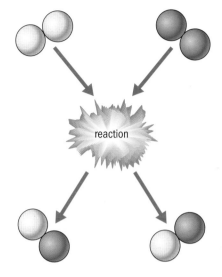

Two molecules collide and react if there is enough energy to allow the atoms to rearrange into products.

Temperature is increased	
	Particles move around faster and collide more often and with more energy
Concentration is increased	
	Particles are closer together and collide more often
Surface area is increased	
	There is more surface for particles to collide with so the particles collide more often with it

Did you know?

Factories making powdered foods need to be careful of explosions! Milk powder and custard powder are so finely divided that if too much powder gets into the air and catches fire it will react so fast that it will explode.

BTEC Assessment activity 1.9 P5 M5

Imagine that you are a chemist working for a company trying to develop a portable oxygen generator for environmental use. Oxygen can be used to react with the pollutants from industry that could contaminate sensitive environmental areas.

You are investigating the possibility of using hydrogen peroxide solution as a source of oxygen. When a catalyst of solid manganese dioxide is added, oxygen is produced. Without a catalyst the reaction is very slow.

1 Write a short report to your manager suggesting what you could do to make the reaction happen as fast as possible. P5

2 Add a short section to explain how the hydrogen peroxide should be stored. You need to make sure that it releases oxygen as slowly as possible to increase its shelf-life. M5

Grading tip

To achieve P5 make sure you clearly identify the factors which you would be changing to increase or decrease the rate.

For M5 remember to use collision theory to explain why these factors affect the reaction in the way you describe.

1.10 Experiments: rates and catalysts

In this section: P5 M5

Key term

Catalyst – a chemical substance which speeds up the rate of a chemical reaction but is not used up and can be re-used again and again.

How can you monitor a reaction?

Watching the cross

If a reaction between dissolved chemicals produces a solid, it will make the reaction mixture go cloudy. You can measure the time taken until you can no longer see through the mixture. A cross drawn on paper underneath a flask is a good way of helping you decide this.

Measuring gases

Some reactions produce gases. The amount of gas that is produced in a given time can be used as a measure of the rate of reaction.

When magnesium ribbon is added to hydrochloric acid it produces hydrogen gas:

magnesium + hydrochloric acid → magnesium chloride + hydrogen

$$Mg + 2HCl \rightarrow MgCl_2 + H_2$$

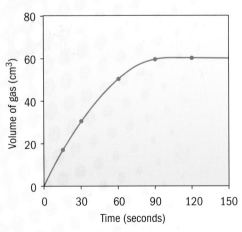

For some reactions the 'cross' method is a good way of studying rate.

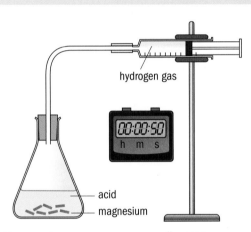

If a reaction produces a gas you can collect it in a gas syringe.

If you measure the volume of gas produced every 10 seconds you can plot graphs to show what happens during the time the reaction is happening. A good way of analysing the graph is to find how long it takes to produce a certain volume of gas.

Plotting a graph of volume against time is an excellent way of helping you analyse your results.

Worked example

A reaction produces 50 cm³ of gas in 20 s. What is the rate of the reaction?

Rate of reaction can be calculated by working out the volume of gas produced every second.

Use the equation rate of reaction $= \dfrac{\text{volume}}{\text{time}}$.

So rate $= \dfrac{50}{20} = 2.5$ cm³/s

Catalysts

Hydrogen peroxide has a chemical formula very similar to water: H_2O_2.

The extra oxygen atom is quite easily lost and so a solution of hydrogen peroxide breaks down into simpler substances:

$$\text{hydrogen peroxide} \rightarrow \text{water} + \text{oxygen}$$
$$2H_2O_2 \rightarrow 2H_2O + O_2$$

If you look at a solution of hydrogen peroxide you will see small bubbles appear from time to time as oxygen is formed. The reaction is happening very slowly.

If a small amount of a chemical called manganese oxide is added, the bubbling becomes more violent, showing the reaction rate has increased.

If the manganese oxide is filtered off, there is still the same mass left and it can be used again. It is called a **catalyst**. Because it is not used up, manganese oxide does not appear in the chemical equation.

How do catalysts work?

A lot of catalysts are solids. The way in which they work can be quite complex but the principle is that some of the reactant particles stick to the solid. When other particles collide with the ones that are sticking to the solid, less energy is needed to break the bonds in the particles and start the reaction

Activity A

Use the graph opposite to find the time taken to produce 50 cm³ of gas.

Did you know?

Living organisms also have catalysts in their cells which can break down hydrogen peroxide. If you put a piece of potato into some hydrogen peroxide, biological catalysts called enzymes will make the hydrogen peroxide fizz, just like the manganese oxide catalyst.

Activity B

How could you carry out an experiment to measure the rate of reaction of the hydrogen peroxide with and without a catalyst?

 BTEC ## Assessment activity 1.10   **P5 M5**

You have carried out some experiments to find out how the rate of a reaction between calcium carbonate and hydrochloric acid is affected by changing the concentration of the acid. You measure the volume of carbon dioxide (CO_2) produced.

You obtain the following results:

Experiment 1: acid concentration of 15 g per litre of solution

Time in s	0	15	30	45	60	75	90
Volume of CO_2 (cm³)	0	20	35	44	46	47	48

Experiment 2: acid concentration of 30 g per litre of solution

Time in s	0	15	30	45	60	75	90
Volume of CO_2 (cm³)	0	39	76	90	94	96	97

1 Plot graphs of volume (on the vertical axis) against time (on the horizontal axis) for each experiment. **P5**

2 Use the graph to find out the volume of gas given off after 20 s. Calculate a value for the rate of reaction in cm³ per second. **P5**

3 What can you conclude about the effect of doubling the concentration on the rate of reaction? **M5**

Safety and hazards

Hydrogen peroxide is harmful, even the dilute solutions you may use in the laboratory. It can irritate eyes, and when the concentration is high it is corrosive and will cause skin burns. You must wear eye protection at all times when you are using hydrogen peroxide.

HARMFUL

Grading tip

When you draw a graph you need to:

- label the vertical and horizontal axis
- choose suitable scales to make your plotting easy
- try and draw smooth curves to show the pattern of the points.

1.11 Industrial chemistry

In this section:

Key term

Yield – the yield of a chemical reaction tells you how much of the reactants are converted into products.

Ammonia is a very important chemical. It can be made into ammonium nitrate, which is used as a fertiliser.

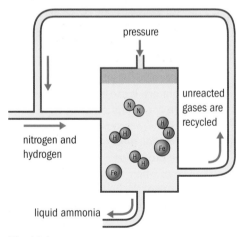

The Haber process

 Did you know?

Chemists measure pressure in 'atmospheres' (abbreviated to atm). A pressure of 200 atmospheres is 200 times greater than the normal pressure of the Earth's atmosphere.

Making ammonia: the Haber process

In 1909 the German chemist Fritz Haber invented a way of making ammonia cheaply. He used nitrogen from the air and hydrogen from natural gas:

nitrogen + hydrogen → ammonia

$$N_2 + 3H_2 \rightarrow 2NH_3$$

To make sure the reaction is fast enough to make ammonia quickly and cheaply:

- a temperature of 400°C is used
- a high pressure of 200 atmospheres is used making the gas molecules squash close together
- a catalyst is used to speed up the reaction, otherwise it would take months for any ammonia to be detected at all in the reaction.

Nitrogen and hydrogen are continuously fed into the reactor. Any unreacted gases are recycled. The reactor has very thick walls to withstand the high pressures used.

 Activity A

Use particle collision theory to explain why the reaction becomes faster at a high temperature.

Why these conditions?

Rate

If you look at the table below, you will see that increasing the pressure and the temperature makes the reaction go much faster.

	100°C	400°C
100 atm	Slow	Fast
200 atm	Fast	Very fast

The reaction would be even faster if the temperature and pressure were increased even more. So why isn't the Haber process operated at, say, 1000°C and 500 atmospheres?

- It might increase the chance of explosions – high pressures and temperatures can be very dangerous.

- Very high temperatures and pressures require lots of energy from burning fuel so, although the reaction is faster, it would cost much more to make a tonne of ammonia.

- The reactor would need to be very strong with very thick walls to withstand the pressure. This would cost a lot more money to build.

- The **yield** of the reaction would be lower.

Yield

The yield of a chemical reaction tells you how much of the reactants are converted into products. A yield of 100% would mean that all of the reactants are converted into products. Chemists can carry out experiments to measure the yield at different conditions.

Atom economy

The Haber process is an example of an industrial reaction which has a good **atom economy**. This means that all of the atoms in the reactants end up in the ammonia, rather than in other waste products.

Activity B

With a partner discuss why 400°C and 200 atmospheres are chosen for the Haber process. (Hint: think about both cost and rate.)

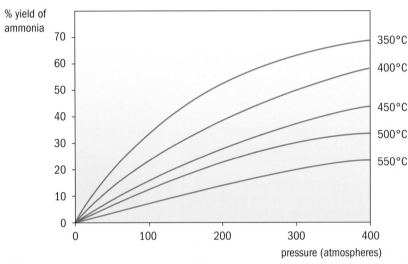

This graph shows you that the yield in the Haber process gets bigger when the pressure is increased, but it gets smaller when the temperature is increased.

The production of bromine doesn't have a very good atom economy:

$$Cl_2 + 2NaBr \rightarrow Br_2 + 2NaCl$$

The sodium and chlorine atoms in the reactants end up in the waste product sodium chloride.

Activity C

(i) What is the yield at the normal operating conditions of 400°C and 200 atmospheres?

(ii) The yield would be bigger if the reaction was done at 350°C and 200 atmospheres. Use ideas about rate to explain why this wouldn't actually be a good idea.

BTEC Assessment activity 1.11 D3 M5

Imagine that you are a chemist working for a company manufacturing ammonia. Your line manager wants to cut the cost of the Haber process by doing the reaction at 200°C and 100 atm and also use less catalyst. Write a report or give a PowerPoint presentation to explain why this would not be a good idea. D3 M5

Grading tip

To achieve **M5** explain, using collision theory, how the different factors will affect the rate.

To meet **D3** you need to be able to analyse how these changes would affect the yield. You should use a graph to help you. Find similar graphs to the one on this page or on the Internet. The BBC Bitesize website is a good source of information – try searching using 'BBC Bitesize Haber Process'.

Activity D

Explain why the production of ethanol (C_2H_5OH) from ethene and water has a good atom economy:

$$C_2H_4 + H_2O \rightarrow C_2H_5OH$$

1.12 Changing Earth: natural factors

In this section:

Key term

Tectonic plates – pieces of the Earth's crust which float on the mantle.

Safety and hazards

Being a volcanologist – a scientist who studies volcanoes – is an exciting but dangerous job. Volcanologists need to get very close to active volcanoes to study them, and if a volcano erupts the gas can travel at speeds of hundreds of miles an hour so it may be impossible to escape. Modern instruments can make earlier predictions of when a volcano is about to erupt, giving scientists more of a chance to flee.

Did you know?

There are about 1500 active volcanoes in the world. The biggest volcanic eruption in the last 1000 years was probably one in Indonesia in 1815. The dust in the atmosphere made the weather in Europe and America that year so cold that it is was known as 'the year without a summer'. Snow fell in June and crops failed.

Activity A

Make a table to list the negative and positive effects of a volcanic eruption to humans living on the Earth.

The explosion of Mount St Helens in 1980 released millions of tonnes of dust and gas into the atmosphere.

Volcanoes

One morning in May 1980, the landscape of Washington State, USA, changed forever. Mount St Helens volcano erupted, blowing away half of the mountain and forming a huge cloud of dust and ash which spread out into the atmosphere. The eruption of a volcano can have a massive impact on the Earth and its atmosphere.

- When volcanoes on the seabed erupt, so much molten rock is released that new islands are formed.

- Many of the world's highest mountain ranges – like the Andes and the Rockies – are chains of old volcanoes, most of which are no longer active.

- The gases released by volcanic eruptions have changed the atmosphere over the past few billion years. Sulfur dioxide (which forms acid rain) and carbon dioxide (a greenhouse gas) are both released in volcanic eruptions.

- Volcanic eruptions may also cause the Earth to cool slightly for a few years after the eruption. The dust from the eruption reflects some of the sunlight and stops it from heating the Earth.

- The soil around volcanoes is often very fertile because the lava and ash from volcanic eruptions contains many nutrient elements, such as phosphorus.

Earthquakes

You will have seen pictures on the television of the damage which earthquakes can cause to towns and cities.

Earthquakes have also changed the natural landscape. They can cause huge landslides and alter coastlines. In 1958 at Lituya Bay in Alaska, USA, an earthquake caused a colossal landslide into the sea. This generated a mega-tsunami with a wave over 500 m high which destroyed the shoreline of the bay.

Tectonic plates

Only certain parts of the world are at risk of severe earthquakes or volcanic eruptions. Scientists now understand this because they have discovered that the Earth's crust is made up of giant plates of rock, called **tectonic plates**.

The tectonic plates move slowly at about 2 cm per year. Earthquakes happen when two plates meet and have to slide past each other. Plates colliding together or moving apart can also cause volcanoes to be formed.

When tectonic plates move together and crumple up the crust, mountains can be formed.

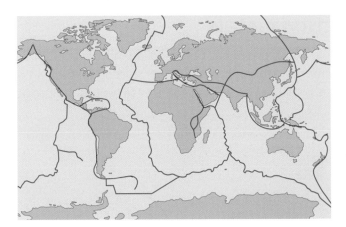

The tectonic plates which make up the Earth's crust.

Case study

Earthquake prediction

Sofia works as a geologist. She travels to parts of the world where severe earthquakes happen and monitors small movements of the ground to try and predict whether one is likely to happen in the next few years.

Use the map of tectonic plates to suggest some places where Sofia might need to go to. You may want to use an atlas to help you identify the countries.

Grading tip

For **P7** see if you can find maps or pictures to show how the landscape has changed after the eruption. This will allow you to show that volcanoes are one factor which changes the surface. You will also need to make a comment about how the atmosphere might have changed.

For **M7** make sure that you describe how the movement of tectonic plates over millions of years has caused mountains to be formed. You should find some clear diagrams of tectonic plate movements to illustrate your presentation.

BTEC Assessment activity 1.12   **P7** **M7**

1 **(a)** Find out about a volcanic eruption which has happened in the last 150 years. Where did it happen? Why do you think it happened? How did it affect the landscape and the atmosphere? Some examples are Krakatau or Pinatubo.

(b) Make a poster with maps and pictures to show what you have found out. **P7**

2 Find out about two mountain ranges which have been formed by the movement of tectonic plates. Good examples could be the Andes and the Himalayas. Make a short presentation to explain how they were formed

A good website for detailed information is **The Story of Plate Tectonics** **M7**

Functional skills

ICT

This assignment will give you the chance to demonstrate your ICT skills of finding and selecting information.

1.13 Changing Earth: human activities

In this section: P6 M6

Key terms

Ore – a rock from which metals can be extracted.

Greenhouse gas – a gas which is causing global warming by absorbing heat emitted from the Earth.

The Bingham Canyon is over a kilometre deep and was made by humans.

Unit 11: You can learn more about the effects that humans have on the environment on pages 214 – 215.

Activity B

Copper sulfide, copper carbonate and copper oxide are some of the ores of copper used in copper extraction. Which of these ores could produce sulfur dioxide during extraction processes?

Mining for metals

The Bingham Canyon in Utah, USA, is the biggest hole in the ground in the world. It has been dug by a company which needs copper **ore** to use in its copper extraction plant.

Many people think that huge mines like this are very ugly to look at. Another problem is that the mine produces a quarter of a million tonnes of waste rock each day. This rock must be disposed of.

Activity A

Why is copper such a valuable metal? You may want to look back at earlier pages in this unit to find out.

Extracting metals

Even when the copper ore has been successfully dug out of the ground, the environmental problems are only just starting. Extraction processes produce lots of damaging waste products as well as the copper metal

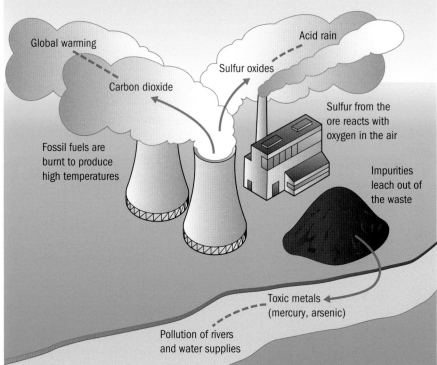

The environmental problems caused by the extraction of metals from ores.

Acid rain

When acidic gases like sulfur dioxide dissolve in rain water they form acids, such as sulfuric acid. When the rain falls it finds its way into lakes and the soil, killing fish and plants. It can also react with rocks such as limestone and cause stonework to erode.

Sulfur dioxide is also produced from volcanic eruptions, so some acid rain may be natural.

Global warming

Any process that involves burning something containing carbon to provide heat or energy gives off carbon dioxide.

Carbon dioxide is an example of a **greenhouse gas**. It acts a bit like a blanket around the Earth, stopping heat escaping into space. When there is more carbon dioxide in the atmosphere the Earth gets warmer.

Scientists predict that humans might double the amount of carbon dioxide in the atmosphere by 2050. This means the temperature of the Earth may increase significantly – maybe by up to 6°C.

Extracting materials from the sea and air

Lumps of metal ore called nodules can be found on the seabed. Scientists think that there might be 500 billion tonnes of these nodules, including valuable metals like manganese and nickel. But if the seabed was dredged to collect these nodules it would destroy the fish and other creatures that live on the seabed.

The air is a mixture of gases, including oxygen and nitrogen and the unreactive gases helium and argon. If air is liquefied, these can be boiled off at different temperatures and collected. Argon is used in welding and oxygen is used in hospitals and by scuba divers. Although we will never run out of air, energy is needed to separate the gases from air. This energy is obtained by burning fossil fuels, which releases carbon dioxide into the atmosphere, adding to global warming.

Did you know?

Carbon dioxide is emitted from volcanoes. Some scientists think that if there was a lot of volcanic activity when the Earth was very young it could have helped to warm the Earth up enough for life to become possible.

Oxygen, extracted from air, is used by pilots who fly at high altitudes.

BTEC Assessment activity 1.13   P6 M6

1 A chemical company plans to dig a small mine on local farmland and then build a chemical plant to extract the metal nickel from its ore.

Write a letter to the local paper to explain why you don't want this plan to go ahead. **P6**

2 In a group think of as many ways as you can in which human activities are affecting the environment. Start off by listing them, but then try and put them into a spider diagram with lines pointing to the different kinds of effect, such as: global warming, habitat destruction, killing of wildlife, etc. To help you, look at Unit 11 as well as websites such as **Young People's Trust for the environment. M6**

Grading tip

To meet **P6** you will need to list some of the ways in which this plan will affect the environment. You should include ways in which it will affect wildlife as well as the people living near the mine and chemical plant.

Make sure that you include activities which involve extracting substances from the sea and air as well as the ground to obtain **M6**.

PLTS

Creative thinkers

These activities will help you develop your skills as creative thinkers.

1.14 Solving pollution problems

In this section:

Key term

Chemical scrubber – a piece of equipment added to a chimney that removes some of the polluting gases passing up it.

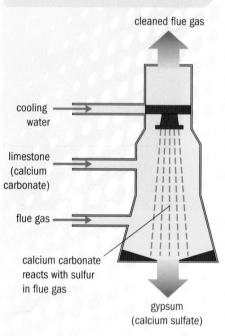

cleaned flue gas

cooling water

limestone (calcium carbonate)

flue gas

calcium carbonate reacts with sulfur in flue gas

gypsum (calcium sulfate)

Chemical scrubbers are used in industrial chimneys to remove harmful sulfur dioxide gas. The sulfur dioxide reacts with limestone (calcium carbonate) to produce gypsum (calcium sulfate)

Solving the problems of pollution caused by industry is a top priority for scientists.

Aluminium collected from recycling bins can be melted down and used again.

Reducing acid rain

Metal extraction can cause acid rain (see pages 28–29). The burning of fossil fuels can also add to the problem because fuels like coal often have sulfur impurities that form sulfur dioxide when burned. Clouds are made up of water droplets. When sulfur dioxide dissolves in the water droplets, the result is acid rain. Between 1985 and 1990, governments in Europe and the USA agreed to try and reduce the amount of sulfur dioxide and other acidic gases released by industries in their countries. UK industry now emits about 70% less sulfur dioxide than in 1980. This was achieved by introducing new technology, such as **chemical scrubbers**, in factory chimneys.

Recycling

It's not just industry that needs to take responsibility for caring for the environment. We can all help to reduce pollution. Recycling metals like aluminium is one easy way.

Every can which is recycled means that less aluminium needs to be mined. Less fuel is needed to melt old aluminium than to extract new aluminium from ore.

Activity A

What other items are you encouraged to recycle? How does recycling them help the environment?

Case study: Saving the planet

Emily works for the government. She is putting together a website to encourage people to take action at home and in their workplaces to reduce the impact they have on the environment.

What sort of things might she suggest that people should do? How do they help the environment?

The carbon dioxide problem

It is much more difficult to solve the problem of carbon dioxide (CO_2) that causes climate change. Whenever fossil fuels are burned, CO_2 is released, and most industry, transport and electricity generation rely on burning fossil fuels.

Unit 2: You can learn more about nuclear power and renewable energy sources on pages 58–61.

What can be done?

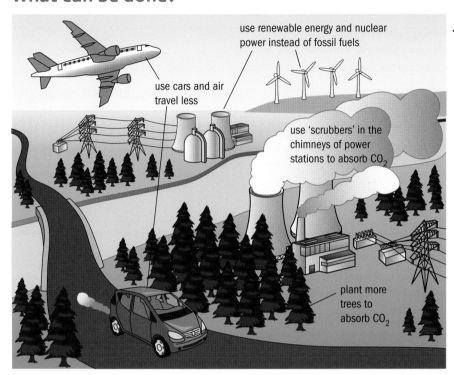

use renewable energy and nuclear power instead of fossil fuels

use cars and air travel less

use 'scrubbers' in the chimneys of power stations to absorb CO_2

plant more trees to absorb CO_2

The different ways in which global warming could be tackled.

Activity B

Why does planting trees help to reduce the amount of CO_2 in the atmosphere? (Hint: you might need to think about an important chemical process which happens in all plants.)

Did you know?

Climate change could mean that in the future much of the world would be too hot or dry to grow crops. What is happening now is that polar ice-caps are starting to melt and sea levels are rising, leading to frequent flooding of low-lying countries such as Bangladesh.

Why aren't things changing?

Many of the suggestions for reducing pollution are quite expensive and involve a big change in the way things are done. Countries like India and China are now using a lot of fossil fuel energy to develop their industry and economy. Some people want governments to agree to spend money on making sure that less fossil fuel is burned. But so far not everyone is in agreement that this would be the right thing to do.

Arguments for	Arguments against
If we don't take action the Earth will heat up to dangerous levels	We can't be sure how much the Earth will really heat up
We already have the technology to reduce the amount of CO_2 released into the atmosphere so it should be used	It will cost far too much money for some countries
If we don't act together then nothing will happen	Each country should make up its mind what to do

BTEC Assessment activity 1.14

Write a newspaper report about the choices which the world must make about climate change. Explain some of the possible solutions and how they could help solve the problem. At the end, give your opinion of what you think the world should be doing. **D4 M6**

Grading tip

To meet **M6** you need to give examples of how human decisions affect the environment. To achieve **D4** you will also need to explain how the various methods of reducing emissions would work.

1.15 The world of tomorrow

In this section: **D**4

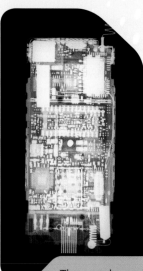

The metals used in mobile phones may run out in your lifetime.

Try and imagine life without the substances produced by the chemical industry. These could be the polymers in your clothes, the lithium metal in your mobile phone battery, the petrol and diesel which powers the cars and buses we travel in.

All of these substances are produced from finite resources. This means that the resources will eventually run out because there is only a certain amount that we know of left. Scientists have worked out how long these reserves could last, unless we take some action.

Resource	Used for	How long could it last?
Crude oil	Burning as fuel, making polymers	about 100 years
Metal ores	Manufacturing metals	cadmium: 70 years silver: 29 years aluminium: 1000 years

Sustainable development

Sustainable development is a way of making sure that we can carry on meeting our needs without exhausting the Earth's resources. It also means not polluting the environment so that future generations can survive and prosper.

Recycling and reducing our use of fossil fuels are two ideas which are important in sustainable development.

Organic farming

Using fertilisers and pesticides makes food crops cheaper to grow and gives better yields to the farmer. But they can cause the death of other organisms in the environment and affect ecosystems. To avoid damaging the environment some farmers prefer to farm without using chemicals.

Activity A

Why do you think it is difficult to be sure about how long the reserves of crude oil and metal ores could last?

Science focus

Biofuels

Crops like maize can be used to produce a fuel called biodiesel, and sugar cane can be used to produce ethanol which can replace petrol. Crops are renewable resources because they can be replaced as fast as they are used up.

BTEC Assessment activity 1.15 **D**4

Many businesses are now under pressure to show that they are committed to sustainable development. Imagine that you are a manager of an international manufacturing company. Write a short leaflet for your staff to explain what sustainable development is and suggest ways in which the business could operate in a sustainable way.

Grading tip

To achieve **D**4 you need to explain each problem caused by human activities and then go on to give the solution to the problem.

WorkSpace

Joanne Beech
Research Scientist, molecular modelling department

I am part of a research team looking into new compounds that I hope can be used to extract metals from solutions. I use 3D computer molecular modelling to predict whether the molecules will do what I want them to do.

I spend a lot of time getting data from other universities and programming this into the 3D molecular modelling program on the computer.

I talk to scientists in laboratories where I work and also in other universities, as I hope they will make the molecule that I design. Once they have made the correct molecule, I carry out tests on it with a team of scientists in a laboratory to see whether my hypothesis is correct.

Sometimes it is not, but the molecule works well for a different use instead, which is good.

The best thing about my job is that it is very interesting, as no one has made these molecules before. I get to travel around the world presenting my research findings and meeting other scientists working in this field.

Molecules similar to those I have been working on are now being used to extract toxic and radioactive metals from waste water. I feel proud that I was part of a worldwide team researching these new molecules, and now they are cleaning up pollution and helping people. Scientists all around the world are using these molecules to try to find cures for cancer, as industrial catalysts, as molecular switches and for many more uses.

Think about it!

1 If you could design a molecule for a particular purpose what would it do?
2 Would you prefer to be the person in the laboratory synthesising the molecule or the person designing it using a computer?

33

Just checking

1   What is the difference between a mixture (like air) and a compound (like water)?
2   List four physical properties which you could measure or find out for a substance.
3   How many electrons can normally fit into (a) the first shell (b) the second and third shell?
4   What happens to the reactivity of elements as you go down the group in (a) group 1 and (b) group 7?
5   Give three examples of useful compounds formed from group 1 or group 7 elements and describe why they are useful.
6   What happens to the rate of a chemical reaction when you heat it up? Use collision theory to explain this.
7   Describe three human activities which have affected the Earth or its atmosphere.
8   List two changes to Earth which might be caused by the movement of tectonic plates.

Assignment tips

To get the grade you deserve in your assignments remember the following.

• Be very careful how you choose your words. You need to use the correct chemical words to describe the substances you are writing about.

• When you are writing reports on experiments, think about how you are going to present your results. A table is a very clear way of doing this but you will need to think very carefully about what headings you use.

• When you are using resources on the Internet, think very carefully about exactly what information you need to put in your report. Don't be tempted to just cut and paste large chunks – you should try and put as much as possible into your own words.

Some of the key information you'll need to remember includes the following.

• The difference between physical and chemical properties.

• The trends in the periodic table.

• The factors affecting rates of reactions – remember that you can use collision theory to explain all of these factors.

• Knowing that humans are affecting the atmosphere and the Earth by their activities. Volcanoes and plate movements have also caused huge changes in the atmosphere and the surface of the Earth.

You may find the following websites useful as you work through the unit.

For information on...	Visit...
elements in the periodic table	Online periodic table
rates of reactions and reactions in industry	Reaction rates from S-Cool
tectonic plates, earthquakes and mountain building	The Story of Plate Tectonics

2 Energy and our Universe

Most of the appliances we use at home and at work use energy from sources that are running out. If we are not careful we won't have any energy to do the things that we take for granted. By understanding energy better, we can plan for the future by designing and building new technology that lets us derive energy from sources that will not run out.

In this unit you will learn how energy is transferred and used along with the different sources of energy and how they can be used to generate electricity. You will investigate how we can make better use of the energy we use at home and in the workplace. You will also have the opportunity to carry out practical work, for example investigating how to minimise energy loss at home.

You will also learn about different types of light and radiation and how they can be used in our everyday lives and in the world of work, such as the use of gamma radiation to treat cancer patients.

Finally you will learn about the Universe and our place in it. You will have the opportunity to investigate the origin of the Universe and our Solar System and discover theories that astronomers have to explain how the Universe is changing.

Learning outcomes

After completing this unit, you should:

1 be able to investigate energy transformations

2 know the properties and applications of waves and radiation

3 know how electricity that is produced from different sources can be transferred to electric circuits in the home and industry

4 know the components of the Solar System and the way the Universe is changing.

Assessment and grading criteria

This table shows you what you must do in order to achieve a **pass**, **merit** or **distinction** grade, and where you can find activities in this book to help you.

To achieve a **pass** grade, the evidence must show that the learner is able to:	To achieve a **merit** grade, the evidence must show that, in addition to the pass criteria, the learner is able to:	To achieve a **distinction** grade, the evidence must show that, in addition to the pass and merit criteria, the learner is able to:
P1 Carry out practical investigations that demonstrate how various types of energy can be transformed **See Assessment activities 2.1, 2.2, 2.3 and 2.4**	**M1** Describe the energy transformations and the efficiency of the transformation process in these investigations **See Assessment activity 2.2**	**D1** Explain how energy losses due to energy transformations in the home or workplace can be minimised to reduce the impact on the environment **See Assessment activity 2.4**
P2 Calculate the efficiency of energy transformations **See Assessment activity 2.5**		
P3 Describe the electromagnetic spectrum **See Assessment activity 2.6**	**M2** Describe the uses of ionising and non-ionising radiation in the home or workplace **See Assessment activity 2.8**	**D2** Discuss the possible negative effects of ionising and non-ionising radiation **See Assessment activity 2.8**
P4 Describe the different types of radiation, including non-ionising and ionising radiation **See Assessment activity 2.8**		
P5 Describe how waves can be used for communication **See Assessment activity 2.7**	**M3** Explain the advantages of wireless communication **See Assessment activity 2.7**	**D3** Compare wired and wireless communication systems **See Assessment activity 2.7**
P6 Describe how electricity can be produced **See Assessment activities 2.9, 2.10 and 2.11**	**M4** Compare the efficiency of electricity generated from different sources **See Assessment activities 2.11 and 2.12**	**D4** Assess how to minimise energy losses when transmitting electricity and when converting it into other forms for consumer applications **See Assessment activity 2.12**
P7 Describe how electrical energy is transferred to the home or industry **See Assessment activity 2.12**		
P8 Describe the use of measuring instruments to check values predicted by Ohm's law in given electric circuits **See Assessment activity 2.9**		
P9 Describe the composition of the solar system **See Assessment activity 2.13**	**M5** Describe the main theory of how the universe was formed **See Assessment activity 2.14**	**D5** Evaluate the main theory of how the universe was formed **See Assessment activity 2.14**
P10 Identify evidence that shows how the universe is changing. **See Assessment activity 2.14**	**M6** Explain how the evidence shows that the universe is changing. **See Assessment activity 2.14**	**D6** Evaluate the evidence that shows how the universe is changing. **See Assessment activity 2.14**

How you will be assessed

Your assessment could be in the form of:

- an observation sheet e.g. from your practical investigations into energy
- a report or presentation e.g. on the properties and applications of waves and radiation
- a leaflet e.g. on the production and distribution of electricity
- a presentation e.g. on the Solar System and the changing Universe.

Tariq, 18 years old

I enjoyed this unit and I particularly liked the section on the Solar System as looking at the night sky fascinates me. It is amazing how we can see objects that are millions of miles away from us. Our class took a trip to the National Space Centre, in Leicester, which was fantastic and it brought this unit together.

I found the section on using light in communication very useful as it showed me that there are lots of technologies, some better than others. We experimented with laser light, which I found really interesting.

Energy issues are always on the news and the section on energy allowed me to be part of this debate. I feel that completing this unit has improved my practical skills and made me more aware of the world we live in.

Catalyst

Have you got the energy?

Imagine our lives without energy. How could we work without eating or drinking? How could a bus move from one bus stop to another without the fuel its engine needs? How could your mp3 player work without the electrical energy it requires to power it up?

In small groups discuss some other things you have used recently that require energy. In your groups work out what type of energy has been used.

2.1 Understanding types of energy

Types of energy

Energy is vital to everyday life and we use it to do all sorts of things. The table below shows some examples of different types of energy.

Sound energy is used to test metals in the aerospace and automotive industries. Cracks, or weak areas, reflect the energy.

Type of energy	What is it?	Example
Potential (e.g. elastic, gravitational, chemical)	Stored energy that has the potential of doing work	
Kinetic	Movement energy	
Light (electromagnetic)	Bright objects give out light energy	
Sound	Things that vibrate give off sound energy	
Thermal (heat)	Energy that is transferred from a hot region to a cold region	
Electrical	Flow of charge in an electric circuit	

Grading tip

Part of meeting the **P1** criteria is to list types of energy. When doing your assignment, make sure you give all the different types of energy given in the table.

Energy transformations

We need energy to do all sorts of things. Running, reading and even sleeping require energy. Energy can be transformed (changed) from one form to another. Anything that takes in energy must also give out energy. Here are some examples.

- A girl running gets her energy from the food she eats. The energy is then transformed to movement (kinetic energy), sound and heat energy.

- The light bulb that lights your room gets its energy from electricity. The energy is then transformed to light and heat energy. Remember: don't touch a lit bulb – it will burn.

Activity A

Write down three ways in which you have experienced energy being transformed today.

Using energy at home and in the workplace

We use many different appliances at home and at work that convert energy from one form into others.

Activity B

For each thing pictured on the right, write down the type of energy that is going into it and the types of energy that it is giving out. (Hint: remember that most things give out more than one form of energy.)

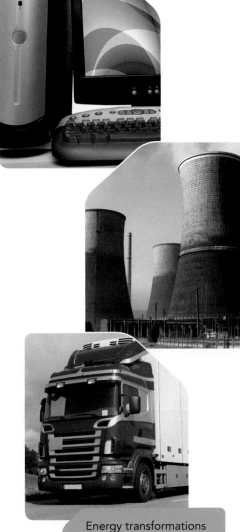

Energy transformations are all around us.

Assessment activity 2.1

You are a food scientist working for a supermarket, looking at energy in food.

1 Find out how much energy is stored in a can of drink (any type). This value will be marked clearly on the label. **P1**

2 What form of energy is in this drink? **P1**

3 Investigate what happens to this drink as it goes into your body. **P1**

Grading tip

Remember, everything requires energy – even sleep! This means that you should be able to find enough types of energy to cover the content for **P1**.

 PLTS

Independent enquirers and Self-managers

When you carry out your investigation you will be learning to enquire independently as well as developing your self-management skills.

2.2 Describing energy

In this section: P1 M1

Key terms

Energy block diagram – shows the forms of energy going into and out of a system.

Sankey diagram – shows how much energy is going into and out of a system.

Conservation of energy – tells us that energy is transformed to various forms and is not destroyed.

When engineers and designers create the appliances we use in everyday life they need to know how much energy is transformed to useful forms and how much is wasted. They can then improve their designs by trying to reduce the amount of wasted energy. For instance, we now have more efficient 'energy-saving' light bulbs in our homes.

Case study: Energy-efficient flight

Jenny is a trainee engineer working for an aerospace company. She is working with other engineers to design a more efficient engine for the planes. They want the engine to transform as much energy as possible into useful forms and to reduce the amount of energy that is wasted.

Which types of energy given out by the engine are wasted?

Investigating energy

You need to be able to describe energy changes that take place in everyday situations. It is helpful to break the problem down. This example shows you how you could do this:

- Consider a ball on a work bench. What kind of energy does the ball have? (Hint: what energy is related to having the potential to do something?)

- Now consider what happens as the ball falls. What energy is being transformed? (Hint: which energy is related to motion?)

- As the ball hits the ground and then rebounds, does it reach the height it fell from? Explain your answer in terms of how energy is transformed.

Engineers design aeroplanes to be as energy efficient as possible.

Tracking transformations

We use **energy block diagrams** to understand how energy is transformed. This block diagram shows the energy transfers that occur in a moving lorry.

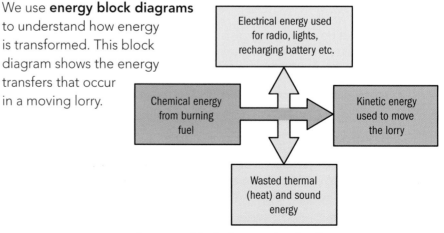

An energy block diagram showing the energy transfers that occur in a moving lorry.

The block diagram shows you that the lorry is powered by chemical energy in the form of fuel. The chemical energy is transformed into:

- kinetic energy that moves the lorry
- electrical energy that powers the lights, radio, recharges the battery etc.
- sound energy
- thermal (heat) energy.

The heat and sound energy are transferred to the surroundings as wasted energy.

Useful versus wasteful energy

It is useful to know how much energy is actually transferred into useful energy and how much into wasteful energy. You can show this by constructing a different type of block diagram called a **Sankey diagram**. A Sankey diagram for the moving lorry is shown on the right.

In a Sankey diagram the energy flow is shown by arrows. Broad arrows show large energy transfers. Narrow arrows indicate small energy transfers. We say that the width of the arrow is proportional to the energy.

The total amount of energy that comes out of the lorry is equal to the total amount of energy that goes in. We say that the energy is conserved. Physicists call this the law of **conservation of energy**.

Activity A

Draw a block diagram to show the energy transformations for someone using a hair dryer. (Hint: energy comes out of the hair dryer in more than one form.)

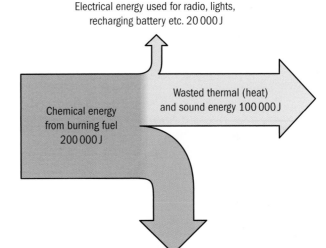

Electrical energy used for radio, lights, recharging battery etc. 20 000 J

Chemical energy from burning fuel 200 000 J

Wasted thermal (heat) and sound energy 100 000 J

Kinetic energy used to move the lorry 80 000 J

A Sankey diagram showing the size of the energy transfers for a moving lorry.

BTEC **Assessment activity 2.2** P1 M1

You are working for a leading IT company. Your manager wants you to look into energy-efficient computers. To start, you investigate the energy used by one of the company's existing computers.

1 State in words the types of energy involved when the computer is in use. **P1**

2 Draw a block diagram to show the energy transformations. **M1**

3 350 J of electrical energy is supplied to the computer. In the process 65 J is used to generate light energy, 190 J is transformed into thermal (heat) energy and 95 J is transformed into sound energy.

   Draw a Sankey diagram to show the energy transfers. **M1**

Grading tip

Remember that when you draw a Sankey diagram, the amount of energy leaving the system must be the same as the energy that enters it.

2.3 Understanding thermal energy

In this section: **P1**

Key terms

Free electrons – electrons within the atom of a metal that are shielded from the nucleus and are free to move.

Density – the amount of matter that occupies a specific volume; something heavy that takes up a small space has a higher density than something that weighs the same but takes up more space.

Vacuum flasks keep liquids hot by minimising heat loss due to conduction, convection and radiation.

When you touch a metal gate on a winter morning it feels cold. This is because the thermal (heat) energy from your hand is being transferred to the metal.

Scientists need to understand how thermal energy is transferred so that they can design useful products. For example, a vacuum flask is used to keep liquids hot (or cold) by preventing heat transfer. Saucepans are made out of stainless steel so that they transfer heat quickly from the cooker to the food inside the pan.

Thermal energy can be transferred in three ways: **conduction**, **convection** and **radiation**.

Conduction

You know that all substances consist of atoms. In a solid, the atoms are close together; in a liquid, the atoms are more spread out; and in a gas, they are very far apart.

Unit 1: Page 6 shows the structures of solids, liquids and gases.

The atoms in substances vibrate. When a substance is heated, its atoms vibrate more. If one end of a metal bar is heated, the other end eventually gets hot. You may have noticed this if you've used a metal spoon in a saucepan. This is because the heat is transferred from atom to atom through vibrations; this is called conduction. Solids conduct thermal energy better than liquids or gases because the atoms are closer together in solids. Metals are the best conductors of heat because they also have **free electrons** that transfer thermal energy.

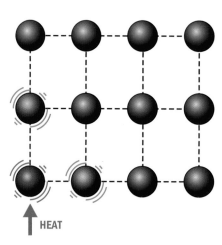

HEAT

A non-metal transfers heat through the vibration of its atoms. These are poor conductors of heat but good insulators.

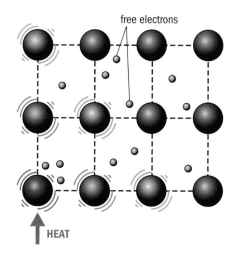

free electrons

HEAT

A metal transfers heat through the movement of free electrons as well as through the vibration of its atoms.

Activity A

Imagine heating up some baked beans in a metal saucepan. You stir the beans with a metal spoon. Using the idea of conduction, explain why the spoon gets hot.

Convection

The atoms in liquids and gases are free to move around because they are joined by only weak forces. Thermal energy can be transferred because of the movement of these atoms. This is called convection.

Convection allows a radiator to heat a whole room rather than just the air immediately surrounding it. This is shown in the diagram on the right.

Activity B

Now imagine heating up some soup. Even if you don't stir it the whole pan of soup eventually heats up. Use the idea of convection to explain why.

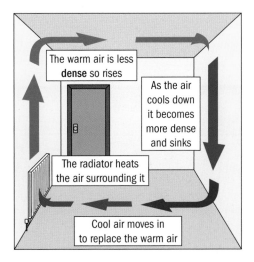

The warm air is less **dense** so rises

As the air cools down it becomes more dense and sinks

The radiator heats the air surrounding it

Cool air moves in to replace the warm air

This room is being heated by a radiator; the convection current is shown by arrows.

Radiation

Radiation is the third way of transferring thermal energy. The heat is transferred by infrared light waves. It does not involve atoms. Radiation is absorbed by dark dull objects and is reflected by shiny substances such as metals. You may have seen an athlete wrapped in a shiny blanket after a race – this prevents the body temperature from dropping too quickly.

Did you know?

The warmth that we get from the Sun is from infrared radiation, coming from the Sun almost 92 million miles away.

 BTEC ## Assessment activity 2.3

1 Explain why the whole of a pan of soup gets hot, even if you don't stir it. **P1**
2 Work in groups of three. Produce a leaflet showing different ways that we use heat transfer in the home and the workplace. **P1**

Grading tip

Remember that solids transfer thermal energy by conduction and liquids and gases by convection. Radiation is light and doesn't need a medium to transfer thermal energy.

 PLTS

Team workers and Self-managers

Producing the leaflet in your groups will help you develop team-working and self-management skills

2.4 Catch that energy

In this section: P1 M1 D1

The cost of energy is going up and our non-renewable energy resources are going down. Minimising loss of energy is becoming important for all of us. Also, in generating the energy that we use, carbon dioxide gas is given off, which is thought to be responsible for making the Earth warmer. This means that reducing energy loss is good not only for our pockets but also for our planet.

The red areas in this thermal image of a house show where most heat energy is being lost.

Activity A

Look at the thermal image of the house. Identify which areas of the house are losing energy.

Heat is lost from our houses mostly through the walls and roof, and to a lesser extent through the doors, floor and windows. The diagram below shows how energy can be saved.

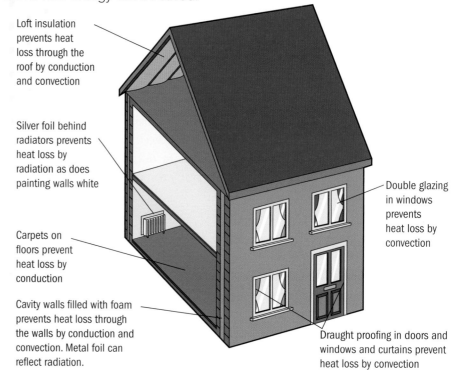

Loft insulation prevents heat loss through the roof by conduction and convection

Silver foil behind radiators prevents heat loss by radiation as does painting walls white

Carpets on floors prevent heat loss by conduction

Cavity walls filled with foam prevents heat loss through the walls by conduction and convection. Metal foil can reflect radiation.

Double glazing in windows prevents heat loss by convection

Draught proofing in doors and windows and curtains prevent heat loss by convection

Methods of insulating a house.

Grading tip

Remember that the criteria P1, M1 and D1 need to relate to each other. So when you are planning the investigation P1, make sure that it relates to minimising energy in the home or workplace. Make sure you include experiments on conduction, convection and radiation.

BTEC Assessment activity 2.4 P1 D1

1 Investigate your own house. List the methods that are used to minimise energy loss and explain how they do this. P1 D1

2 What else might you do to minimise energy loss from your house? D1

   Hint: Think about the different ways heat can be transferred.

Kevin Wright

Principal Manufacturing Engineer, Astrium Ltd

I am an engineer working in the UK's Space Industry and I'm involved with production of electronic circuits which will be fitted in a satellite to work in space.

My responsibilities include:

- creating manuals for production personnel to follow as they build the electronic circuits, including instructions on using equipment safely
- making sure any chemicals we use have an up-to-date COSHH certificate (Control of Substances Hazardous to Health). We have to tell people if a material is hazardous and what to do if they come into contact with it
- making sure the circuits we produce meet the high standards necessary for them to be used in space.

I like working with end products which will actually go into space. I enjoy my work because it can affect our everyday lives. Our satellites can help climatologists better understand our environment by observing climate change, or can help improve global communications and the quality of television broadcasts from space.

Our work has to meet high standards set by external bodies such as the European Space Agency.

When we make electronic circuits we use very thin gold wires to electrically connect microchips to the rest of the circuit. These wires are thinner than human hair but have to be strong enough to survive huge forces and vibrations during rocket launch once the circuit is inside a satellite.

Each wire is tested by pulling it with a special machine to make sure that it won't break.

Think about it!

- A new machine arrives from America but is only wired to connect to their 120 V mains supply. What would you do?
- You have installed a new component-cleaning process but the chemical it needs doesn't have a COSHH certificate. What would you do?

2.5 Efficiency

Key terms

Input – the energy that goes into a system.

Output – the energy that goes out of the system.

Tungsten filament light bulb – the standard type of light bulb in which the filament (the tightly curled wire that glows) is made out of the metal tungsten.

Often, a lot of the energy that goes into a system is wasted, mainly as heat. To save energy and money, electronics manufacturers are developing appliances that make better use of energy and therefore waste less. We describe these as energy efficient. One successful example is the energy-saving light bulb.

The energy that is usefully used by an appliance is given by the **efficiency**, which we can calculate using this equation:

$$\text{efficiency (\%)} = \frac{\text{useful energy output from the system}}{\text{total energy input to the system}} \times 100\%$$

The efficiency is usually given as a percentage, so it varies from 0 to 100%. Maximum efficiency is indicated by 100%, meaning that all the energy **input** is converted to useful energy **output**.

For example, petrol engines in cars transfer only 30% of the chemical energy in the fuel to kinetic energy used to move the car. Electric cars are more efficient.

Energy-saving bulbs waste up to 75% less energy through heat than standard tungsten filament light bulbs.

Activity A

Write down the ways in which a petrol-engine car wastes energy.

Worked example

The energy input per second to a desk lamp with a standard tungsten filament light bulb is 100 J and the output light energy (useful energy) is 5 J. Energy expressed as joules per second is actually the power, which has the unit of watts and the symbol W.

How efficient is the lamp? Give your answer as a percentage.

Using the equation above:

$$\text{efficiency} = \frac{5}{100} \times 100\% = 5\%$$

This lamp is only 5% efficient. Where do you think the other 95% of the energy goes?

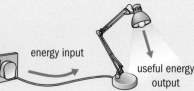

energy input

useful energy output

Activity B

Work out the efficiency of a fluorescent lamp if the useful energy given out each second is 60 J. Assume that it has the same energy input as the tungsten filament lamp of 100 J.

Which is more efficient?

Saving our world's energy resources

There are many sources of energy. They can be divided into two types: **renewable** and **non-renewable**. Renewable energy sources are sources that will never run out if we continue to replenish them. Non-renewable energy sources cannot be replaced once they have been used.

Did you know?

UK businesses waste £8.5 billion worth of energy every year.

Type	Source	Energy	Uses
Non-renewable	Fossil fuels (remains of dead plants and animals that died millions of years ago)	Thermal energy obtained by burning oil, natural gas and coal	Powering vehicles, heating homes, generating electricity
Non-renewable	Nuclear	Thermal energy given off during the splitting of atoms	Generating electricity
Renewable	Wind	Kinetic energy transferred to wind turbines	Generating electricity
Renewable	Biofuels	Crops are fermented to make ethanol. This is burned to give thermal energy	Powering cars
Renewable	Sun	Thermal energy captured by solar panels	Heating water in homes, generating electricity

Energy assessors calculate an energy rating of your home – you need this if you want to sell.

BTEC Assessment activity 2.5

1 An electricity company has designed a power station using the potential energy in water from hill reservoirs. The average input is 800 MW and the average output is 200 MW. What is the efficiency? **P2**

You are a member of a committee set up by the government to investigate options for different energy sources.

2 Work in groups of four with each person choosing a different energy source. Then undertake research to find out the efficiency, cost, amount of energy that can be produced and the advantages/disadvantages of each energy source. **M4**

3 Present your findings to the rest of the group, then discuss which energy sources are most suitable to meet the country's energy needs. **M4**

4 Produce a leaflet that outlines your recommendations with the reasons why. **M4**

Functional skills

ICT

You could use your ICT skills when making your leaflet.

Grading tip

To meet **P2** you will need to calculate the efficiency of the energy transformations you investigate in **P1**. Remember that the useful energy output will always be less than the input energy.

2.6 Understanding waves

In this section: **P3**

Key terms

Displacement – how far the wave is disturbed from its rest position.

Oscillation – a complete to and fro movement; this could be going up and down, or sideways.

We are surrounded by waves, but mostly invisible ones. What other waves can you think of?

A beach is an obvious place to see waves in the sea. But this isn't the only place you'll find waves – they are all around us. You are using waves to read this sentence. Light waves are reflected from the book into the retinas of your eyes, where the information is turned into electrical signals which are sent to your brain from your eyes. Sound waves carry music from a radio to your ears.

Activity A

List three examples of waves that you have used today.

What is a wave?

The diagram on the left shows a wave. The properties of a wave are described using the terms **amplitude**, **wavelength**, **frequency**, **period** and **speed**.

The amplitude of a wave is the maximum **displacement** from its fixed position. This is also called its **equilibrium position**. The wavelength of the wave is the distance between two identical points on the wave as it repeats itself. The **period** is the time for one complete **oscillation**.

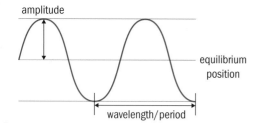

Diagram of wave showing amplitude, wavelength and period.

Frequency and speed

The frequency of a wave is the number of complete oscillations it makes in one second. The unit of frequency is the hertz (Hz). Because many waves oscillate very quickly, frequency is often given in kilohertz (kHz), which means 1000 waves in one second, or even megahertz (MHz), which means one million waves in one second.

The frequency and period of a wave are related by the equation:

$$\text{period} = \frac{1}{\text{frequency}}$$

so the period decreases as the frequency increases.

Worked example

The frequency of microwaves used by a microwave oven is 2000 MHz. What is the period of the microwaves?

First remember to change the frequency prefix to a standard number. 2000 MHz is 2 000 000 000 Hz.

$$\text{period} = \frac{1}{\text{frequency}} = \frac{1}{(2\,000\,000\,000)}$$

period = 0.5×10^{-9} seconds (half of a billionth of a second)

Did you know?

Light travels at a speed of approximately 300 million metres per second. This value is true for all types of light. We write this as 3×10^8 m/s. Sound waves are much slower – in air they travel at about 330 m/s.

The speed of a wave, which is how quickly it travels along, depends on both the frequency and wavelength. It is given by the equation:

$$\text{speed} = \text{wavelength} \times \text{frequency}$$

The speed will be in metres per second (m/s), wavelength in metres (m) and frequency in hertz (Hz).

Case study: Keep your distance

Alan works as an engineer for a car company. He is helping to design a safety system that uses light waves to work out how far away the car in front is. If you are too close to the car in front, the system slows your car down automatically. In an emergency it would automatically apply the brakes for you.

Can you think of another situation in which this technology would be useful?

 ## Assessment activity 2.6

1   In groups, discuss how you could model the movement of a wave.
2   Construct your model or role play it to the other groups.

Grading tip

Remember that the longer the wavelength the smaller the frequency. When calculating the speed, make sure that you change the prefixes (e.g. the 'M' in MHz) to numbers, otherwise your answers will be wrong!

Light can be used to sense the distance between a car and other objects.

2.7 Understanding the electromagnetic spectrum

In this section: P3 P5 M3 D3

Key term

Electromagnetic spectrum – the different types of electromagnetic radiation, arranged in the order of frequency and wavelength, from radio waves to gamma rays.

The electromagnetic spectrum

Electromagnetic radiation is a wave. The colours of the rainbow are just the small range of radiation that our eyes can detect as visible light. Electromagnetic radiation outside this range is invisible to humans. All of the different wavelengths and frequencies of radiation, from radio waves, through visible light to X-rays and gamma rays, form the **electromagnetic spectrum**.

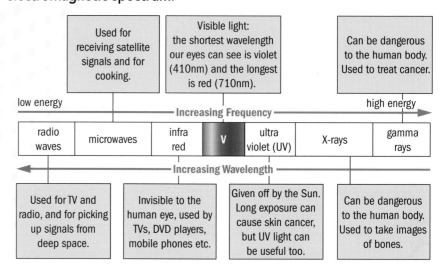

radio waves	microwaves	infra red	V	ultra violet (UV)	X-rays	gamma rays

low energy ← Increasing Frequency → high energy

← Increasing Wavelength

Used for receiving satellite signals and for cooking.

Visible light: the shortest wavelength our eyes can see is violet (410nm) and the longest is red (710nm).

Can be dangerous to the human body. Used to treat cancer.

Used for TV and radio, and for picking up signals from deep space.

Invisible to the human eye, used by TVs, DVD players, mobile phones etc.

Given off by the Sun. Long exposure can cause skin cancer, but UV light can be useful too.

Can be dangerous to the human body. Used to take images of bones.

The electromagnetic spectrum and some of its applications.

Activity A

Put these types of electromagnetic radiation in order of increasing wavelength: visible green light, X-rays, microwaves, ultraviolet.

Visible light is measured in nanometres. A nanometre is a billionth of a metre. All radiation that makes up the electromagnetic spectrum travels at a speed of about 300 million metres per second.

Activity B

Give one application each for microwaves, gamma rays and infrared light.

Understanding waves in communication

Many electronic devices use electromagnetic waves in some way. Some require wires to work, some don't. The table on the next page shows some examples of wireless and wired communication.

The colours of a rainbow are just a small part of the electromagnetic spectrum.

Method of transmission	Examples	Advantages	Disadvantages
Wires	Cable TV, Internet and phone calls; infrared is sent through optical fibres	Excellent picture quality Can only be intercepted by physical access	Difficult to use as and where you want Cables must be laid
Wireless	Wireless keyboards, mice and remote controls; all using infrared	TV and games consoles can be controlled from a distance Keyboards/mice can be placed in a suitable place without having to rearrange wires	Phones, keyboards, mice: heavy battery use
	Wireless phones, laptops; all using radio waves	Laptops can be used in different parts of the house	Laptops: signals can be intercepted remotely
	Satellite; use of microwaves to transmit TV and mobile phone communication	Can cover large distances Can carry a lot of TV stations; TV, radio and Internet can be accessed in remote areas	There is a delay in communication Very expensive to set up

Did you know?

The honey bee can see ultraviolet light. Snakes such as the viper can see infrared.

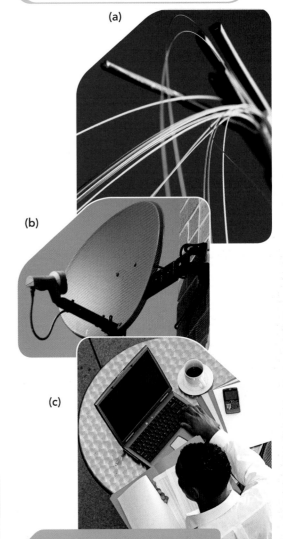

(a)

(b)

(c)

Communication:
(a) cable using optical fibres,
(b) satellite dish,
(c) Wi-Fi wireless connection.

BTEC Assessment activity 2.7

You have just started work as a salesperson at a telecommunication company. You are researching the market.

1 Find out which parts of the electromagnetic spectrum are used for communications. **P5**

2 Think of two types of communication devices that you have used today that rely on electromagnetic radiation to work. **P5**

3 Working in pairs, one of you should take the role of trying to sell 'with-wire' technology, using a specific example from the table. The other should try to sell wireless technology, using a different example. After the role play summarise what you have found out by writing an advert. **M3 D3**

Grading tip

In order to meet **P3**, make sure that you can describe all the areas of the spectrum that are covered on these pages. To meet **P5** you need to describe both wireless and wired communication.

2.8 Understanding radiation

In this section: P4 M2 D2

Key terms

Nucleus – the inner part of the atom, where protons and neutrons are found.

Radiation – energy spreading out, as carried by electromagnetic radiation, or carried by a particle.

Ionising radiation – radiation that can remove electrons from atoms, causing the atom to become positively charged.

Non-ionising radiation – radiation that does not remove electrons from atoms, e.g. microwaves or infrared.

When we think of radiation, we usually think of things like nuclear bombs and radiation leaks, which are uncontrolled radiation and are extremely dangerous. However, medical physicists can use controlled radiation to kill cancer cells in tumours.

A stable **nucleus** has the right number of protons and neutrons so it does not break apart. If the number of protons and neutrons changes, the nucleus becomes unstable and emits **ionising radiation**. This **radiation** has three types: alpha, beta and gamma. They differ in how ionising and how penetrating they are, and how they react to magnetic or electrical fields.

Non-ionising radiation is radiation from the low frequency end of the electromagnetic spectrum: radio, microwave, infrared and visible light.

Alpha (α) radiation

- Alpha radiation consists of α particles. These are helium nuclei, each having two protons and two neutrons, and a charge of +2. When α particles hit another substance, e.g. air, they knock electrons off the particles they hit. This leaves the particles with a positive charge; the particles have been ionised. Alpha radiation is *highly ionising*. (If you swallow α particles, they cause serious damage because they ionise DNA.)

- α particles are large compared with electrons and protons so they cannot penetrate far into a material. For example, a few centimetres of air or a sheet of paper will stop α particles. α particles are *weakly penetrating*. This means there is little chance of α particles getting into the human body through the skin.

- Because α particles have a positive charge, they will be attracted to a negatively charged plate.

Activity A

Describe alpha radiation.

Beta (β) radiation

- Beta radiation consists of fast-moving electrons that have been given off (emitted) by unstable nuclei. If they collide with an atom, they can knock off an electron and ionise the atom. Because β particles are small, they don't ionise as much as α particles. They are *moderately ionising*.

- Because they are less strongly ionising, β particles can travel further than α particles. They can travel through a few millimetres of aluminium before they are stopped. They are *moderately penetrating*. This makes them dangerous if they come into contact with living things.

- Because β particles are electrons, which have a negative charge, they will be attracted to a positively charged metal plate. They are deflected more than α particles because they are lighter.

Gamma (γ) radiation

Gamma radiation is high-energy electromagnetic radiation. It has a very short wavelength and is emitted from unstable nuclei.

- Electromagnetic radiation does not have charge so it is difficult for γ radiation to ionise particles. But because it has very high energy it can still ionise matter. It is *weakly* ionising.

- Because γ radiation is weakly ionising it can travel large distances. It can pass through aluminium and even several centimetres of lead. It is *highly penetrating*. This means that γ radiation is extremely dangerous, both outside and inside the human body.

- Because it does not have a charge, it is not deflected by electric or magnetic fields.

Safety and hazards

We are exposed to tiny doses of radiation in our everyday lives. This is called background radiation. Some of this radiation comes from the food we eat, in the form of radioactive potassium.

Wherever there is a danger of being exposed to higher levels of radiation, especially in the workplace, you will see this symbol.

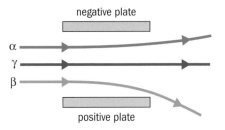

The effect of an electric field on α, β and γ radiation.

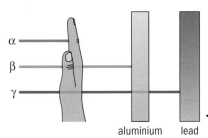

Penetration of α, β and γ radiation.

Activity B

Describe *three* useful applications of radiation.

Using ionising radiation

Alpha radiation is used in smoke alarms. A weak alpha source ionises the air and causes a small current to flow. If smoke gets into the detector, the current reduces and the alarm sounds.

Beta radiation is used to control the thickness of paper during production in a paper mill. A Geiger counter measures how much radiation passes through the paper. This is used to control the pressure on the rollers that the paper passes through.

Gamma radiation is used to kill bacteria in food so that the food does not go bad. It is also used in the treatment of cancer.

BTEC Assessment activity 2.8 P4 M2 D2

You are working for a science charity to produce a poster or presentation on radiation.

1 Describe two properties each of α, β and γ radiation and give an application of each. **P4**

2 Describe the applications and dangers of radiation. One of you should investigate applications and the other should investigate the possible dangers in using these applications. Use a computer to prepare a poster or some presentation slides. **M2 D2**

Grading tip

For **P4**, make sure you can describe the nature of the different types of radiation and their absorption properties. For **M2**, don't forget to include applications of non-ionising radiation (pages 50–51). To get **D2**, make sure that you relate the ill effects to each type of ionising and non-ionising radiation.

2.9 Understanding electricity

Key terms

Series – in a series circuit the components are connected in a line, end to end, so that current flows through all of them one after the other.

Parallel – in a parallel circuit the components are in separate paths and the current is split between the paths.

Imagine your world without electricity: no lights, no television, no central heating, no shower. It would be a strange place.

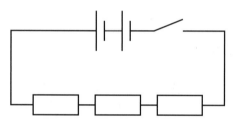

Series circuit.

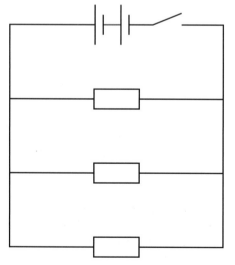

Parallel circuit.

What is electricity?

Electricity is the flow of electrical charge. The charge could be positive and negative ions, as inside the battery of your mobile phone, or negatively charged electrons, as in the wire of your DVD player. When charge flows we say there is a **current**. Electrical energy allows a current to flow in a circuit. For example, when your DVD player is connected to the mains, it forms a circuit. A measure of the energy carried between two points in a circuit is called **voltage** or **potential difference (pd)**. The two points could be each end of the bulb in the circuit shown.

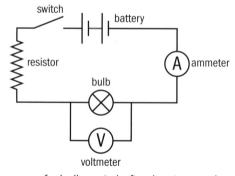

An electric circuit diagram of a bulb, switch, fixed resistor, voltmeter and ammeter.

We use a **voltmeter** to measure voltage and an **ammeter** to measure current. The way we connect the meters is important. A voltmeter is always connected in **parallel**. An ammeter is connected in **series**. The picture above shows a typical circuit diagram of a light bulb with the symbols of the different components.

Case study: Fault finder

Sophia is a technician at an electronics company. Today she is repairing a DVD player that seems to have no power. She wants to measure the voltage and find out if there is a break in the circuit.

How could she do this?

Ohm's law

Ohm's law describes how a current and voltage behave in metals. This law can be written as:

$$\text{voltage} = \text{current} \times \text{resistance}$$
$$V = I \times R$$

In a practical you can use an ammeter and a voltmeter to check the values you calculate for a circuit using Ohm's law.

High levels of current can be dangerous. In the laboratory we use only low levels such as a thousandth of an amp (mA) or a millionth of an amp (µA). All electrical devices, such as televisions, hairdryers and light bulbs, have **resistors**. These limit the current that flows through the components, as they could be damaged if too much current flows through them.

Worked example

1 **What is the voltage across a 300 W speaker if the current flowing is 0.01 A?**

voltage = current × resistance
    = 0.01 × 300 = 3 V

2 **If the voltage across the speaker was 9 V, what current would be flowing?**

voltage = current × resistance

so current = $\dfrac{\text{voltage}}{\text{resistance}}$

= $\dfrac{9}{300}$ = 0.03 A

Activity A

What meter is used to measure current? How should the meter be connected in order to measure the current through the circuit? Draw a diagram to show this.

Safety and hazards

Electrical current is dangerous as it could cause the heart to stop working. You can also get burns from where the current enters and leaves the body. Before working with electrical equipment make sure you ask for a safety briefing from your supervisor.

Table: Units and symbols of electrical properties

Electrical property	Unit	Symbol
Voltage	volt	V
Current	ampere or amp	A
Resistance	ohm	Ω (Greek symbol omega)

Functional skills

Mathematics

Correctly obtaining the value of the current involves identifying the problem and selecting the correct mathematical method.

Grading tip

Make sure that when you perform electrical calculations you change the prefix (e.g. 'm' in mA) to numbers.

BTEC Assessment activity 2.9 P6 P8

You are an electrician. Part of your work is to make sure that electrical circuits are working correctly. To do this you must understand Ohm's law and how to use measuring instruments.

1 Draw the symbols for a voltmeter and an ammeter. **P6**

2 This question uses Ohm's law. If a resistor in a circuit is 1500 Ω, what is the current through it if it is connected across a 1.5 V supply? **P6**

3 Using a circuit diagram, show how you could confirm the current and voltage readings in question 2 by using the correct measuring instruments. **P8**

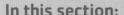
2.10 Producing electrical energy – batteries

Activity A

Write down three appliances you have used today that are powered by batteries. Were the batteries rechargeable or non-rechargeable?

CAUTION DO NOT DISASSEMBLE
ATTENTION NE PAS DEMONTER
BATTERY TYPE / TYPE DE BATTERIE
LITHIUM-ION 3.7V / 700mAh
MADE IN INDONESIA V30145-K1310-X250

You have probably used something powered by a battery today – your alarm clock or watch, mp3 player or a remote control.

Science focus

A battery produces electricity by the chemical reactions that take place inside it. The chemical inside a battery is called an **electrolyte**. Batteries can be rechargeable or non-rechargeable.

If you look at a battery you will see two terminals. One is a positive terminal, called the **anode**. The other is a negative terminal, called the **cathode**. In some batteries, such as AA, C and D batteries, the ends form the terminals.

Table: Examples of different types of batteries and where we use them.

Appliance	Battery material	Battery type
Mobile phone	Lithium ion	Rechargeable
Modern car	Lithium acid	Rechargeable
Very old car	Lead acid	Rechargeable
Laptop	Lithium ion	Rechargeable
Television remote control	Alkaline	Non-rechargeable
Watch	Lithium-iodide	Non-rechargeable

The electricity produced in batteries is described as **direct current (dc)**. Direct current flows in one direction and does not change direction.

Non–rechargeable batteries

A battery is made up of a number of **cells**. For example, the popular AAA battery is a single cell (although we call it a battery) that supplies 1.5V. The flat PP3 is a battery that consists of six 1.5V cells connected in

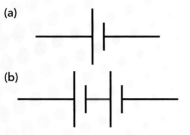

(a)

(b)

Symbols for **(a)** a cell and **(b)** a battery.

series. It therefore supplies 9V. Non-rechargeable batteries contain what are called dry cells. A dry cell is shown on the right.

A chemical reaction takes place between the electrolyte and the anode which produces electrons at the anode. These electrons want to flow towards the cathode where there aren't many electrons, but the salt bridge is in the way. When a wire is placed across the electrodes, the electrons flow through it from the anode to the cathode generating current. The chemicals are gradually used up, until there are none left to produce charge. The battery then stops working.

We use non-rechargeable batteries for items that need little current, such as remote controls, or for things that we don't use often, such as an emergency torch. These batteries are cheap and don't lose their energy (called self-discharge) as quickly as rechargeable batteries. However, they do contain chemicals that are harmful to the environment if they go into landfill.

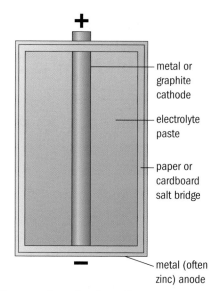

metal or graphite cathode

electrolyte paste

paper or cardboard salt bridge

metal (often zinc) anode

Cross-section of a dry cell.

Activity B

What kind of electricity is produced by a battery? Why does it have this name?

Rechargeable batteries

Cells in rechargeable batteries are called **secondary cells**. These batteries are mostly used in portable items that are used regularly, such as mobile phones and laptop computers. The chemical is used up as the battery is used, but in this case the process is reversible. The battery can be recharged by applying an electric current to it, which reverses the chemical reactions that take place during its use.

Safety and hazards

Dead batteries must be disposed of safely. Some batteries contain toxic mercury that may leak into the environment. Leaking batteries may also cause burns if the acid inside comes into contact with skin. In some areas of the UK, all types of battery can be recycled.

 Assessment activity 2.10

1 Explain the difference between a rechargeable and a non-rechargeable battery. Give five examples of uses of each.

2 Draw a labelled diagram of a primary cell. **P6**

3 Discuss with a partner the advantages and disadvantages of rechargeable and non-rechargeable batteries. **P6**

Grading tip

To meet part of the grading criterion for **P6**, make sure that you include a diagram for the primary cell. To get all of the **P6** criterion you need to also describe another way of generating electricity.

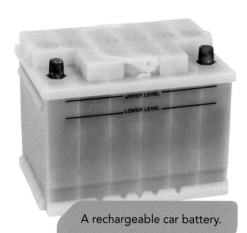

A rechargeable car battery.

2.11 Producing electrical energy – non-renewable sources

In this section: P6 M4

Key terms

Non-renewable energy sources – energy source that we cannot replace, for example, fossil fuels.

Mains electricity – electricity that comes into our homes and places of work. The voltage is normally 230V and the frequency is 50Hz.

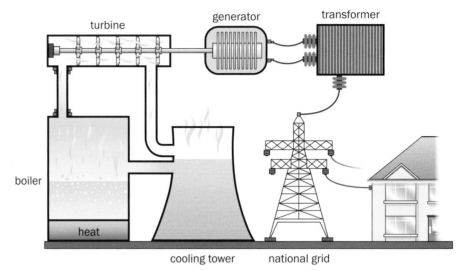

Electricity generation and distribution.

Most power stations produce electricity by heating water to create steam. This steam is used to turn turbines which then rotate a generator to produce electricity. The electricity is then sent to our homes via the national grid.

The water is often heated by **non-renewable energy sources**.

Fossil fuels

In many power stations the non-renewable sources of energy are in the form of fossil fuels: oil, coal or gas. The efficiency of most fossil fuel power stations is only about 30%, although the efficiency of newer ones may be as high as 50%. When fossil fuels burn, carbon dioxide (CO_2) is given off, which is a form of air pollution.

Unit 2: See page 46 for how efficiency is calculated.

Nuclear power

In a **nuclear** power station, energy given out during nuclear reactions is used to heat water to create the steam. No burning of fuel takes place. Electricity generated by nuclear power plants does not create CO_2 and is relatively cheap to produce. It does produce radioactive waste.

Producing electricity – ac generators

Electrical generators use induction to supply electricity. The turbine that is turned by the steam created in the power station boilers then rotates a generator which is a large coil of wire between magnets. The magnetic field induces a current in the coil.

The diagrams (next page) show a simple ac (**alternating current**) generator and the output produced (compared with a direct current).

Safety and hazards

Nuclear power stations generate nuclear waste, which is radioactive. It is very dangerous and needs to be stored safely for thousands of years until it is no longer radioactive. People living near nuclear reactors also worry about radioactive leaks that may occur in the running of the plants.

Did you know?

In the UK, almost 79% of the electricity generated comes from fossil fuels and about 5% is generated by nuclear energy. Nuclear power stations are about 30% efficient – similar to those that use fossil fuels.

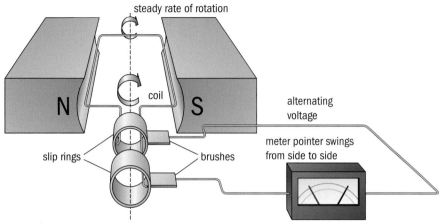

A simple ac generator.

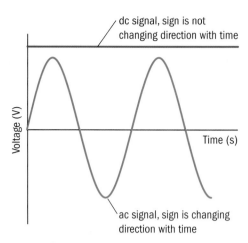

Output from ac and dc generator.

The current generated by the coil is delivered to the circuit via springy metal contacts called brushes which rest on the slip rings. The brushes and slip rings allow constant contact with one side of the coil even though it is rotating. The alternating current is due to the sides of the coil moving through the magnetic field in opposite directions.

Activity A

Write down the name of the device that produces alternating current.

The **mains electricity** supply in our homes is an alternating current with a frequency of 50 Hz. This means the current changes direction 50 times every second.

Case study: Let's get efficient

Mary is a trainee engineer working for an electricity company. Part of her job involves investigating ways to make the electricity generators more efficient.

Make a list of all the areas in a power station where energy may be wasted and the ways that these losses may be reduced.

BTEC Assessment activity 2.11   P6 M4

You must produce a report on nuclear power for an electricity company.

1 Draw a pie chart to show the percentages of electricity that are produced in the UK from fossil fuels and nuclear power. **P6**

2 Discuss with a partner the advantages and disadvantages of fossil fuel power and nuclear power. Which is more efficient and what are the effects on the environment? Put these arguments, along with your pie chart, into a report. **M4**

Science focus

The current produced by an ac generator can be increased by:

- using stronger magnets
- rotating the coil faster
- increasing the number of turns of wire on the coil
- making the coil thicker.

Activity B

List four ways that alternating current can be increased.

Functional skills

English

In discussing nuclear energy and fossil fuels as a way of producing electricity, you will develop both speaking and listening skills, as you present your arguments and listen to the views of others.

Grading tip

When you draw a pie chart, the total must add up to 100%. You will need to include electricity generated by alternative methods, which you can label as 'Other'.

2.12 Producing electrical energy – renewable sources

In this section: P6 P7 M4 D4

Key term

Transformer – a device that changes the voltage of an alternating current without changing its frequency.

The previous pages describe non-renewable energy sources. We can also generate electricity using natural energy such as solar power from the Sun, wind or water power. These sources are described as renewable because they do not run out if we continue to replenish them.

Hydroelectric power

Hydroelectric power stations are one example of the use of renewable energy sources. Water is stored behind a dam, often high up in the mountains. The height of the reservoir provides a source of **potential energy**. When the water is released and flows downhill, the potential energy is converted to **kinetic energy** and the **energy transfer** turns the turbines to generate electricity. Because no heating is required, there is no pollution. Hydroelectric power is thought to be the most efficient method of generating electricity, with nearly 90% efficiency.

Unit 2: Energy transformations are described on pages 38–39.

Activity A

List the energy transformations that take place in a hydroelectric power station.

Hydroelectric power stations are expensive to build but cost little to run and do not cause any pollution.

Building a hydroelectric power station is expensive and some people are also concerned that they cause flooding and spoil the natural beauty of the area.

Wind power

Wind turbines use the kinetic energy of the wind to turn the turbines to produce electricity. No pollution is produced, but people worry that the wind turbines spoil the view of the countryside and about the noise the turbines produce. Although wind is free, wind turbines are expensive to set up and electricity generation depends on the wind – if there is no wind, no electricity is generated. When they do operate, the efficiency is reported to be 35–60%.

Solar power

Solar power can be harnessed using solar cells called **photovoltaic cells**. When the Sun shines on these cells they emit electrons which form a current. The solar panels are expensive, but the energy source is free and no pollution is produced. However, the amount of electricity generated depends on how bright the sunshine is. As with wind power, the efficiency of energy conversion varies. It is reported as 12–25%.

Did you know?

Most of Norway's electricity is produced by hydroelectric power, possible because of all its lakes and mountains. In the UK, only about 5% of electricity is produced in this way.

Activity B

Compare the efficiencies of three renewable methods used to generate electricity and list their advantages and disadvantages.

Getting electricity to our homes and factories

The UK has a network grid of pylon towers linked by copper cables that transfer electrical energy to our homes. The voltage produced at the power station is about 25 000 V. Engineers then increase this voltage to 400 000 V using a **step-up transformer**. Transferring electricity through the national grid at a higher voltage reduces energy losses during the transfer. The higher the voltage, the lower the current becomes so the lower the energy loss. Using thick cables also reduces energy loss because it decreases resistance.

Unit 2: Ohm's law, which describes the relationship between current and voltage, is described on page 55.

We use 230 V mains in our homes and up to 11 000 V in some factories. The voltage from the national grid is reduced using a **step-down transformer**.

Assessment activity 2.12 (P7) (M4) (D4)

1 Explain why the voltage used in the home is different from that used to transmit electricity over the national grid. (P7) (D4)

2 Which equipment is used to increase and decrease the voltage as electricity is transferred from the power station to our homes and factories? (P7)

3 Which type of renewable energy power station would you recommend to be built near your community? Prepare a presentation that describes the advantages and disadvantages including the impact on the local environment. (M4)

Grading tip

For (P7), ensure that you include each stage of electrical generation. Using a diagram will make your description clear. For (M4) consider the efficiencies for both non-renewable energy, such as fossil fuels or nuclear generation, and renewable energy, such as hydroelectric and solar power. For (D4) remember to include "consumer products" in the discussion. These are products such as TVs, washing machines etc.

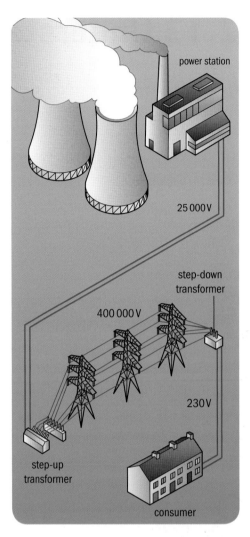

The national grid at work, showing transmission lines and transformers.

2.13 Understanding our Universe

In this section: P9

Key term

Orbit – the path of an object moving through space, such as the path of the Earth as it goes round the Sun.

Did you know?

Stars are so far away that their distances from us are measured in 'light years'. As the name suggests, a light year is the distance travelled by light in 1 year. Light travels 300 million metres in 1 second so a light year is about 10^{13} kilometres.

Case study

Probing Near-Earth Objects (NEOs)

Rachael is a technician at the European Space Agency. She is part of a group of scientists and engineers who are designing the next generation of space probes. Some of these probes will be used to collect samples from space objects close to the Earth, for example asteroids and comets.

What could we learn from analysing these samples?

Stars being born. Each small bulge will eventually form into a collection of planets the size of our Solar System.

The Solar System

The **Solar System** consists of the Sun and all objects that are attracted to the Sun by gravity. These include the eight planets and other objects such as asteroids and meteoroids. The Sun is the brightest star and is the centre of the Solar System. It contains almost 99.9% of all the mass in the Solar System. Because the Sun is so huge, its gravity holds the planets in their **orbits** around it.

Objects in the night sky

If you look at a clear night sky, you will see that it is filled with various objects. With the naked eye you can see the light of thousands of stars, which seem to be arranged in patterns, called constellations. Because the Earth rotates, the stars rise and set like the Sun. You will also see planets. These don't shine their own light but reflect light from other sources, such as the Sun. Because they are close to the Earth, they shine brightly and do not twinkle like stars.

If you are lucky you may see an object with a bright tail. This is likely to be a **comet**. Comets are made from rock, dried ice and frozen gases such as carbon dioxide and methane. They come from outside our Solar System. You may also see 'shooting stars', which are **meteors**. These are bits of dust and rock that enter the Earth's atmosphere. Astronomers have also discovered hundreds of stony objects called **asteroids**, which are also in orbit around the Sun.

Activity A

Which objects in the night sky don't shine with their own light?

The Earth's moon is clearly visible and its appearance changes through the month as it orbits the Earth. With a good telescope you can see that other planets also have moons. Jupiter has 63 moons. One of these, called Io, has active volcanoes on its surface.

Origin of the Solar System

Astronomers believe that the Solar System was formed when clouds of gas and dust collided, because of some sort of explosion that happened in space. Eventually our Sun was formed, together with other objects such as planets. Asteroids and **meteoroids** are believed to be the remains of that cloud.

Geologists have investigated **meteorites** (meteors that have landed on Earth) and estimate they are 4.5 billion years old. The effect of meteoroids and asteroids that hit the surface of the moon is clearly seen as craters, even with the naked eye. Some astrophysicists believe that the Earth was formed by collisions of asteroids and meteoroids.

BTEC Assessment activity 2.13 P9

You are an astronomer working for an observatory. You are invited to a primary school to describe our Solar System to young children.

1   Working in groups of three, construct a model of the Solar System showing the distances of the eight planets from the Sun.

2   In your groups, investigate the theory described above of how the Solar System was formed. Present your results in the form of a poster. P9

Grading tip

For P9: When describing the Solar System, make sure you include objects other than planets; there are many other objects out there apart from planets.

PLTS

Creative thinkers and Team workers

Producing a model of the Solar System will develop your creative skills and presenting your work will help you develop team skills.

The Solar System is made up of the Sun, eight planets, the dwarf planet Pluto, asteroids and comets. Previously Pluto was thought to be a planet.

2.14 Understanding our Universe – how did it all happen?

Key terms

Red shift – light from stars that are travelling away from us comes from closer to the red end of the electromagnetic spectrum than light from the Sun.

Big Bang theory – the theory that the Universe began with an explosion.

Cosmic background radiation – electromagnetic energy that comes from all directions in space and is predicted to have come from the Big Bang.

The Universe is made up of many different objects.

Our Universe

So far we have looked at our Solar System, but our Sun is not the only star in the region. It is one of about 100 billion stars in our **galaxy**, which is called the **Milky Way**. The Milky Way is a spinning spiral disc. On a clear dark night without any pollution or street lights, you can see it as a 'milkyish' light band.

Our Solar System is on the edge of the Milky Way. It takes about 220 million years for our Solar System to orbit the Milky Way, even though it is estimated to be travelling at 100 000 miles per hour. This means that it has just completed one orbit since the first creatures appeared on Earth.

Our galaxy is the second largest in a group of seventeen galaxies. The nearest galaxy to the Milky Way is called M31, the Andromeda galaxy. Beyond this are other clusters of galaxies, with their own stars and planets. These clusters form a shape which is a little like a honeycomb. All these clusters make up what we call the Universe. Astronomers believe that there are about 100 billion galaxies in the Universe.

Activity A

What is the name of our galaxy? Name one other galaxy.

Our Milky Way galaxy: its shape is a spiral and the Sun is near the edge, as shown in this representation.

Expansion of the Universe

Many astrophysicists believe that the Universe is expanding. You can imagine this as bread with raisins in it rising: the raisins represent the galaxies, moving away from each other as the bread rises. Light coming from galaxies has provided evidence for this expansion.

Light forms a spectrum of wavelength and frequency. The visible part of the spectrum starts with violet and ends with red. The further you go towards red, the longer the wavelength.

Astrophysicists have found that light coming from distant galaxies is shifted towards the red end of the spectrum. The more distant the galaxy, the bigger the shift is. They call this a **red shift**. (It is also known as the **Doppler effect**.) A possible explanation for this red shift is that the galaxies are moving away from us. This suggests that the Universe is expanding.

Unit 2: The electromagnetic spectrum is described on page 50.

The Big Bang theory

According to the **Big Bang theory**, galaxies and indeed the Universe were once a fixed point that then exploded. The theory also suggests that radiation was given off during this explosion, and that this radiation should still be detected today. This radiation is called cosmic background radiation and it was detected in the 1960s. NASA confirmed this discovery in 1992, using its newly built satellite called Cosmic Background Explorer (COBE).

So what next for the Universe? Cosmologists believe that the Universe could follow one of the following paths.

- It could continue to expand for ever.

- The expansion will slow down, but won't quite stop.

- The expansion could come to a complete stop and then reverse until the Universe collapses into a massive black hole (singularity).

Unit 18: See page 316 for more information about black holes.

Did you know?

In the centre of our galaxy (the Milky Way) there is a black hole that is 4 million times bigger than our Sun.

Activity B

What does the Doppler effect tell us about our Universe?

BTEC Assessment activity 2.14 P10 M5 M6 D5 D6

You are being interviewed for a job at a space technology company. You must produce a presentation on space. In your presentation:

1 List the evidence that suggests that the Universe is changing. P10

2 Explain how the evidence indicates that the Universe is changing. M6

3 Evaluate the strengths and weaknesses of this evidence. D6

4 Describe the Big Bang theory of how the Universe was formed M5 ; how sure are you that this theory is correct? D5

Grading tip

To obtain P10, make sure you include the red shift and the COBE as evidence that the Universe is expanding. In attempting M6, remember that you need to *explain* how the evidence you identified for P10 suggests a changing Universe. For D6 you need to evaluate the evidence for and against stating that the Universe is changing.

Just checking

1 What is the difference between rechargeable and non-rechargeable batteries?
2 With the aid of a diagram, describe how an ac electrical generator works. Sketch a graph showing the electrical current that is produced.
3 Sketch the current provided by dc supply.
4 Describe how electricity is brought to our homes.
5 List three ways that heat is lost from a house.
7 What is the name of our galaxy?
8 How many planets are there in our Solar System? Name these planets.
9 Name three types of radiation and give an application of each.

Assignment tips

To get the grade you deserve in your assignments remember the following.

- Make sure that your assignments are written as clearly as possible. Always read them through when you have finished.

- Make sure that, when you plan experiments, you have thought about what kind of results you expect to get and have prepared a table for them. What kind of apparatus is likely to be available? Make sure you plan well, allowing yourself plenty of time.

- Don't forget to include the correct units when solving numerical problems. Always check your calculations before you hand in your work for assessment.

Some of the key information you'll need to remember includes the following.

- Knowing the difference between energy block diagrams and Sankey diagrams – remember the width of each arrow in a Sankey diagram corresponds to the value of the energy transferred.

- Knowing the difference between ionising and non-ionising radiation – remember ionising radiation knocks electrons out of the atoms it comes into contact with.

- When doing work on the electromagnetic spectrum, remember the smaller a wavelength is, the higher the frequency.

- Renewable sources of energy are those that don't run out, for example wind energy and solar energy. Non-renewable sources of energy are ones that will run out, for example fossil fuels and nuclear fuels.

- You may find the following websites useful as you work through this unit.

For information on...	Visit...
the different types of energy, its transfer and uses	Energy revision
the full range of electricity generation technology	Electricity generation
energy-saving measures	Saving energy

3 Biology and our environment

Biology covers all areas of the natural world, which consists of millions of living organisms of all shapes, sizes and functions.

This unit will allow you to understand how these organisms interact and depend on each other. You will learn how evolutionary processes have allowed them to adapt to their environment in order to survive.

You will also be assessed on the effects of our own impact on the environment as humans and the factors that affect our health.

There are many different roles within industry that use biology. For example, a science technician, a medical researcher, a student doctor, a conservation worker, a food scientist, a beauty therapist or a horticulturalist. All these jobs require a range of skills including researching, carrying out practical investigations, reporting and presenting information in a variety of ways.

This unit will also provide you with the knowledge and understanding to be able to complete Unit 6 'Health Applications of Life Science'.

Learning outcomes

After completing this unit you should:

1 be able to investigate the functioning and classification of organisms

2 be able to investigate the impact of human activity on the environment

3 know the factors which can affect and control human health.

Assessment and grading criteria

This table shows you what you must do in order to achieve a **pass**, **merit** or **distinction** grade, and where you can find activities in this book to help you.

To achieve a **pass** grade, the evidence must show that the learner is able to:	To achieve a **merit** grade, the evidence must show that, in addition to the pass criteria, the learner is able to:	To achieve a **distinction** grade, the evidence must show that, in addition to the pass and merit criteria, the learner is able to:
P1 Describe how the functioning of organisms relates to the genes in their cells **See Assessment activities 3.1, 3.2 and 3.3**	**M1** Describe how variation within a species brings about evolutionary change **See Assessment activities 3.2 and 3.4**	**D1** Explain how genes control variation within a species using a simple coded message **See Assessment activities 3.2 and 3.3**
P2 Construct simple identification keys to show how variation between species can be classified **See Assessment activity 3.5**	**M2** Explain how organisms within an ecosystem interact over time **See Assessment activity 3.6**	
P3 Describe the interdependence and adaptation of organisms **See Assessment activity 3.6**		
P4 Carry out an investigation into the impact of human activity on an environment **See Assessment activity 3.7**	**M3** Describe how to measure the effect of human activity on an environment **See Assessment activity 3.7**	**D2** Explain how the environmental effect of human activity might be minimised in the future **See Assessment activity 3.7**
P5 Describe the effect of different internal and external factors on human health **See Assessment activities 3.8, 3.9, 3.10 and 3.11**	**M4** Explain how selected medical, social and inherited factors disrupt body systems to cause ill health. **See Assessment activities 3.8, 3.9, 3.10 and 3.11**	**D3** Describe the social issues which arise as a result of the selected medical, social and inherited factors and the illnesses they cause. **See Assessment activities 3.8, 3.9, 3.10 and 3.11**
P6 Identify the control mechanisms which enable the human body to maintain optimal health. **See Assessment activities 3.12 and 3.13**		

How you will be assessed

Your assessment could be in the form of:

- a leaflet e.g. explaining how genes control cell function
- classification keys
- a flow diagram e.g. explaining interdependence between organisms
- a presentation e.g. on the factors that affect human health.

Liam, 17 years old

I have been studying BTEC First Applied Science at my local sixth form college.

I think this unit looks good as it covers evolution, how we affect the environment and how different factors affect us all. It includes everything you need to help you pass the unit and there is information on the merit and distinction criteria.

There are opportunities to do some practical work. This includes using microscopes, which I like, and I think that I would enjoy this quite a lot. It also gives me an insight into the different jobs and careers that are available in biology when I complete my course.

Catalyst

How much do you know about biology and the environment? Discuss what you know about the following.

1 How many different kinds of organisms are there on Earth and how do they interact with each other?

2 What activities do we as humans do that can affect the delicate balance of the Earth and its atmosphere?

3 What causes illness and disease and how can they be cured or prevented?

3.1 Cells and cell function

Key terms

Tissue – a group of similar cells acting together to perform a particular function.

Epithelial cells – one of several cells arranged in one or more layers that form part of a lining or covering of the body.

Unit 3: See pages 72–73 for more about DNA and chromosomes.

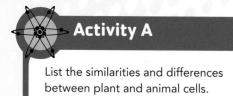

Activity A

List the similarities and differences between plant and animal cells.

How you are made

All life that exists on Earth is made up from **cells**. Cells that make up the body can be compared to bricks that make up a house. Cells are very small so you will need a microscope to see them.

Features and functions of a cell

The table below shows the function of specific features that make up animal and plant cells. Some of these features are common to animal cells and plant cells but others are unique to plant cells.

Cell feature	Function	In animal cell?	In plant cell?
Cell membrane	Partially **permeable** so allows the movement of substances into and out of the cell	Yes	Yes
Nucleus	Contains DNA arranged in chromosomes Controls the activities of the cell, e.g. maintenance, repair and growth	Yes	Yes
Cytoplasm	Jelly-like substance Where chemical reactions of the cell occur	Yes	Yes
Cell wall	Made from **cellulose** Prevents the cell from bursting and gives the cell shape	No	Yes
Vacuole	Contains cell **sap**	No	Yes
Chloroplast	Contains **chlorophyll** Where glucose is made in **photosynthesis**	No	Yes
Mitochondrion	Releases energy for the cell	Yes	Yes

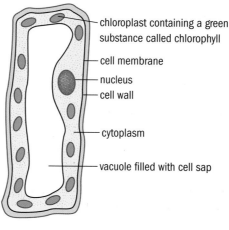

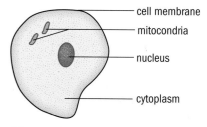

(a) Typical plant cell.

(b) Typical animal cell.

Specialised cells – cells with a difference

As cells develop, they can become specialised to perform specific functions. This is called **differentiation**. Some examples of specialised cells can be seen below:

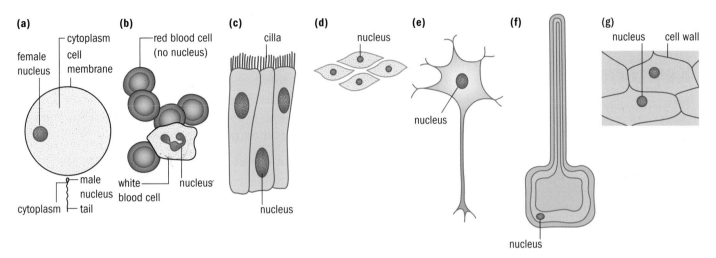

Specialised cells: **(a)** egg and sperm, **(b)** blood cells, **(c)** ciliated epithelial cell, **(d)** muscle cell, **(e)** nerve cell, **(f)** root hair cell, **(g)** onion cell.

Tissues

Tissues are made up from a group of cells which have a similar structure and do the same job in the body. Some examples of tissues in the body are:

- nervous tissue found in the nervous system

- epithelial tissue found in the skin and linings of the inside of the body – different **epithelial cells** perform different functions, e.g. secretion or absorption; some cells have tiny hairs (cilia)

- muscular tissue, which makes up the muscles involved in movement, in the heart or which also line the insides of the body.

Did you know?

Stem cells are cells which are undifferentiated. They have the ability to grow into any type of specialised cell under the right conditions.

Activity B

Give three examples of tissues and where they are found in the body.

Grading tip

You need to use your knowledge of cells to help you to complete **P1**.

BTEC Assessment activity 3.1 **P1**

1 Which part of a cell controls the activities of the cell and what does it contain? **P1**

2 Research further into cell structure and variation between cells using the Internet and books. Make notes on your findings. **P1**

Functional skills

ICT and English

You can use your researching skills in ICT and reading in books to search for more information about cells and their function.

3.2 Our genetic make-up

Key terms

DNA – deoxyribonucleic acid is a nucleic acid that contains the genetic instructions used in the development and functioning of all known living organisms.

Nucleotide – consists of a sugar, a base and a phosphate group; many join together to form DNA.

Genetic code – the instructions in a gene that tell the cell how to make a specific protein.

Genes allow different cells to perform specific functions and are responsible for the variation between all living organisms.

Our genetic make-up

DNA is a chemical found in the chromosomes within the nucleus of nearly all cells. DNA is a string of genes, a bit like the beads on a necklace. Each gene codes for a particular characteristic such as hair colour or the colour of your eyes. The diagram below shows the difference between chromosomes, genes and DNA.

Unit 3: To learn more about genetic coding see pages 74–75.

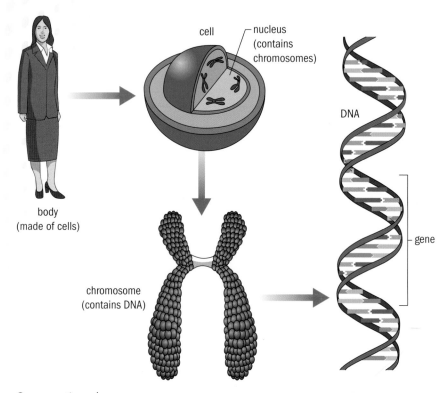

Our genetic make-up.

Did you know?

You have one gene that codes for the eyes but the colour depends on which alleles you inherited from your parents. If you have blue eyes then a recessive allele has coded for them, and if you have brown eyes a dominant allele has coded for them.

 Activity A

Where is DNA found in a cell?

The structure of DNA

DNA is a double-stranded molecule made up of **nucleotides** which then join together to make a polynucleotide. Each nucleotide contains a base of **adenine** (A), **thymine** (T), **guanine** (G) or **cytosine** (C) plus a sugar and a phosphate group.

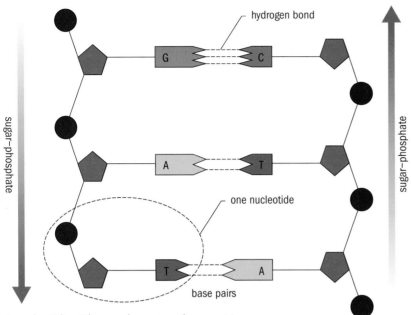

hydrogen bond

sugar–phosphate

sugar–phosphate

G — C

A — T

one nucleotide

T — A

base pairs

A nucleotide with complementary base pairing.

The diagram shows that A pairs with T and that C pairs with G. This is due to a base pairing rule which allows only these two sets of pairings. This is called **complementary base pairing**. The pairs are held together by **hydrogen bonds**.

Different forms of the same gene are called **alleles**. You inherit one allele for each gene from your father and one from your mother. Many characteristics you have are due to alleles. Alleles determine different characteristics and can be **dominant** or **recessive**.

Case study: Working with DNA

Davina works as a public liaison officer for a research institute that is part of the Human Genome Project. She is writing an article for the institute's website. The article is to describe the basics of DNA, chromosomes and genes. It also needs to explain why the genetic code is so important. She wants to show that the goal is knowing where certain genes are and on which chromosome they are found. Scientists and geneticists can then hopefully start to cure or prevent many hereditary genetic diseases.

How would you write the article? Discuss this in small groups.

Grading tip

When you are working towards **P1**, **M1** and **D1**, avoid getting confused with the structure of DNA and the new key words by reading up thoroughly on this tricky topic. You should also practise drawing and labelling DNA to help your understanding.

BTEC **Assessment activity 3.2**   **P1** **M1** **D1**

Work in pairs to complete the following activities.

1   Discuss with your partner the structure of DNA and then write down three key points.

2   Prepare a short presentation on the relationship between DNA, chromosomes and genes. **P1** **M1** **D1**

Functional skills

ICT

You could use a computer to prepare a slide show as the basis of the presentation.

3.3 Genes control the activities of cells

Key terms

Protein synthesis – is the process whereby new proteins are made in living things according to codes (instructions) given by DNA.

Substrate – the material or substance on which an enzyme acts.

Messenger RNA – the form of RNA that carries information from DNA in the nucleus to the ribosome sites for protein synthesis.

Science focus

Enzymes will bind to their **substrate** only if they are the correct shape or, in other words, **specific** to the substrate. This works like a key fitting into a lock. It has to fit in order to open the door just as the substrate needs to fit into the enzyme's binding site for a reaction to occur.

Activity A

Why is it so important that different proteins are made using the instructions from DNA?

The genetic code

The genes on the DNA are like instruction manuals to make proteins for the body and to control cell functions. The order of the nucleotide bases on the DNA molecule forms a genetic code for the order of **protein synthesis**. The table below shows some examples of proteins and their functions.

Protein	Function
Intracellular enzymes	Control processes within the cell by speeding up reactions (catalysing) and breaking larger molecules into smaller ones
Extracellular enzymes	Control process outside the cell in the same way as intracellular enzymes
Structural proteins	Make cell membranes, hair, nails, muscles and haemoglobin (this is the chemical that binds to red blood cells and allows oxygen to be transported around the body)
Hormones	Bring about changes in the body to keep it functioning and regulate bodily processes, for example regulation of blood glucose levels
Antibodies	Help the body defend itself against invading microorganisms

Protein structure

Proteins are made from **amino acids**. Amino acids exist in long chains. These are put together in a different order to make different proteins according to the instructions from each gene on the DNA molecule. The bases of nucleotides code for an amino acid, and it takes three nucleotides to code for each amino acid. This is called a **triplet code**. Each triplet codes for one particular amino acid. For example, the nucleotide base CCT codes for the amino acid called proline.

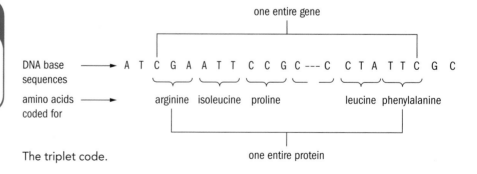

The triplet code.

All organisms on Earth have the same triplets of bases that code for amino acids. That is why we share a high percentage of our genes with other organisms including bananas (50%) and chimpanzees (more than 98%).

How DNA makes proteins

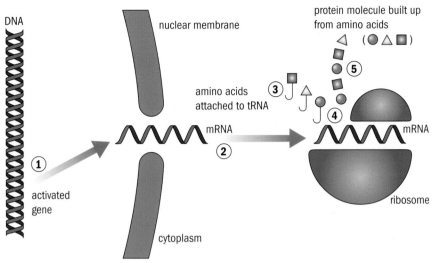

Protein synthesis.

DNA is in the cell nucleus but proteins are made in the cytoplasm. So the DNA has to send a messenger to the cytoplasm to tell it what proteins to make (1). This messenger is called **messenger RNA** (mRNA) (2). The production of mRNA is called **transcription**. The mRNA travels into the cytoplasm to the **ribosomes** where the proteins are then made. Another kind of RNA, called **transfer RNA** (tRNA) (3), then puts the amino acids into the correct order to make a specific protein (4, 5). This process is called **translation**.

Activity B

Working in groups, discuss what you think could happen if something went wrong in the production of these proteins.

Once the protein has been made it folds itself into a shape. The shape allows it to perform a particular function. This is very important as enzymes (large protein molecules) have a specific **active site**. The **substrate** must fit into the active site of the enzyme to allow it to catalyse the reaction.

 Assessment activity 3.3

For each of the following questions, write a list of three key points. Then discuss your answers with a partner.

1 How does a gene code for a protein to be made? **D1**

2 Why do proteins need to be made for the body? Give examples. **D1**

3 How does an enzyme allow organisms to function? **P1**

Did you know?

There are 20 amino acids but they have the ability to make thousands of different proteins. Think about how many keys a piano has and how many tunes can be played. There are lots of combinations. Amino acids act in the same way to make lots of proteins.

Grading tip

For **P1**, you need to discuss briefly how the production of new proteins, coded for by DNA, allows different bodily functions to happen.

For **D1**, you need to research to find out more about the genetic code and protein synthesis. You also need to explain in detail how this leads to differences between individuals.

PLTS

Effective participators

You may find this aspect of genetics difficult so you can be an effective participator, discussing issues of concern with your work and seeking resolutions where needed.

3.4 Variation and evolution

In this section: **M1**

Key terms

Mutation – occurs when a DNA gene is damaged or changed in such a way as to alter the genetic message carried by that gene.

Evolution – the change in the genetic material of a population of organisms from one generation to the next.

Adaptation – changes to an organism that make it better equipped to survive in a particular environment.

Natural selection – the process by which evolutionary changes occur over many years so only organisms which are best adapted to the environment survive.

Variation

Genes are responsible for how each individual is different from everyone else, but some characteristics can be affected by environmental factors. Some characteristics can be affected by both genes and environmental factors.

Influenced by environment	Influenced by genes and environment	Influenced by genes
language	skin colour	eye colour
religion	height	blood type
	weight	

Activity A

Think of some other characteristics that you have inherited from your parents. Have you changed any of them?

Mutations

Mutations are genetic changes in the DNA. These alter the instructions for how the gene makes proteins. The DNA then codes for a different set of instructions involving different amino acids. This means it will make a different protein or even a protein that doesn't function at all. Some gene mutations are harmful and kill the cell. Other mutations may cause the cell to divide uncontrollably, as in **cancer**, causing a tumour to develop.

Unit 3: Find out more about how the structure of different proteins is determined on pages 74–75.

Mutations can happen in individual genes or in a whole chromosome. They usually occur spontaneously. Some environmental factors can also alter your **genotype** (the type of genes you have) to cause mutations to occur. These include:

- ionising radiation, for example X-rays

- ultraviolet (UV) light, for example from the Sun or sunbeds

- chemicals (**carcinogens**) such as benzene in cigarette smoke.

Mutations can occur in a number of ways, as shown in the table on the left.

Mutation	Description
Deletion	Nucleotide is lost
Substitution	One nucleotide is exchanged for another
Duplication	Nucleotide is inserted twice

Table: The three main types of genetic mutation.

Benefits of mutations

Mutations aren't always a bad thing. Sometimes mutations may improve a characteristic and this forms the basis of **evolution**. For example, people who carry the gene for sickle cell anaemia but don't suffer from the disease are more resistant to malaria.

Unit 3: See page 88 for more about sickle cell anaemia.

Adaptation, natural selection and evolution

Many living things change to adapt to their environment. This helps them to survive and breed, passing on their genes to their offspring. As a result, **adaptations** that help the organism to survive become more common in the population. This process is known as **natural selection**. The genetic inheritance of changed characteristics can lead to new species. This is called evolution but it doesn't happen overnight. It involves a gradual change in the features or characteristics of a species and may take thousands of years to happen. For example, icefish have evolved anti-freeze genes and Galapagos finches have evolved different shaped beaks as they have different diets of seeds, flowers, insects etc.

MRSA and natural selection

Superbugs such as MRSA bacteria can evolve very quickly. MRSA is very good at changing to become **resistant** to the antibiotics that are used to kill it or stop it reproducing. By becoming resistant, bacteria are able to survive and reproduce and cause more harm until scientists produce a new antibiotic.

Did you know?

Peppered moths can be either pale and speckled, or dark in colour. They feed during the night and rest on tree trunks in the day but are eaten by birds, if the birds see them. Before 1850 the pale variety was most common; by 1895, after industrial activity made tree bark dark with soot, nearly all of the moths were dark. The pale variety was easily seen by birds and eaten but the dark variety was not seen and survived. So the 'dark' genes were passed on and the species evolved.

Peppered moths.

BTEC Assessment activity 3.4 · M1

Use the information on these two pages and further research using the Internet and books to help you to answer the following questions.

1 Working in pairs, discuss how changes (variation) in DNA can occur. Then make three important statements that summarise your discussions. **M1**

2 What can mutations do for an organism in a beneficial way? How can this help it? **M1**

3 What is natural selection? Give some examples of this. **M1**

Grading tip

For **M1**, use referenced or hand-drawn diagrams to help you show different aspects of variation and adaptation.

Functional skills

English and ICT

You can use your ICT skills to search for diagrams and information on this topic, and use websites and books to practise your reading and selecting information skills.

3.5 Classifying organisms

In this section: **P2**

Key terms

Classification – the systematic grouping of different organisms.

Organism – a living thing that has, or could develop, the ability to act or function independently.

Classification

Biologists use **classification** to sort different **organisms** into groups depending on individual characteristics as shown in the table.

Organism	Characteristics
Viruses	Non-living These are the smallest microorganisms (20–250 nanometres) Have a protein coat and carry genetic information Once inside living cells viruses reproduce using the cell's own mechanisms. They make multiple copies of themselves, causing illness
Bacteria	Sizes vary but very small (up to 10 micrometres in length) Reproduce asexually Can release harmful toxins that make you feel ill
Protoctists	Single-celled or small multi-cellular creatures Larger than bacteria but still microscopic
Fungi	Multi-cellular Most live in or on their food Consist of hyphae (fine threads) which penetrate food and secrete digestive enzymes to digest food that they then feed off
Plants	Multi-cellular Cells have rigid cellulose walls around them Make their own food through photosynthesis
Animals	Multi-cellular No cell walls Highly organised organs, tissues and systems

Identifying organisms

Scientists use **identification keys** to classify and sort organisms with similar characteristics into groups so that they can easily identify them.

Activity A

Look at the examples of identification keys on the page opposite and use them to identify each animal below.

(a) 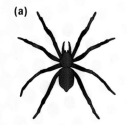 (b) 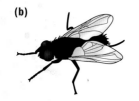 (c)  (d)  (e)

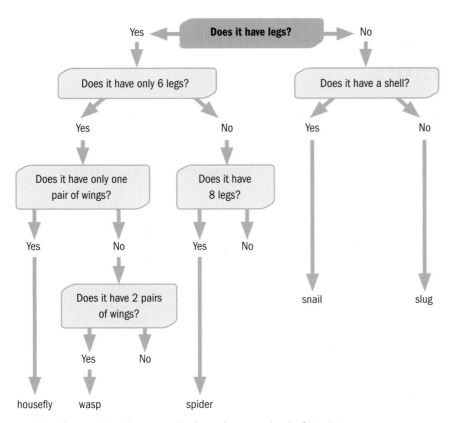

1.	Has legs	Go to 2
	Has no legs	Go to 4
2.	Has 6 legs	Go to 3
	Has 8 legs	Spider
3.	Has 1 pair of wings	Housefly
	Has 2 pairs of wings	Wasp
4.	Has a shell	Snail
	Has no shell	Slug

A numbered identification key.

An identification key that uses the branching method of Yes/No answers to identify organisms.

Activity B

Working in pairs, discuss which you think is the best method of identification to use and why.

Case study: Crime scene investigation

Daniel is a forensic science technician who often works at crime scenes. Part of his job is to collect and identify insects. The insects found on a corpse at the crime scene can help to tell how long the body has been dead. They can also indicate where an item of clothing has been. Daniel carefully collects the insects and takes them back to the laboratory where he identifies them. He can then tell whether their natural habitat is the crime scene or whether they have come from somewhere else.

How could knowing if the clothing has been somewhere else help in a crime scene investigation?

Grading tip

In order to construct your own identification key for **P2** , choose a variety of organisms with different characteristics. Plan out how to construct your identification key using a rough sketch and then draw up your final version.

 BTEC **Assessment activity 3.5** **P2**

1 Select five organisms of your choice and plan how you could create an identification key to group these organisms using:

(a) the Yes/No answer method

(b) a numbered identification key. **P2**

 ## PLTS

Creative thinkers

This activity allows you to generate ideas and explore possibilities in order to construct classification keys.

3.6 Interdependence and adaptation of living organisms

In this section: P3 M2

Key terms

Parasite – an organism which receives benefits from another by causing damage to it.

A tapeworm.

Science focus

If living organisms don't adapt to the environment as it changes then they will more than likely die. It is only the ones that change and adapt that will survive and pass on their genes to their offspring.

Interdependence

Organisms that live within certain ecosystems depend on each other to survive. This process is called **interdependence**. Plants depend on carbon dioxide and water in the atmosphere to produce oxygen and glucose by photosynthesis. If there were no plants then all living organisms including ourselves would die. Interdependence between different organisms within certain ecosystems can involve the following relationships.

- **Parasite and host** – parasites such as a tapeworm depend upon host animals to survive.

- **Predator and prey** – predators such as lions depend on there being enough antelope for them to survive.

Food chains and food webs

Food chains and webs show the organisms present in an ecosystem and how the different organisms are interdependent. If the numbers of one species in the food chain decreases, for example by disease or hunting, the species that normally feeds on that particular organism will need to find food from another source or it may also die.

The organisms within an ecosystem interact over time. For example, after a volcanic eruption the land is initially populated by lichen and moss. This will support a variety of insect life. As the lichen decays it will provide nutrients for larger plants. These will in turn allow larger animals to move into the ecosystem.

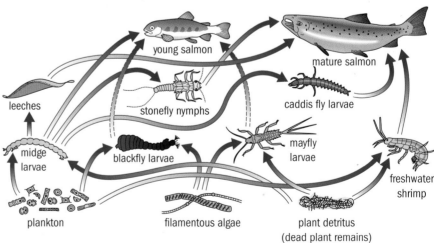

The food web of a salmon. The arrows show the energy flow between organisms and what they eat. The producer is always at the bottom of the food chain or web.

Case study: Conservation and interdependence

Jake works for a conservation agency. Part of his job is to monitor the numbers of species within different ecosystems. Any changes, for example due to disease, hunting or lack of food, may affect the food chain or food web. By noticing these changes, he will be able to identify species which may need protecting. This will help to conserve the whole ecosystem.

If the numbers of certain species started to decrease, what action would Jake need to take and why?

Adaptation

Different organisms have adapted in order to survive in their particular environment by having particular characteristics. The polar bear is one example.

The thick fur of the bear's winter coat and thick layer of fat under the skin provide excellent insulation against the low temperatures

Large feet act like snowshoes and spread the weight of the bear

Small ears further reduce the surface area through which heat can be lost

The fur is covered with oils, which makes it water-repellent and so keeps the skin dry when the bear is swimming in the Arctic seas

The polar bear is a large animal. (A large male, standing on its hind legs, could look in through the upstairs window of a house!) This gives it a relatively small surface area compared with its volume and helps to conserve heat

How a polar bear has adapted to increase its chances of survival in the Arctic.

Activity A

What other animals or plants can you think of that have particular characteristics or **adaptations** that allow them to survive in their environment?

 BTEC ### Assessment activity 3.6

1 Using books and/or the Internet, research how different organisms that live together in a rainforest are dependent on each other to survive. For example, you might look up an animation of a rainforest food web using the Internet to help you. **P3**

2 Create a flow diagram to demonstrate your findings. This could be in a similar format to the food web seen above. **P3**

3 Using the Internet, research how the environment around Mount St. Helens was recolonised following the volcanic eruption in 1980. Draw a flow diagram to show your findings. **M2**

Grading tip

For **P3**, 'describe' means to show what you know about interdependence and adaptation by presenting a number of labelled drawings, pictures or flow diagrams, for example.

For **M2**, annotate the flow diagram to include the reasons why the ecosystem evolved.

 ### Functional skills

ICT

You can use your ICT skills to create a flow diagram once you have researched a food web of your choice.

3.7 Humans and the environment

In this section: P4 M3 D2

Key term

Monoculture – farming only one type of crop over a wide area of land.

Did you know?

100 animal and plant species from the rainforest become extinct each day due to deforestation.

A monoculture of palm oil trees.

Grading tip

For **M3**, think about how human impact has brought about a change in the environment. This can be measured by looking at the changes in the number of species, soil or air quality, or weather patterns over a period of time in a particular place or habitat.

For **D2**, 'explain' requires use of scientific knowledge to provide a detailed account of how environmental damage can be minimised.

The environment is changing and human impact is a major factor contributing to this. The human population is increasing due to better medical care and standards of living. This means that more strain is being put on the planet. More land is needed for farming so natural habitats are lost. More factories and power stations are built so there is more pollution. Other indicators of human impact include living factors, e.g. **monoculture**, and non-living factors, e.g. erosion and acid rain.

Advantages of monoculture	Disadvantages of monoculture
Cheap	Removal of hedges when clearing fields
Uses modern farming methods	If one plant gets a disease the rest will
Mass production of crops to sell	Vital habitats are destroyed as water is drained from fields

Activity A

List five ways you could reduce your own impact on the environment.

Acid rain

Limestone statues and stone walls can become worn by the acid rain produced from sulfur dioxide emissions and nitrogen oxides produced by burning fossil fuels and exhaust emissions. This acid rain also damages some plants and trees.

There are other human factors which have a *positive* impact on the environment, like recycling and conservation.

BTEC Assessment activity 3.7   P4 M3 D2

1 Working in pairs, create a questionnaire that investigates human impact on the environment. What do you think are the four most important questions? Give your questionnaire to 20 people to complete and then discuss the findings. **P4**

2 Using books and the Internet, research how environmentalists work to assess, measure and minimise damage to the environment. Explain how the actions they take are scientifically based on the measurements they have made. **P4 M3 D2**

Unit 11: For more help on answering this question, turn to pages 214–215.

Meg Woodward

STEM
AMBASSADORS
ILLUMINATING
FUTURES

Nationally coordinated by STEMNET

Laboratory Technician, Analytical Testing Ltd

I work for a water-testing company. With a small team of technicians I analyse water for quality and pollution indicators.

Samples come into the laboratory from field study officers who take them from rivers. Large companies, for instance plastics manufacturers, also send us samples to check whether their waste water can safely be released into nearby waterways.

One test we perform is called biochemical oxygen demand (BOD) test. BOD tests are carried out to assess the quality of waste water and water in our rivers and waterways.

Oxygen levels in the water are measured twice a few days apart. A high BOD indicates that there are large amounts of aerobic (oxygen-requiring) microorganisms in the water sample, which means a high level of pollution.

Another test we carry out is the chemical oxygen demand (COD) test. This test is carried out on effluent waters; these are waste waters from industrial processes. Companies have to comply with strict legal restrictions about releasing effluent into the environment.

A COD test requires a sample of the water to be added to a chemical oxidant, for instance potassium dichromate, and a strong acid, such as sulfuric acid. This mixture is then incubated at a set temperature for a set time period to see how much oxidant is used up. The test uses specialist laboratory equipment and requires exacting laboratory technique and keen analytical skills to obtain accurate results. Any mistakes could result in polluted water being released into our national waterways and thus damaging valuable ecosystems.

Think about it!

1 How could a company treat waste water samples found to have a high BOD?

2 What kinds of organisms would you expect to find in a river with a low BOD?

3.8 Social factors that affect human health – drugs

In this section: P5 M4 D3

Key terms

Dependence – the state of relying on or being controlled by something.

Addiction – being abnormally tolerant to and dependent on something that is psychologically or physically habit-forming (especially alcohol or narcotic drugs).

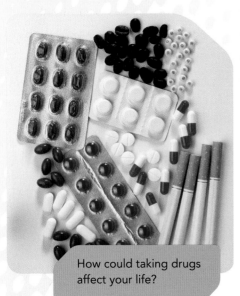

How could taking drugs affect your life?

Safety and hazards

- If several drugs are mixed this can lead to severe side effects and even put the drug user in a coma or lead to death.
- There is a high risk of getting hepatitis and HIV through sharing dirty needles.

There are many different types of drugs. It is very important to know what different types of drugs do to the body, how they are used and misused and the social implications of drug taking.

The use and misuse of drugs

A drug is a chemical substance. It alters how your mind or body works. Drugs can be used as medicines to treat an illness, disease or infection. They are usually prescribed by a doctor or bought over a counter at the chemist. When drugs are abused, they are used for the effects they have on the mind and body and may be bought illegally. Drug abuse can lead to physical, economic and social harm. It can involve using drugs to:

- improve sporting performance, e.g. steroids and amphetamines
- make the person feel good or 'high' – this can lead to **dependence** and **addiction** and have serious effects on your overall health
- enhance the mood, e.g. nicotine, alcohol and caffeine – these are used socially and are legal but can also be addictive and can cause serious health problems.

Activity A

What are the disadvantages of taking illegal drugs? Create a spider diagram or poster to show your findings.

Nearly all drugs induce a level of **psychological dependence** where the user feels like they can't live without it.

Drug	Effect on the body	Long-term health problems
Heroin	Good pain killer and provides a mood 'high' Extremely addictive causing serious social problems	**Physical dependence** – if you stop taking the drug you will get severe withdrawal symptoms including shaking and vomiting You need to take more as time goes on to get the same effect
Cannabis	Depressant hallucinogen	Can cause mental health problems such as schizophrenia and paranoia
Alcohol	Depressant	Kidney damage, stomach ulcers and sclerosis of the liver

Drug	Effect on the body	Long-term health problems
Nicotine	Mild stimulant	Increased risk of lung cancer, emphysema and bronchitis from smoking
Caffeine	Mild stimulant	Can cause physical dependence
Solvents	Depressant	Can include liver, brain and kidney damage if the toxic effects don't kill you first whilst inhaling solvents

Social factors associated with taking illegal drugs

- Losing your job
- Stealing goods or money to pay for your drug addiction
- Losing your family and friends through antisocial behaviour
- If you are pregnant, giving birth to a low birth-weight baby who also has a dependence on the drug
- Risk of an early death due to suicide, infections or an overdose

Diet and exercise

What you eat and drink is important; your diet provides the nutrients and vitamins needed to keep healthy. Exercise is also essential to maintain a steady weight, keep you strong and supple, and prevent diseases including heart disease and some cancers. If you don't balance your calorie intake with what your body needs, you will gain or lose weight.

Other factors that affect human health

- **Sunbathing** without a high factor sunscreen can cause sunburn and lead to skin cancer.
- **Loud noises** can damage your hearing, for example standing too close to the speakers at a live concert or using machinery without wearing ear-defenders.

BTEC Assessment activity 3.8   P5 M4 D3

1   Create a table with two columns. In the first column, list ten different factors that can affect our health. In the second column, show how each factor affects our health and if it is positive or negative. **P5 M4**

2   Discuss the social implications of a close friend taking drugs. How would it affect them? How would it affect their relationship with you? **P5 M4 D3**

Case study

Investigating the misuse of drugs

Lauren is a researcher at the World Health Organization. She is investigating the illegal use of drugs in society and the effects that it has on the individual, their work commitments and their families. Lauren must decide where the study will be based, who will take part and what information she should gather.

List five tips you could give Lauren to help with her research.

Grading tip

For **P5**, a list of the main factors and how they affect health is adequate.

For **M4**, you need to explain how the factor affects our health not just list it.

For **D3**, for three chosen factors that affect health, discuss the social impact for the person involved and implications for society.

Functional skills

English

You can practise your extended writing when explaining the health impact of social factors.

3.9 Microorganisms and their effect on health

In this section: P5 M4 D3

Key terms

Microorganism – any organism of microscopic size.

Pathogen – an agent that causes disease, especially a living microorganism.

Vector – an organism that carries disease from one animal or plant to another.

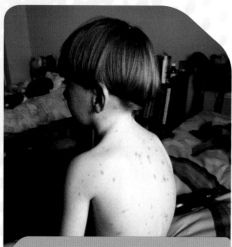

If you have been ill you have probably been affected by microorganisms. Microorganisms can be useful but many make us ill by causing diseases.

Did you know?

Mammals can be vectors for rabies, but the UK is currently free from rabies. Rat fleas are vectors for the bubonic plague. Mosquitoes are vectors for malaria, passing the protoctists that cause the disease from one human to another.

Microorganisms

Microorganisms, or microbes, are microscopic living organisms made up of a single cell. They can form colonies of millions of individual cells gathered together which can be seen with the naked eye. There are four types of microorganisms: **bacteria**, **fungi**, **viruses** and **protoctists**. Many microorganisms are helpful, for example they can aid digestion, but others can cause harm. Microorganisms that cause disease are called **pathogens**. Microbiologists work to identify and provide the correct treatment. They may also treat plants affected by certain conditions caused by microorganisms.

The spread of pathogens between humans

Once a person becomes ill from a microorganism it can be passed on to infect other humans. The ways in which microorganisms spread include:

- **in the air:** coughing, sneezing or breathing releases thousands of tiny water droplets containing the pathogen into the air

- **direct contact:** e.g. shaking hands, kissing or unprotected sex

- **infected food and water:** the presence of sewage in water, especially in poorer countries, or lack of hygiene when preparing food

- **vectors:** organisms that can spread disease, e.g. houseflies can spread salmonella, typhoid or cholera; tsetse flies spread sleeping sickness; and mosquitoes spread malaria.

Activity A

List the four main types of microorganisms and the ways in which they are spread.

Type of microorganism	Disease it can cause	Symptoms
Viruses Reproduce inside the cells of living things	Influenza (flu)	Blocked and runny nose, headache and fever
	HIV	Lethargy, profound sweating, weight loss, diarrhoea, dry cough and swollen lymph nodes
Bacteria Reproduce rapidly in ideal conditions and produce poisonous toxins	Cholera Food poisoning	Both cause extreme diarrhoea and sickness due to toxins produced by large numbers of bacteria
	Tuberculosis (TB)	Coughing, fever, weight loss Damages many organs especially the lungs which can become so scarred a person cannot breathe
	Meningitis	Head and neck ache, rash, vomiting and fever
Protoctists Live in or off humans or animals in return for food and shelter	Malaria	Flu-like symptoms
	Dysentery	Severe diarrhoea, vomiting and ulcers
Fungi Live in or on their food	Athletes foot *Candida* (thrush)	Both cause itching, soreness and inflammation

Safety and hazards

Sexually transmitted diseases, e.g. syphilis, gonorrhoea, HIV and genital herpes, can be passed from human to human during unprotected sex.

Activity B

Use the Internet to find images for each type of microorganism listed in the table. Produce a health information poster displaying the images and the symptoms of the associated disease.

Functional skills

English

You can use your reading and selecting information skills to research and select the information you require and use your writing skills to communicate this.

 BTEC ## Assessment activity 3.9

1 How do microorganisms spread disease? List the various ways and give examples. **M4**

2 How could being affected by a pathogen cause social problems? **P5** **M4** **D3**

Grading tip

For **P5**, **M4** and **D3**, select at least three different pathogens to include in your assignment. Then you should explain how each causes disease and how it affects the sufferer in work, at home and socially.

PLTS

Creative thinkers

You will be able to use your creative thinking skills to ask questions to extend your own thinking about how organisms function.

3.10 Inherited and autoimmune effects on health

In this section: P5 M4 D3

You may have some features that remind you of your parents. You have inherited these through their genes being passed on to you. Genes can also pass on **inherited diseases** and conditions which affect your health. Geneticists work to find cures for many inherited conditions while health professionals treat the existing symptoms.

Life–threatening inherited conditions and diseases

Sickle cell anaemia

Red blood cells contain haemoglobin that carries oxygen around the body. In sickle cell anaemia there is a problem with the haemoglobin causing red blood cells to be in the shape of a sickle instead of the normal disc shape (see photo). The sickle cells form clumps and become trapped in the blood vessels. This means they cannot deliver oxygen around the body. This can cause pain, serious infection and damage to the organs.

Cystic fibrosis

Cystic fibrosis (CF) causes large amounts of thick sticky mucus to build up in the airways and digestive tract. Symptoms include a persistent cough, regular chest infections, long-lasting diarrhoea and poor weight gain due to difficulty in digesting food. CF also affects the pancreas, which produces enzymes to help break down food molecules. This can lead to malnutrition where the body doesn't receive enough nutrients for good health.

Physiotherapy and medication help to remove the mucus from the lungs and prevent it from causing infections which can damage the lungs.

Key terms

Inherited disease – disease or disorder that is inherited genetically.

Autoimmune disease – illness that occurs when the body tissues are attacked by its own immune system.

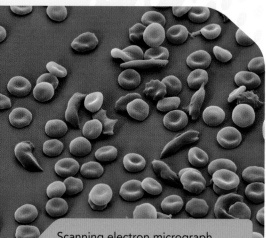

Scanning electron micrograph showing normal red blood cells (rounded) and abnormally shaped red blood cells caused by sickle cell anaemia.

 Did you know?

Cystic fibrosis affects over 8000 people in the UK. It is the UK's most common life-threatening inherited disease which affects vital organs in the body. Multiple sclerosis is the most common progressive and disabling condition of the nervous system in young adults. We do not know the cure for any of the diseases and conditions mentioned in this section.

 Activity A

Imagine living with a painful and life-threatening disease like cystic fibrosis or sickle cell anaemia. What treatments are available now? Would you want geneticists and other scientists to do research to discover a cure?

Autoimmune diseases

Some hereditary diseases cause the immune system to attack the body's own tissues. These are called **autoimmune diseases**. There is no cure for these diseases; only treatment is available. Some examples of these are listed in the table.

Disease	Description	Symptoms and treatment
Multiple sclerosis	A life-threatening condition that damages the myelin surrounding each nerve – this interferes with the electrical pathway for messages between the brain and the body	Symptoms vary but can include loss of balance, bladder problems, fatigue, problems remembering and thinking, changes in mood, depression, pain, muscle spasms and problems with speech, swallowing and vision Treated with drugs and physiotherapy
Rheumatoid arthritis	A chronic inflammatory disease of the joints where the immune system attacks the cells that line the joints	Treated with drugs to numb the pain and reduce inflammation in the joints Complementary therapies such as acupuncture and hydrotherapy can sometimes help
Crohn's disease	An inflammatory bowel disease – it can affect the whole of the internal linings of the digestive system from the mouth to the anus but usually attacks the intestines	The inflammation can affect the wall of the stomach and cause lesions. Symptoms include diarrhoea, abdominal pain, fever and tiredness Treatment includes drugs and changes in lifestyle

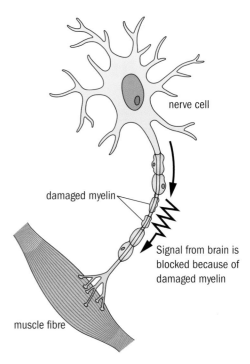

Damaged myelin in multiple sclerosis.

Unit 22: To learn more about nerves in the human body, see pages 364–365.

Social implications of inherited diseases and conditions

The impact of an inherited condition on a person's social life varies depending on the disease and how it is being controlled. It may affect the sufferer's ability to work, their self-confidence and feeling of wellbeing. The sufferer could also have financial difficulties and problems with socialising due to treatment.

Activity B

Working with a partner, choose one of the autoimmune diseases and describe how it would affect your daily routine while you are studying.

 **Assessment activity 3.10**

1 Choose two inherited diseases, one of which should be an autoimmune disease, and research each further using books and the Internet. **P5 M4 D3**

2 What are the social implications for each of the diseases you have chosen for Question 1? **D3**

Grading tip

For **P5**, **M4** and **D3**, select two inherited diseases, one of which should be an autoimmune disease, to include in your assignment and explain how they cause disease and how it affects both the sufferer in work, at home and socially.

PLTS

Reflective learners

You can be a reflective learner in the way you communicate your learning on the factors that affect health.

3.11 The control of health with vaccinations

In this section:

Key terms

Vaccine – prepares the body to fight against certain infectious diseases in case we come into contact with them in the future.

Antibody – a large protein produced in response to an antigen on the surface of a pathogen.

An immunologist at work making a vaccine.

A child suffering from mumps.

A **vaccine** is used to prevent you getting a particular illness. Many diseases spread easily between humans. This spread needs to be prevented to stop an epidemic happening. Many people who suffer from a disease make a full recovery but sometimes complications lead to serious damage or even death.

Unit 6: For more on vaccinations and the immune system, see pages 150–151.

How vaccines work

Vaccines stimulate our immune system to produce **antibodies** without us having to be infected with the disease. Immunologists use actual **pathogens** to make vaccines but they have to make sure that the vaccine doesn't trigger the disease itself.

Producing vaccines

*   Using a weakened or attenuated strain of the pathogen by repeatedly growing it to select a strain that doesn't cause the actual disease

*   Extracting the part of the pathogen that causes the immune response

*   Killing the pathogen by heating it or adding a chemical

To make the vaccine the pathogen is combined with stabilisers, preservatives to prevent contamination and an **adjuvant**. This is a substance which increases the body's immune response.

Activity A

Working in pairs, explain to your partner what a vaccine is.

Immunisation programmes

Immunisation programmes provide people with a vaccination to try to prevent the onset and spread of many diseases. Some currently offered by the NHS include MMR, HPV and BCG vaccinations.

Measles, mumps and rubella (MMR)

MMR is a combined vaccine to protect against measles, mumps and rubella (German measles). It is usually given to 3–5-year-olds and needs two injections.

Measles	One of the most infectious diseases known. Measles causes a high fever, a rash and a feeling of being unwell. Complications can cause chest infections, fits, swelling of the brain, brain damage or death
Mumps	Symptoms include headache, fever and painful swollen glands in the face, neck and jaw. Mumps can lead to deafness, swelling of the brain and infertility in men
Rubella	If a pregnant woman gets rubella it can lead to significant defects in the baby, which can include deafness, eye abnormalities and heart disease

Did you know?

In a UK study, 40% of cervical smears from 20–24-year-old women were positive for HPV DNA. This showed they were infected with human papilloma virus. This proves how important it is to use protection during sex.

Human papilloma virus (HPV)

You can get HPV by having unprotected sex with someone who has the virus. Most women who have sex will get infected by HPV at some time in their lives. The virus is usually harmless or can cause genital warts. If the infection lasts for a long time it can damage cells and lead to cancer of the cervix. The HPV vaccine protects against some strains of cervical cancer, but not all. This means women still need regular screening using smear tests throughout their life. The vaccine's protection lasts for up to 6 years.

Activity B

List all vaccinations you have had. Ask your parents/guardian about any you had as a small child.

Grading tip

For **P5**, **M4** and **D3**, select at least three different diseases which are treated with vaccinations to include in your assignment. Explain how each disease is caused and how it affects the sufferer in work, at home and socially.

Bacillus Calmette–Guérin (BCG)

This immunisation protects against tuberculosis (TB), which is highly contagious and easily spread between humans through coughing or sneezing. It used to be given to school children at 13 years of age. Now it is only offered to groups of people at high risk, usually babies or people living in crowded unhygienic conditions.

Functional skills

ICT

You will be able to use your ICT skills to research the information you need here.

Unit 3: See page 87 for more information on TB.

BTEC Assessment activity 3.11   **P5** **M4** **D3**

1. Why do we need vaccinations? Create and complete a table to show this. **P5** **M4**
2. Using the Internet and books, do some further research into the diseases above. List three points for each disease explaining how they make you ill. **P5** **M4** **D3**
3. Choose two diseases from the list above. How could family and friends be affected by knowing someone with these diseases? **D3**

PLTS

Independent enquirers

You can be an independent enquirer as you explore issues from different perspectives when considering the effect of different internal and external factors on human health.

3.12 Homeostasis

In this section: **P6**

Key terms

Homeostasis – how the body maintains a constant internal environment.

ADH (anti-diuretic hormone) – a hormone released by the pituitary gland in the brain to reabsorb water back into the body's cells.

Feel the heat.

 ### Did you know?

When you sweat, especially during prolonged, heavy exercise, the body loses water and salts. Sports drinks are designed to replace the water and salts in the correct balance.

The environment in which we live is always changing but inside our bodies it stays relatively constant. Our temperature, blood glucose levels and water levels all stay relatively stable under different environmental conditions such as exercise or illness.

What is homeostasis?

Homeostasis keeps a constant internal environment inside the body. The internal environment is the surroundings of cells such as the blood and tissue fluid. The tissue fluid contains salts and water to allow cells to function properly. Large changes to the fluid can kill or damage the cells. Homeostasis works to keep steady the following in the body:

- blood glucose levels
- water levels
- salt (ion) levels
- pH of blood
- temperature
- oxygen
- carbon dioxide levels and urea excreted as waste products.

Large changes away from the normal levels for any of these can have a damaging effect on the body as a whole and may even lead to death.

 ### Activity A

Working with a partner, discuss in which situations the factors listed above may change and may need to be regulated by homeostasis.

The organs involved in homeostasis

Homeostasis relies on feedback from the body if it is to keep the internal environment constant. This involves the cooperation of many organs as shown in the table.

Lungs	Excrete carbon dioxide from, and get oxygen into, the body
Liver and pancreas	Maintain glucose levels in the blood
Liver	Involved in temperature regulation
Kidneys	Remove waste products such as urea and maintain salt and water levels
Skin and liver	Help keep the body temperature at a constant level
Blood	Help transport materials around the body
Brain: hypothalamus	Thermostat to monitor temperature, monitor for concentration of blood chemicals and gases such as carbon dioxide

Brain: pituitary gland	Releases hormones, such as **ADH**, to regulate water content
Muscles	Help maintain a stable body temperature – muscular activity and shivering generate heat

Activity B

Produce some fact cards using the information in the table. Then test your partner on what they know about the organs involved with homeostasis.

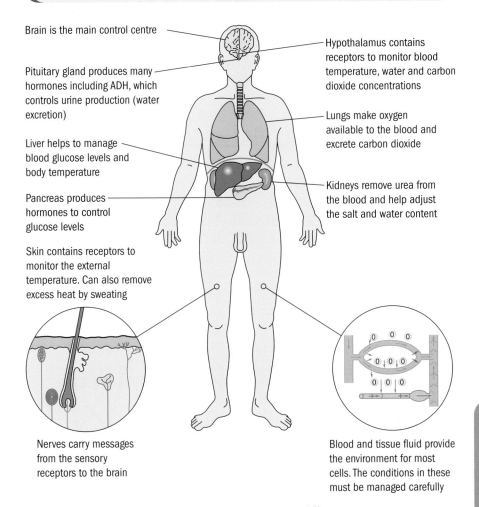

Brain is the main control centre

Pituitary gland produces many hormones including ADH, which controls urine production (water excretion)

Liver helps to manage blood glucose levels and body temperature

Pancreas produces hormones to control glucose levels

Skin contains receptors to monitor the external temperature. Can also remove excess heat by sweating

Hypothalamus contains receptors to monitor blood temperature, water and carbon dioxide concentrations

Lungs make oxygen available to the blood and excrete carbon dioxide

Kidneys remove urea from the blood and help adjust the salt and water content

Nerves carry messages from the sensory receptors to the brain

Blood and tissue fluid provide the environment for most cells. The conditions in these must be managed carefully

Homeostasis is a whole body process involving many different organs.

Grading tip

For **P6**, you need to use this information to help you identify how blood glucose and temperature are regulated.

BTEC Assessment activity 3.12 **P6**

1 List the organs which are involved in homeostasis and state the function of each. **P6**

2 Explain why homeostasis is so important to maintain a constant environment within the body. **P6**

PLTS

Self-managers

You can show that you are a self-manager as you work towards your goal, showing initiative and commitment to complete the tasks.

3.13 Maintaining a steady state

In this section: P6

Key term

Thermoregulation – regulation of the body's temperature by homeostasis.

Unit 10: For more about enzymes, read pages 196–197.

The role of the nervous system in regulating temperature

The normal body temperature is 37°C and it doesn't usually change by more than 1–2°C. The body works to keep the temperature at 37°C using **homeostasis**. Regulation of temperature, or **thermoregulation**, also involves the nervous system.

Why the body needs to stay at 37°C

Enzymes are important catalysts of many reactions in the body. The warmer the conditions, the faster reactions can take place. But if the body temperature rises above 40°C then enzymes and other molecules can start to change shape so they don't work so well. A temperature of 37°C ensures all reactions in the body are fast enough.

How the body regulates temperature

The hypothalamus in the brain acts as a thermostat to detect changes in body temperature above or below 37°C. Temperature receptors in the hypothalamus regulate internal temperature; temperature receptors in the skin monitor external temperature. The receptors send messages via the nervous system to bring about changes to keep the body temperature at 37°C. These changes are shown in the picture.

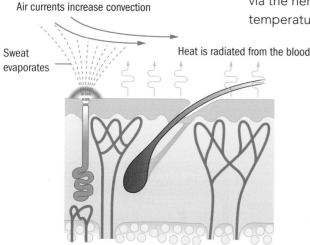

Air currents increase convection

Sweat evaporates

Heat is radiated from the blood

No sweat is produced

Air currents cannot get near the skin

If you live somewhere cold for a long time you will deposit fat under the skin

When you are too cold

- The blood capillaries in your skin get narrower (constrict) so less blood passes through them. Less heat is lost by radiation.
- Sweat production stops.
- Hairs stand on end preventing air currents getting near to the skin surface*.
- Your skeletal muscles start involuntary contractions causing shivering. This produces heat.
- Behavioural responses such as putting more clothes on and taking warm drinks.
- Metabolic rate is increased generating more heat.

* Body hair is more important in other mamals. Head hair prevents heat loss from the head.

When you are too hot

- The blood capillaries in your skin get wider (dilate) so more blood can flow through them.
- Your sweat glands make sweat and deposit it on the skin surface. As this evaporates it cools the skin.
- Hairs lie flat allowing air currents near to the skin. More heat is lost by convection*.
- Your breathing rate increases so more heat is lost in expired air.
- Behavioural responses such as wearing loose clothing and taking cold drinks.
- Metabolic rate is reduced so less heat is produced.

How the skin regulates body temperature when it is too hot or too cold.

The role of the endocrine system in regulating blood glucose levels

After eating a large meal the amount of glucose in the blood rises above the normal safe level. If you go without food for a long period of time the blood glucose levels in your body fall below the normal safe level. The body needs glucose for the process of respiration but in the right amounts. Homeostasis returns blood glucose levels back to normal. The liver and pancreas do this with the help of some hormones which form part of the endocrine system.

The **pancreas** monitors and controls the concentration of glucose in the blood. It produces two hormones.

- **Insulin** causes glucose to move out of the blood to be stored in the liver and muscles as glycogen.

- **Glucagon** causes glycogen to convert back into glucose and enter the blood stream to increase the amount of glucose in the blood.

The **liver** stores excess glucose as glycogen to lower blood glucose levels. It then releases glucose when blood glucose levels are low.

Case study: Using stem cells to treat diabetes

Jade is a molecular biologist who works in a scientific research laboratory where they perform clinical trials. She has to organise a group of people with type 1 diabetes to trial a new method of treating diabetes. This involves taking stem cells from their bone marrow. These are cells which have not yet specialised. Her research involves trying to get the stem cells to produce insulin. If this were possible then the stem cells could be injected back into the person to help their pancreas produce insulin again. The person will then no longer have to inject insulin daily. This would be a significant benefit to their daily life.

Use the Internet to find out the other sources of stem cells that are being used in research to find a cure for diabetes. Discuss with a partner how this work is affected by the laws governing stem cell research in different countries.

 BTEC ## Assessment activity 3.13 **P6**

1 Discuss how and why the body maintains a steady internal environment by homeostasis. Then write a list of five key points. **P6**

2 Create flow diagrams or a poster to show how the following are regulated:

   (a) temperature by the nervous system **P6**

   (b) blood glucose levels by the endocrine system. **P6**

 ## Activity A

Working in pairs, discuss why it is so important that the body temperature stays very close to 37°C.

 ## Did you know?

In type 1 diabetes, blood glucose levels do not return to normal safe levels after eating. This can be due to the pancreas not producing enough insulin to move glucose out of the blood. If blood glucose levels remain too high this can make the person very ill. The sufferer needs to monitor their glucose levels and inject insulin to keep their blood glucose at a safe level.

 ## Activity B

Working in groups, create a role play to show how hormones regulate blood glucose levels in the body.

Grading tip

For **P6**, you will need to use these examples and research further using books, the Internet and by practical investigations. Keep it simple and avoid long descriptions which may confuse you and the person assessing the work.

Remember, when enzymes are overheated they become **denatured**. Do not say they are killed because enzymes are not alive in the first place.

 ## Functional skills

English and ICT

You can use your ICT and reading skills to research and create a flow diagram to show how homeostasis regulates the body.

Just checking

1 Draw a simple plant and animal cell. Show where the DNA and genes are.
2 Explain how genes lead to variation in groups.
3 How and why do we need to classify organisms?
4 How does human activity impact on our environment?
5 What factors can make us ill? Give examples.
6 What is there to prevent and stop us from being ill?

Assignment tips

To get the grade you deserve in your assignments remember the following.

- Show evidence that you have chosen the information needed for each of the pass, merit and distinction criteria for the assignment.

- Demonstrate that you have used a variety of different resources to complete your assignment. Check that you have referenced all your work correctly.

- Use a variety of ways to present your material. For example, as a leaflet, classification keys, flow diagrams, surveys, reports or presentations.

- If you are presenting information in a leaflet, remember who is going to be reading it and present the information clearly so it is easy for them to understand.

Some of the key information you'll need to remember includes the following.

- The passing down of genes from one generation to the next involves passing on physical characteristics. It can also pass on medical conditions and diseases.

- Mutations in DNA are not always harmful to an organism. They can be beneficial.

- Organisms can adapt to survive. This is linked to natural selection where organisms best adapted to their environment are more likely to survive and pass on their genes to the next generation.

- Humans can impact on the environment to cause long-term changes. The impact needs to be monitored to help to identify changes and prevent further damage.

You may find the following resources useful as you work through this unit.

For information on...	Visit...
topics in this unit including homeostasis	Homeostasis
current issues and research in biology	New Scientist
biological facts and explanations	Roberts M and Ingram N, *Nelson Science – Biology, 2nd Edition* (Nelson Thornes, 2001) ISBN 9780748762385

Credit value: 5

4 Applications of chemical substances

Chemistry is a fast-moving and exciting subject. Chemists and chemical engineers are designing and producing completely new substances to meet the demands of today's society.

From nanotechnology in your mobile phone to 'smart materials' used in medicine, chemistry is creating solutions to problems which didn't even exist 20 years ago.

To do this, chemists need to understand how substances bond together and how the different kinds of bonds affect the properties of the substances.

In this unit, you will be using some of the knowledge and skills you built up in Unit 1 independently. Then you will make sense of some of the energy changes that happen when chemical bonds are broken and formed. You will look in detail at some of the substances which can be made from crude oil; finally, you will find out how today chemists are working on a tiny scale to produce exciting new substances.

Learning outcomes

After completing this unit you should:

1 be able to investigate chemical substances with different types of bonding

2 be able to investigate exothermic and endothermic reactions

3 be able to investigate organic compounds that are used in society today

4 know about specialised materials and their applications.

Assessment and grading criteria

This table shows you what you must do in order to achieve a **pass**, **merit** or **distinction** grade, and where you can find activities in this book to help you.

To achieve a **pass** grade, the evidence must show that the learner is able to:	To achieve a **merit** grade, the evidence must show that, in addition to the pass criteria, the learner is able to:	To achieve a **distinction** grade, the evidence must show that, in addition to the pass and merit criteria, the learner is able to:
P1 Carry out experiments to identify compounds with different bonding types **See Assessment activity 4.1**	**M1** Describe the properties of chemical substances with different types of bonds **See Assessment activity 4.2**	**D1** Explain why chemical substances with different bonds have different properties **See Assessment activity 4.2**
P2 Carry out experiments to investigate exothermic and endothermic reactions **See Assessment activity 4.3**	**M2** Explain the temperature changes that occur during exothermic and endothermic reactions **See Assessment activities 4.3 and 4.4**	**D2** Explain the energy changes that take place during exothermic and endothermic reactions **See Assessment activities 4.3 and 4.4**
P3 Carry out experiments to identify organic compounds **See Assessment activity 4.5**	**M3** Describe the uses of organic compounds in our society **See Assessment activity 4.6**	**D3** Explain the benefits and disadvantages of using organic compounds in our society **See Assessment activity 4.6**
P4 Identify applications of specialised materials. **See Assessment activities 4.7 and 4.8**	**M4** Describe the production of specialised materials. **See Assessment activity 4.7**	**D4** Explain the implications of nanochemistry. **See Assessment activity 4.8**

How you will be assessed

Your assessment could be in the form of:

- a presentation of results e.g. from a practical showing how the bonding in substances may be identified by the properties

- a report on a practical investigation e.g. on the energy changes in chemical reactions

- an article, video or podcast e.g. on the uses of nanochemistry and new materials.

Hayley, 18 years old

I thought I might find this unit quite difficult because it's all about things that you can't see, like atoms and chemical bonds. But our teacher used models and computer animations and I really started to understand it.

We did a lot of experiments to measure energy changes in chemical reactions. I enjoyed burning some fuels to see which ones produced most energy. It was interesting to see how quickly water can be heated up by burning really small amounts of fuel.

It's amazing to see how chemists have produced all of these new molecules, like nanotubes. We spent quite a lot of time using the Internet to find out about these new substances and then we made presentations on them. I've become a lot better at deciding which websites are going to be the most helpful for my assignments.

Catalyst

Chemists design sportswear!

Think about a sport you play or a hobby you have which needs special clothes or equipment, such as Lycra shorts for cycling or protective equipment for hockey players. Tell your partner about the clothes or equipment and discuss the following questions.

- Why are special clothes or equipment needed?

- What properties do they need to have?

- Are there natural substances which can do the job or have chemists been involved in designing suitable alternative materials? You might need to use the Internet for this question.

4.1 Types of bonding

In this section: **P1**

Key terms

Bonding – the way in which particles are held together in chemical substances.

Ionic bonding – bonding between positive and negative ions.

Covalent bonding – bonding between atoms which are sharing one or more pairs of electrons.

Dative covalent bonding – covalent bonding in which both electrons in a shared pair come from the same atom.

Metallic bonding – bonding between metal ions which are surrounded by a 'sea' of moving electrons.

Ion – an atom (or group of atoms) with a positive or negative charge. Examples Na^+, SO_4^{2-}.

Molecule – two or more atoms bonded together by covalent bonds.

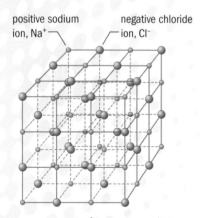

positive sodium ion, Na^+ — negative chloride ion, Cl^-

The arrangement of ions in a sodium chloride lattice.

Activity A

Iron oxide is a black substance with **ionic bonding**. Which ion is positive and which is negative?

Different substances, different bonding

The chemical industry handles a huge range of different substances. Sometimes the substances may look very similar; if you see a pile of black powder it could be iron filings used in fingerprint detection, covalent nanoparticles used in new medical treatments or ionic iron oxide used in black paints. Each substance needs to be handled in different ways depending on how the particles are bonded.

Ionic bonding

The **bonding** in substances like iron oxide or sodium chloride is ionic. In sodium chloride, for example, the sodium atoms have lost an electron to become a positively charged sodium **ion**. Chlorine atoms have gained an electron to become a negatively charged chloride ion. Both ions now have full outer shells (eight electrons).

The positive and negative ions attract each other, and they are held together in a regular arrangement called an ionic lattice.

When metals bond to non-metals they usually form ionic bonds. Metal atoms form positive ions and non-metals form negative ions.

Ionic compounds often dissolve well in water and the solution which is formed conducts electricity – this is important in electrical cells and batteries.

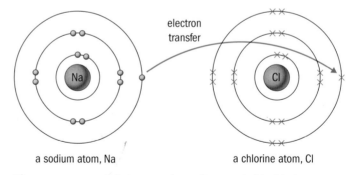

electron transfer

a sodium atom, Na   a chlorine atom, Cl

The arrangement of electrons in sodium and chloride ions.

Covalent bonding

Covalent bonding normally happens when non-metal atoms bond together. The atoms do not form charged ions. Instead the atoms *share* one or more pairs of electrons. The negative charge of the electrons attracts the positive nucleus in each atom and holds them together.

(a) 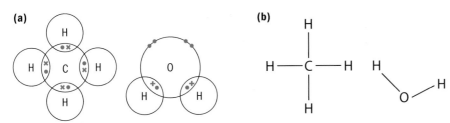 (b)

(a) The electron arrangement of some covalent **molecules**. Notice that the outer shell of each atom is now full. (b) The lines on these diagrams of molecules show where the electrons are being shared.

Simple covalent substances include water and methane. Unlike ionic substances there are no charged ions in a covalent substance, so they do not conduct electricity even when dissolved. In fact, most covalent substances don't dissolve in water, although they dissolve quite well in organic solvents.

 ## Activity B

Ammonia (NH_3) contains a nitrogen atom forming three covalent bonds to hydrogen atoms. Draw a dot and cross diagram to show the electron arrangement.

Metallic bonding

It is normally easy to recognise metals by their shiny appearance. The bonding in metals is not covalent or ionic – it is a third type of bonding called **metallic bonding**.

The metal atoms are arranged in layers. Each atom loses some electrons to become a positive ion. A 'sea' of moving negative electrons surrounds the positive ions. The electrons and ions are attracted together.

Because the electrons in the lattice can move, metals conduct electricity even when solid.

BTEC ## Assessment activity 4.1 **P1**

Three black substances are being tested in a laboratory to find out what type of bonding they have.

- Substance A: composed of quite large black lumps that do not dissolve in water; conducts electricity.
- Substance B dissolves in water to give a deep purple solution which conducts electricity.
- Substance C doesn't dissolve very well in water, although it dissolves in the organic solvent hexane to give a purple solution. When this was tested it did not conduct electricity.

Using the evidence, decide what type of bonding is found in substances A, B and C. **P1**

Science focus

Although covalent bonds always involve a pair of electrons being shared, the two electrons don't have to come from different atoms. Look at the dot and cross diagram of the ammonium ion below. In one of the bonds both of the electrons come from the nitrogen atom. But the bond has the same properties as the other covalent bonds – it has the same strength, for example.

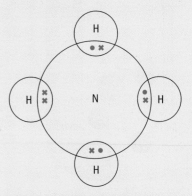

The ammonium ion contains a **dative covalent bond**.

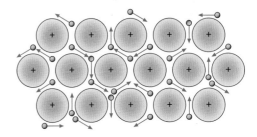

 metal ion · electron

Metallic bonding.

Grading tip

Remember that only some types of bonding cause substances to conduct electricity. Use the information on these pages to summarise what you know about each kind of bonding in a table. It will help you to interpret results like the one in this **P1** exercise.

4.2 The properties of substances

Handling substances safely is vital in the chemical industry. Knowing the bonding type in a substance helps you predict its properties. It is particularly important to know what might happen to the substance in the case of a fire or if it leaks into water. For example, you may need to know whether a substance dissolves in water.

Properties of substances with ionic bonding

Ionic substances normally have the following properties:

Property	Explanation
High melting point	The + and – ions attract each other strongly, so a lot of energy is needed to pull them apart
Conduct electricity when they are dissolved in water or melted	The ions separate in the liquid so they are free to move and carry the electric current
Often dissolve in water	Water is attracted to the charged ions and bonds to them

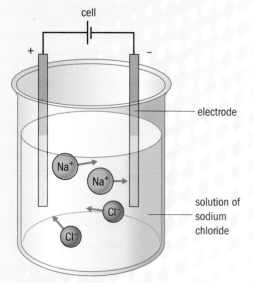

The charged ions in a solution of an ionic compound allow it to conduct an electric current.

Properties of substances with covalent bonding

The covalent bonding in molecules means that the properties of these substances are very different from those of ionic substances.

Property	Explanation
Low melting points and boiling points (sometimes the substances are gases or liquids)	Although the covalent bonds between the atoms are very strong, the forces between the molecules are very weak
Not usually soluble in water (but they may dissolve in organic solvents, like hexane)	The molecules aren't charged so water molecules aren't attracted to them
Do not conduct electricity, even when melted	There are no charged ions in the substance to carry the current. If an ionic substance dissolves in a covalent liquid, the conductivity increases because there will now be some free ions

Burning petrol cannot be put out using water because petrol does not mix with water. It just floats on top of the water, where it continues to burn.

 Activity A

Nitrogen in the air is a molecule with the formula N_2. It contains very strong covalent bonds, which means that it is usually very unreactive. Use the information in the table above to explain why, even though the covalent bonds are strong, nitrogen is a gas in the atmosphere.

Case study: Pure water

Dipak works for a company which makes water purifiers. One of the tests that he does is to measure the conductivity of water (whether it conducts electricity) after it passes through the purifier.

One day Dipak discovers that a sample of water has an unusually high conductivity.

Do you think the water is pure?

What kind of substance might have got into the water?

Properties of substances with metallic bonding

Apart from their shiny appearance, metals have some other important properties.

Property	Explanation
Conduct electricity when solid and molten	The electrons are able to move freely and carry the electric current
Normally have a high melting point and boiling point	The attraction between the ions and the electrons is very strong

Metallic paints on cars look shiny because they have flakes of metal mixed in with the paint.

  Assessment activity 4.2   **M1** **D1**

You are working for a decontamination company which has been called in to advise on the best way of disposing of chemicals at an abandoned chemical plant. The chemicals are in unlabelled containers.

The chemicals are:

- sodium chloride, an ionic compound which can be disposed of safely into the sea nearby
- paraffin wax, a covalent compound which can be buried in landfill
- zinc metal, which must be recycled.

Work in a group to devise some tests which you could do to be certain which chemical is which.

Present your findings as a PowerPoint presentation. To convince your audience that you are right, you should also try and explain how the bonding helps you to explain the property. **M1** **D1**

 Activity B

Can you think of a metal which does not have a high melting point?

Grading tip

For **M1** think about which property is the most helpful to allow you to decide what type of bonding is present.

For **D1** include some diagrams showing ionic, covalent and metallic bonding to help you with your explanations of the properties.

4.3 Energy changes in chemical reactions

In this section: **P2** **M2** **D2**

Key terms

Exothermic – an exothermic reaction gives out energy. Chemical energy in the system is transferred to heat energy in the surroundings.

Endothermic – an endothermic reaction takes in energy. Heat energy in the surroundings is transferred to chemical energy in the system.

The thermite reaction between aluminium and iron oxide releases heat energy.

Did you know?

Most of the reactions which you see in the laboratory are exothermic reactions. Endothermic reactions are much rarer because energy needs to be provided continuously to make them happen.

Exothermic reactions

The thermite reaction is an **exothermic** chemical reaction. Exothermic means that the reaction gives out energy, in the form of heat, light or even electrical energy.

Case study: Blast off

Reisha designs fuels for the rockets used to launch satellites. A rocket fuel works by producing a lot of hot gas when it reacts with oxygen.

The fuel she chooses for the rockets used to launch satellites is a mixture of hydrazine (N_2H_4) and dinitrogen tetroxide (N_2O_4). These react in an exothermic reaction.

Why is it important that the reaction is exothermic? Can you suggest any other important features of a good rocket fuel?

Activity A

Can you think of any other chemical reactions you have seen where heat energy is released?

Endothermic reactions

Not all reactions produce heat. Some actually take in heat – **endothermic** reactions. Think about putting a cold pack on your wrist. Your skin starts to feel cold because heat energy from your skin is transferred to the chemicals in the cold pack. But the cold pack itself doesn't get hotter. Instead the heat energy from your hand is used to make the two substances (ammonium nitrate and water) react. The heat energy becomes stored as chemical energy in the products of the reaction.

Energy transfer – system and surroundings

You know from Unit 2 that energy has to be formed from something. So where does the energy formed in exothermic reactions come from? All chemicals have chemical energy. When chemicals react, some of this chemical energy may be transformed to heat energy – or some heat energy may be transformed to chemical energy.

The chemicals in a reaction are often called the system. Everything else – like the water in a beaker or test tube – is called the surroundings. Using words like system and surroundings helps you to explain exactly where energy comes from or goes to.

Activity B

Copy and complete this description of the energy transfers in an endothermic reaction.

Heat energy in the _____ is transferred into _____ in the system.

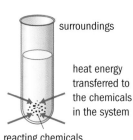
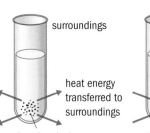

surroundings · surroundings

heat energy transferred to surroundings

heat energy transferred to the chemicals in the system

reacting chemicals (system)

reacting chemicals (system)

Energy transfers in an exothermic reaction.

Energy transfers in an endothermic reaction.

Investigating reactions

Lots of chemical reactions are exothermic. You could investigate some in your laboratory. The easiest way of doing this is to set up an experiment in which the energy from an exothermic reaction is transferred to water. This will cause the temperature of the water to rise. Two ways of carrying out these experiments are shown below.

You may also do some endothermic reactions in solutions. You will notice the water in the solution getting colder: heat energy from the water is transferred to chemical energy in the system.

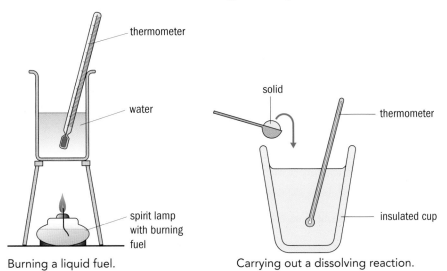

thermometer

water

spirit lamp with burning fuel

solid

thermometer

insulated cup

Burning a liquid fuel.

Carrying out a dissolving reaction.

PLTS

Reflective learners

Carrying out these experiments will help you develop the skills of a reflective learner. You will need to think about what is happening to the energy and then put your thoughts into words to describe the energy transfer processes.

Grading tip

You will need to decide whether the reactions are exothermic or endothermic to achieve **P2**. Remember that if the water surrounding the chemicals in a reaction gets colder, then heat energy has been transferred to the chemicals, making the reaction endothermic. This should also help you with your explanations of temperature changes for **M2**.

To gain **D2** you will need to use words like 'system', 'surroundings', 'heat energy' and 'chemical energy' in your explanation.

BTEC Assessment activity 4.3 **P2 M2 D2**

A group of students carry out some experiments in which they dissolve some ionic solids in water and measure the temperature change.

Substance	Starting temperature	Highest or lowest temperature reached
Calcium chloride	21°C	35°C
Ammonium nitrate	19°C	1°C
Sodium chloride	21°C	20°C

Discuss with other students what energy transfer processes are happening. **P2 M2 D2**

4.4 Explaining energy changes in chemical reactions

In this section: M2 D2

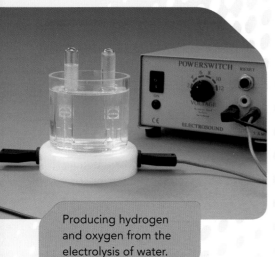

Producing hydrogen and oxygen from the electrolysis of water.

Making and breaking bonds

Breaking bonds – taking in energy

Water (H_2O) can be broken apart to produce hydrogen and oxygen. A lot of energy is needed to do this because breaking up water molecules means breaking chemical bonds. Bond breaking needs energy.

This energy can be provided in the form of electrical energy, such as in the electrolysis of water. This method is used to produce hydrogen for use as a fuel, such as in fuel cells. The oxygen is a valuable by-product and is used in other industries, for example to remove carbon from iron in steel-making.

Making bonds – giving out energy

In a fuel cell, hydrogen and oxygen atoms bond together to form water. Electrical energy is released, because bond forming gives out energy.

Explaining energy changes

When a chemical reaction happens there is a change in the way the particles are bonded. Normally some bonds are broken and others are formed. Bond breaking needs energy, and bond forming gives out energy.

A good way of showing bonds breaking and forming, and the energy changes involved, is as an energy diagram.

When methane is used as a fuel, it reacts with oxygen to form carbon dioxide and water.

Did you know?

In the last few years new types of cars have been manufactured which are powered by hydrogen fuel cells. Many people think that these cars may eventually replace petrol and diesel cars because they are less polluting.

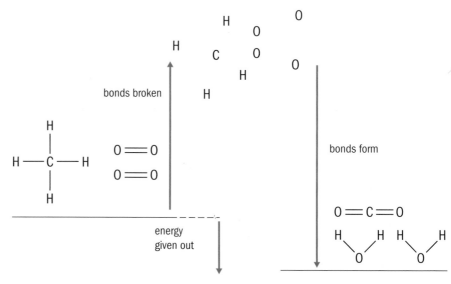

Bonds breaking and forming when methane burns in oxygen.

The diagram for methane burning shows you that quite a lot of energy is taken in to break the bonds inside the methane molecules and the oxygen molecules. But even more energy is given out when new bonds form to make water and carbon dioxide molecules. The extra energy is transferred to heat energy. The reaction gives out heat – it is exothermic.

Instant heat

When calcium oxide is added to water it reacts and dissolves. This is a very exothermic process and it has been used as a way of heating coffee in a can.

Activity A

When a molecule of methane reacts with oxygen, four carbon–hydrogen bonds are broken. List all the other bonds which are broken and formed in the reaction.

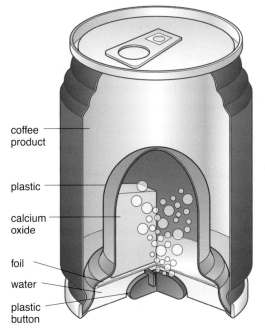

coffee
product

plastic

calcium
oxide

foil

water

plastic
button

When a plastic button is pressed at the bottom of the self-heating can, water and calcium oxide start reacting. Heat energy is transferred to the coffee.

Activity B

It is possible to adapt the can so that it can cool drinks instead of heating them. What kind of reaction would you need to cool a drink?

BTEC Assessment activity 4.4   M2 D2

Imagine that you are working for the company that manufactures the self-heating coffee can.

1 You have been asked to find some alternatives to calcium oxide in the self-heating can. You have also been asked to find out whether there are any suitable substances which can be used to make the can self-cooling.

 Look back at pages 104–105 for ideas.

 Prepare a short written report to recommend which solids would be the best choice for the self-heating and self-cooling cans.

2 Write a short advertising leaflet for the can, targeted at 15-year-old science students, which explains the science behind the way the can works. M2 D2

Grading tip

For M2 you should make sure that you mention bonds in your report and use words like exothermic in your explanation.

For D2 you will need to think about the strength of the bonds which break and the bonds which form. You may find it useful to try and draw a diagram for the dissolving reactions – follow the same style as the diagram showing the burning of methane above.

4.5 Organic compounds

Key terms

Organic compounds – compounds obtained from a source that was once alive, such as crude oil.

Alkanes – molecules containing only carbon and hydrogen atoms with the general formula C_nH_{2n+2}.

Alkenes – molecules containing only carbon and hydrogen atoms with the general formula C_nH_{2n}. Alkenes contain a C=C double bond.

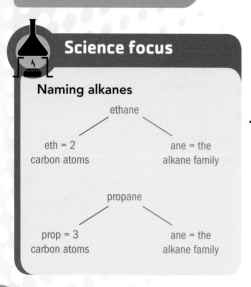

Crude oil is the raw material from which a huge range of chemicals are made.

Science focus

Naming alkanes

ethane

eth = 2 carbon atoms

ane = the alkane family

propane

prop = 3 carbon atoms

ane = the alkane family

Families of molecules

You probably know that many of the fossil fuels which we burn, such as petrol and diesel, come from crude oil. But the molecules in crude oil are also the starting point for the manufacture of a huge range of **organic compounds**. These are used to make things like dyes, detergents, plastics and medicines.

Most of the molecules in crude oil are from a family of organic molecules called **alkanes**, but **alkenes** are also present.

Alkanes

These contain long chains of carbon atoms bonded together by single C–C bonds.

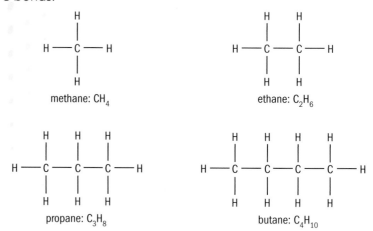

methane: CH_4

ethane: C_2H_6

propane: C_3H_8

butane: C_4H_{10}

The structures, names and formulas of some alkanes.

You will see that there is a pattern in the way these substances are named. The start of the name (the prefix) tells you how many carbon atoms are in the chain, and the end of the name reminds you that the molecule belongs to the alkane family.

The formulas also follow a pattern. You will notice that the number of hydrogen atoms is double the number of carbon atoms plus an extra two atoms. Chemists often say that 'the general formula of an alkane is C_nH_{2n+2}'.

Activity A

Which of these molecules are alkanes? $C_{10}H_{20}$, $C_{12}H_{26}$, C_6H_6, C_7H_{16}.

Other organic molecules

There are many other families of molecules with different atoms or different bonding. An alkene molecule has a double bond between two of the carbon atoms. Alcohols and carboxylic acids contain oxygen atoms as well as carbon and hydrogen.

Other families include molecules which contain nitrogen or halogens (like chlorine) as well as carbon and hydrogen atoms.

Table: The names and formulas of some families of organic molecules.

Family	Name	Formula	Example
Alkenes	End in -ene	C_nH_{2n}	Butene C_4H_8
Alcohols	End in -anol	$C_nH_{2n+1}OH$	Methanol CH_3OH
Carboxylic acids	End in -anoic acid	$C_nH_{2n+1}COOH$	Ethanoic acid CH_3COOH

Recognising organic molecules

If you know the structure of a molecule it is easy to recognise which family the molecule comes from. You can use: the name (look at the ending); the formula (see whether it fits any of the general formulas); the structure.

Practical tests

Table: Laboratory tests for some organic compounds.

Family	Test	Result
Alkenes	Add bromine solution	Colour change from orange to colourless
Alcohols	Add potassium dichromate solution and sulfuric acid, heat in a water bath	Colour change from orange to green
Carboxylic acids	Dissolve in water and add universal indicator	Indicator goes orange or red; pH of 2–4

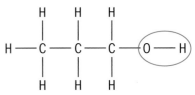

alkenes
contain $C=C$ bonds

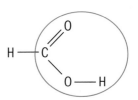

alcohols
contain $O-H$ bonds

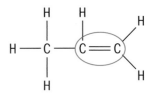

caboxylic acids
contain $C=O$ and
$O-H$ bonds

The structures of some different types of organic molecules.

Activity B

Which family do each of these molecules belong to?

1 (a) pentene   (b) hexanol
  (c) dodecane   (d) butanoic acid

2 (a) C_4H_9OH (b) CH_3COOH
  (c) $C_{12}H_{26}$ (d) $C_7H_{15}OH$

BTEC Assessment activity 4.5 P3

You are working in a laboratory. The labels have come off the bottles of four organic liquids. You know that the four liquids are methanol, octane, ethanoic acid and hexene. You temporarily label the bottles A, B, C and D and then carry out tests to identify the liquids:

- when bromine solution was added to liquid A, the mixture changed colour from orange to colourless
- when universal indicator was added to liquid B, it turned orange-red, suggesting a pH of about 3
- when a mixture of potassium dichromate and sulfuric acid was heated with liquid C, the potassium dichromate changed from orange to green
- the same tests were done on liquid D but no change was observed for any of them.

Using the results of the tests, identify A, B, C and D. P3

Grading tip

Use the information about the results of the practical tests to work out what family each liquid belongs to in order to obtain P3.

4.6 Using organic molecules

In this section: M3 D3

The petrol in our cars and the PVC used to make flooring, drainpipes and other objects do not appear very similar. But both of them are composed of organic chemicals, so they contain chains of carbon atoms bonded together.

Key terms

Polymer – a large molecule made from many small monomer molecules joined together.

Monomer – small molecule that can join together with others to make a long chain called a polymer molecule.

PVC is a polymer made from alkene molecules.

Activity A

(i) Use the word equations on the right to explain why alkanes and alcohols make good fuels.

(ii) Give one possible disadvantage of using alkanes as fuels.

Science focus

PVC can be made into a much more flexible polymer by adding a chemical called a plasticiser. The PVC used in window frames is called uPVC because it is unplasticised PVC. Plasticised PVC can be used to make upholstery and clothing.

Fuels

Alkanes are particularly useful as fuels. They can be gases like methane, which is natural gas used for cooking and heating, or liquids like heptane, which is used in petrol. They react with the oxygen in air in a very exothermic reaction:

> heptane + oxygen → carbon dioxide + water (+ heat)

Alcohols are also useful as fuels. They burn in a similar way to alkanes but produce less soot so they are a cleaner fuel.

> ethanol + oxygen → carbon dioxide + water (+ heat)

Many governments around the world are encouraging petrol companies to add ethanol to petrol in an attempt to reduce the amount of petrol being used. Apart from being cleaner, ethanol can also be made from the sugar in sugar cane – so it is a renewable fuel.

Polymers

If alkene molecules are heated up at a high pressure and with a catalyst, a reaction called **polymerisation** happens. Many alkene molecules join together to form one long **polymer** molecule. The alkenes used in this process are called **monomers**.

Case study: Polymers

Asif works as a development chemist for a company manufacturing polymers. He can change the properties of the polymer by using different monomers. At the moment the company is making polymers for use as plastic sheets to cover garden furniture in bad weather.

What properties does this polymer need to have?

Give the name of a polymer which might be suitable.

Polyethene is the polymer formed from ethene monomers. It is a strong but flexible material that can be used to make plastic bags and bottles. Different alkenes can be used as monomers. One common polymer is PVC (sometimes called polychloroethene). This is made from monomers which contain a chlorine atom. The polymer is stronger and more rigid than polyethene and so it is used in building products, like window frames.

chloroethene molecules

PVC (polychloroethene)

Alkenes react to produce polymers.

Solvents

Most alcohols are liquids. You can dissolve substances in alcohol just as you can dissolve them in water. Ethanol is often used as a solvent. Perfumes are made from very concentrated scent molecules dissolved in ethanol to make a dilute solution. When you dab perfume on your skin, the ethanol evaporates, leaving the scented molecules on your skin.

The dry-cleaning industry uses a lot of organic molecules as solvents to dissolve dirt and grease on clothes.

Food preservation

The food industry uses carboxylic acids in the production of food items that need a long shelf life. Food can be preserved using carboxylic acids, such as citric acid, because most bacteria cannot survive acid conditions.

Uses and risks

Table: Some uses and risks of using organic compounds.

Uses of organic compounds	Risks and disadvantages of using organic compounds
Medicines (e.g. paracetamol – a painkiller)	They are often flammable (can catch fire easily)
Dyes (e.g. mauveine – the first completely synthetic dye)	They are usually made from crude oil, a non-renewable resource
Flavourings for food (e.g. octyl ethanoate, which makes food smell of oranges)	The vapours released from organic solvents can be harmful or even toxic
Cosmetics (e.g. glycerine in moisturisers)	Some organic substances, such as benzene, are now known to cause cancer

Perfume is made up using ethanol as a solvent.

Activity B

(i) Find out and draw the structure of citric acid.

(ii) Which part of the structure makes this compound a carboxylic acid?

Grading tip

You could start your search by deciding the name of an organic substance from each of the four families – use pages 108–109 to help you.

To obtain **M3** you will need to describe the uses you have found out about. For **D3** you will need to go further and explain why this use is important. You will also need to look at the list of risks and disadvantages described in the table and decide whether any of them will be a problem for your chosen substance.

BTEC Assessment activity 4.6 M3 D3

1 Use the Internet to research the use of some organic chemicals. You should include an example of an alkane, an alkene, an alcohol and a carboxylic acid. You may like to work in groups. **M3**

2 Prepare a PowerPoint presentation on the uses of these organic substances. Include any hazards, risks or disadvantages in the way they are used. **D3**

4.7 New materials

In this section: **P**4 **M**4

The polymer Kevlar is used in bullet-proof helmets and clothing.

Activity A

Polymers like Bakelite are electrical insulators. Why was this important in the way that Bakelite was used?

Did you know?

Gecko lizards are able to walk upside down because their feet can stick to just about any surface, but PTFE is the only solid which they cannot stick to.

The chemicals in this thermochromic thermometer change colour when the temperature changes.

Designer polymers

Kevlar

Artificial polymers were first made by chemists over 100 years ago. The first of these artificial polymers was Bakelite, which was used as the casing for radios and telephones and electrical equipment. More recently, chemists have learned to *design* polymers so that they have the special properties they need for specialised uses. Kevlar is a good example. It is made by joining small carboxylic acid molecules with another type of organic molecule called an amine. It is used in bullet-proof and stab-proof clothing worn by soldiers and the police. Kevlar is perfect for this job because it is lightweight and flexible, but also extremely strong.

PTFE

Sometimes strength isn't the most important property. A polymer called PTFE (polytetrafluoroethene) has been developed to be slippery – there is no friction when other substances rub against it. This makes it useful for coating non-stick frying pans and even as an additive to ink to help it flow better.

PTFE can be made into fibres and woven into a sheet of fabric. This is called Gore-Tex™ and is used in making waterproof clothing. The fabric has tiny holes in it that let water vapour from sweat out of the clothing but don't let droplets of rainwater in. You may have used a Gore-Tex™ waterproof coat.

Smart materials

As you have seen, polymer chemists have designed new materials, each with slightly different properties. It is now possible for them to design materials, called smart materials, with properties that change depending on the conditions.

When the temperature rises, bonds in the dye molecules in a thermochromic thermometer break. This causes the dye to change colour. Different dyes change colour at different temperatures.

Other smart materials include:

Type of material	Property
Shape memory polymers	The polymer can be bent into different shapes but changes in temperature cause the polymer to return to its original shape
Photochromic polymers	Changes in the light cause the polymer to change colour
Piezoelectric polymers	Change shape or colour when a voltage is applied

Production of new materials

Monomers react together under suitable conditions to form a polymer. Additives can be put into the polymerisation reactor to give specific properties. The final polymer can then be drawn into fibres, blown into sheets or moulded into particular shapes depending on the way in which it will be used.

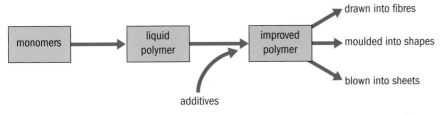

The stages in the manufacture of a polymer.

The specialised properties of polymers can be achieved in several ways.

- Different monomers can be used.

- Additives, such as plasticisers or cross-linking agents, can be added. These can change the strength of the forces between the polymer chains. Dyes might also be added to make the polymer coloured.

- The polymer can be made into fibres, sheets or moulded objects. Fibres, for example, will be needed for making clothing; sheets would be used for making bottles.

Case study

Shape memory

Sajid is a market researcher for a chemical company that is about to start manufacturing a shape-memory polymer.

Before manufacture begins, he needs to find out how big the market is for this substance.

Suggest or research some ways in which this substance could be useful.

What sort of people would Sajid want to talk to in order find out whether there will be a demand for his company's product?

no cross-linking

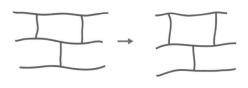

cross-linking

Cross-linking creates bonds between the chains. This makes a polymer stronger by stopping the chains from moving.

BTEC Assessment activity 4.7

1 Unplasticised PVC (uPVC) is a hard material that is water-resistant and remains strong even at temperatures up to 80°C. Find out some uses of uPVC. **P4**

2 Plasticised PVC is much more flexible and is used in a wide range of products from watch straps to medical tubing. Use books or the Internet to find out how PVC is manufactured. You should include details of what monomer is used and how the monomer is made into a polymer. You also need to find out which additives are used in the production. **M4**

3 Design a poster to show the stages of the manufacture. **M4**

Grading tip

To reach **P4** you should be able to find at least two different uses of uPVC.

For **M4** you will need to find out which additives are added and also to describe why the additives are needed.

 ## PLTS

Independent enquirers

This activity will help you develop your skill as an independent enquirer. You will need to be able to select the most suitable websites or books to help you and to choose suitable search criteria or use the index of a book.

4.8 Nanochemistry

In this section: P4 D4

Key term

Nanoparticles – particles that have a diameter of less than 100 nanometres. A nanometre is a billion times smaller than a metre.

In 1985, chemists discovered that carbon atoms could bond together to form nanospheres or nanotubes.

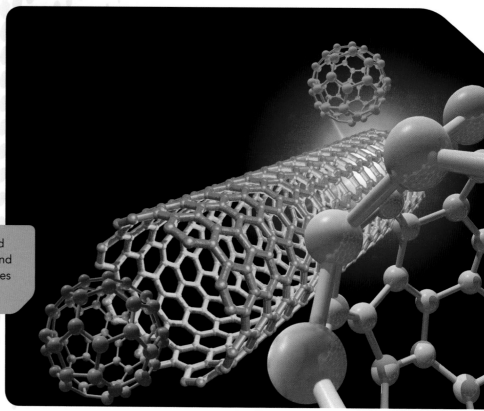

Activity A

Why would a catalyst attached to a nanotube work better than normal catalysts?

Science focus

Carbon nanotubes can be made by vaporising carbon inside a sealed chamber using a high-power laser. For some carbon nanotubes it is important to also vaporise nanometre-sized metal particles. These act as the *seeds* the tubes grow from. The carbon nanotubes can then be collected from the walls of the chamber.

The tiny structures in the picture are examples of **nanoparticles**, which have many special properties which make them useful. Some uses are shown in the table below.

Nanoparticle	Use
Nanotubes (carbon)	These are strong and conduct electricity so they can be used as miniature nanowires for tiny electrical circuits
	A nanotube has a large surface area for its size. Catalysts can be attached to the surface and used to speed up a wide range of reactions
Nanospheres and nanotubes (carbon)	These can be used as 'cages' to hold powerful drug molecules, like those used to treat cancer tumours. The nanoparticles can be designed to release the drug only when it reaches the tumour; this prevents the drug affecting other parts of the body
Titanium dioxide	These nanoparticles are used in sunscreens to block out UV radiation. Normal sunscreen can make the skin look white when it is applied, but using these nanoparticles makes the sunscreen invisible on the skin

Are nanoparticles dangerous?

Nanoparticles also occur as a result of pollution. Soot in the exhaust gases of diesel engines may contain large numbers of carbon nanoparticles. Many scientists think that these nanoparticles may cause lung damage or cancer.

Some people are worried that if we are exposed to more nanoparticles, then we may increase our risk of cancer and other health problems.

This is an **ethical issue**. Our society has to decide whether it is right or wrong to allow industry to manufacture chemicals which may later on prove to be a danger to our health.

BTEC Assessment activity 4.8   P4 D4

1 Make a short news report about nanoparticles for a television station. You will need to explain what they are, why they are useful and what the possible dangers are. **P4 D4**

Use information from the internet to find some up-to-the minute uses of nanoparticles. A helpful website is **Uses of nanoparticles**.

Work in groups of three and present your report as a script, storyboard or video.

Grading tip

For **P4** you will need to find at least two different types of nanoparticles that are being used.

For **D4** you will need to make your report balanced, explaining the importance of nanoparticles as well as the risks.

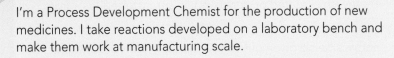

WorkSpace

Colin Benison
Process Development Chemist, AstraZeneca

STEM AMBASSADORS ILLUMINATING FUTURES
Nationally coordinated by STEMNET

I'm a Process Development Chemist for the production of new medicines. I take reactions developed on a laboratory bench and make them work at manufacturing scale.

I design manufacturing processes to produce 10–100 kg of drug compound for use in clinical trials. As we scale the process up towards full-scale manufacture we keep improving it. I work with organic chemists and chemical engineers to ensure the drug is safe for patients and the process is safe for the operators.

Imagine that a clinical trial requires 30 kg of a newly identified potential drug. The organic chemist tells you that the largest amount made so far was 1 kg made for a safety test using a 10 litre laboratory reactor. In the laboratory reaction the reactants were heated to 60°C over 20 minutes, stirred for 2 hours and cooled to 20°C over 20 minutes, and the product filtered off. The reaction produced 250 g of methyl bromide – a toxic gas – as a by-product. The development chemist must work out what will happen when the process is scaled up by a factor of 30.

Think about it!

1 Can it be heated/cooled at the same rate? If not, what happens?

2 How much gas will be given off and how will it be dealt with?

3 How might we purify the product?

Just checking

1 Give one likely difference between the properties of an ionic compound and a covalent compound.
2 The temperature rises during an exothermic reaction. Explain why this happens by using ideas about energy transfers.
3 A student has bottles containing four organic substances: ethanol, butanoic acid, octene and pentane.
  (a) How many carbon atoms are there in each of these molecules?
  (b) What family does each of these molecules belong to?
4 Give a possible use of (a) ethanol and (b) pentane.
5 (a) What is a polymer?
  (b) Give two reasons why polymers are useful substances.

Assignment tips

To get the grade you deserve in your assignments remember to do the following.

- You will need to describe and explain the properties of ionic, covalent and metallic substances. It might be a good idea to make a page about each of these in your notebook. When you come across some information about each type of substance, add it to your page.

- Studying the up-to-date uses of organic and other specialised materials will mean that you will visit lots of websites. If you find a useful website, remember to save the web address so that you can easily go back to it later when you are writing your assignments. You will also need to provide a list of the websites you used as sources.

Some of the key information you'll need to remember includes the following.

- There are three different types of bonding; they all involve attraction between negative and positive charges.

- In exothermic and endothermic reactions, heat energy isn't created or destroyed, it is transferred to or from chemical energy in the system.

- If something feels hot when you touch it, then energy is being given out to your hand. If it feels cold then energy is being taken in from your hand

- Bond breaking takes in energy but bond making gives out energy.

- To help you name or recognise the structure of organic molecules, like propanol, use this rule: the first part of the name tells you the number of carbon atoms and the ending tells you what family it belongs to.

You may find the following websites useful as you work through this unit.

For information on...	Visit...
covalent and ionic bonding	Bonding
energy changes in chemical reactions	Energy changes in chemical reactions
organic chemicals and their uses	Organic chemicals
nanoparticles	Nanoparticles

5 Applications of physical science

Physics touches all aspects of modern life. Maybe you've watched your favourite music video on your mobile today, or maybe you know someone who is thinking about buying a "green" car. Devices that we take for granted are designed and built by scientists and engineers who have an understanding of the physics of these applications.

In this unit you will investigate a range of applications, from how manufacturers are designing safer and more environmentally friendly cars, to how laser technology is being used to revolutionise surgery.

You will learn about forces and how to investigate various applications such as the suspension springs that are found in cars.

You will study different types of motion such as the science behind roller coaster rides, and you will find out about the many ways that light and sound are used in the modern world, including keyhole surgery and ultrasound.

Finally, you will investigate use of electricity such as temperature sensors that can be used to detect temperature changes in a green house.

Learning outcomes

After completing this unit you should:

1. be able to investigate motion
2. be able to investigate forces
3. be able to investigate light and sound waves
4. be able to investigate electricity.

Assessment and grading criteria

This table shows you what you must do in order to achieve a **pass**, **merit** or **distinction** grade, and where you can find activities in this book to help you.

To achieve a **pass** grade, the evidence must show that the learner is able to:	To achieve a **merit** grade, the evidence must show that, in addition to the pass criteria, the learner is able to:	To achieve a **distinction** grade, the evidence must show that, in addition to the pass and merit criteria, the learner is able to:
P1 Carry out an investigation into an application of the uses of motion **See Assessment activities 5.1, 5.2 and 5.3**	**M1** Analyse the results of the investigation into the uses of motion **See Assessment activities 5.1, 5.2 and 5.3**	**D1** Evaluate the investigation into the uses of motion in our world, suggesting improvements to the real-life application **See Assessment activities 5.2 and 5.3**
P2 Carry out an investigation into an application of the uses of force **See Assessment activity 5.4**	**M2** Analyse the results of the investigation into the uses of force **See Assessment activities 5.4 and 5.5**	**D2** Evaluate the investigation into the uses of force in our world, suggesting improvements to the real-life application **See Assessment activities 5.4 and 5.5**
P3 Carry out an investigation into an application of the uses of waves **See Assessment activities 5.6 and 5.7**	**M3** Analyse the results of the investigation into the uses of waves **See Assessment activities 5.6 and 5.7**	**D3** Evaluate the investigation into the uses of waves in our world, suggesting improvements to the real-life application **See Assessment activities 5.6 and 5.7**
P4 Carry out an investigation into an application of the uses of electricity. **See Assessment activity 5.8**	**M4** Analyse the results of the investigation into the uses of electricity. **See Assessment activity 5.8**	**D4** Evaluate the investigation into the uses of electricity in our world, suggesting improvements to the real-life application. **See Assessment activity 5.8**

How you will be assessed

Your assessment could be in the form of:

- experimental results e.g. on motion, forces, waves and electricity
- a company catalogue e.g. giving the applications of waves
- a presentation e.g. on the applications of electricity.

Mohammed, 18 years old

This unit contains lots of things that we use in our lives, and being able to understand these things was really interesting. I enjoyed learning about how air bags and ABS work in cars. I found it fun learning how speeds cameras work too. The fact that this unit contained lots of experiments, including using lasers, helped me to understand the theory behind the applications.

Another section that I found useful was the section on electricity. It made me more aware of its use in hospitals, for example the ECG traces used to check the heart of patients. As I enjoy reading about cars, I was also surprised to learn that there are now some interesting electric cars that have been developed. Basically, it seems that anything new that is happening in the world of science and technology is in this unit!

Catalyst

Opening up the phone

In groups of three, look at a mobile phone. Using the Internet, for example the 'electronics' section of **How stuff works**, think about the following questions:

- How do you think it works?
- What kind of science is involved in a mobile phone?
- What are the different ways that mobile phones can communicate with each other?
- Do you think there are any health risks involved in using a mobile phone?

5.1 Investigating motion

In this section:

Key terms

Displacement – distance from a reference point in a particular direction.

Speed – change in displacement divided by the time taken for the change to occur.

Velocity – speed in a particular direction.

Acceleration – change of velocity divided by the time taken for the change to occur.

Did you know?

In everyday life we tend to measure speed in mph, but for an athlete or an engineer building cars they measure speed in metres per second. We write this as m/s or $m\,s^{-1}$. 1 mile is about 1609 metres.

400 m running track

After one lap the athlete will have travelled a distance of 400 m but his displacement from the start/finish line will be 0 m

start/
finish

The difference between distance and displacement.

Some quantities of motion and their units	
Speed	m/s or $m\,s^{-1}$
Velocity	m/s or $m\,s^{-1}$
Acceleration	m/s² or $m\,s^{-2}$

To understand motion you need to know about distance, **displacement, speed, velocity** and **acceleration**.

Activity A

Describe two things that you did today that involved some sort of motion.

Speed, distance and time

The 2012 Olympic and Paralympic Games, to be hosted by London, will have racing events such as the 5000 m track event. 5000 m is the distance that the athletes will cover over the ground. The winner will be the one who covers the distance in the shortest time. This is the description of speed: the fastest will be the winner. The speed can be written as:

$$speed = \frac{distance}{time}$$

An athlete running the 5000 m will start from rest, accelerate, move at about the same speed for a while before accelerating again at the end. As the speed is changing throughout the race, it is better to take the average speed, which is total distance ÷ total time.

Activity B

A motorist is travelling in a built-up area with a 30 mph speed limit. A speed camera, which takes two photographs in quick succession, shows that he travelled a distance of 10 metres in 0.5 second. What speed, in mph, do you think the camera will record?

Displacement

Displacement is distance from a reference point in a particular direction, so there is a difference between distance and displacement. The athlete running on the track in the diagram shows the difference.

Velocity

Velocity is speed in a specific direction. For example, you may be travelling 10 m/s east (that is velocity) but if you said 10 m/s that would be speed. The unit of velocity is still m/s or mph.

Acceleration

When a car driver presses down on the accelerator, the car speeds up; it accelerates. **Acceleration** is actually the change in velocity, not the change in speed, because acceleration has direction. When the driver applies the brakes the velocity decreases; this is called **deceleration** (negative acceleration). Acceleration has the unit of m/s^2 or m s^{-2}. It can be written as follows:

$$acceleration = \frac{change\ in\ velocity}{change\ in\ time}$$

Activity C

Explain the difference between speed and velocity.

A car accelerates and decelerates many times during a journey.

Worked example

A car speeds up from 1 m/s to 20 m/s in ten seconds. What is its acceleration?

$$acceleration = \frac{change\ in\ velocity}{change\ in\ time}$$

$$acceleration = \frac{(20 - 1)}{10} = \frac{19}{10} = 1.9\ m/s^2$$

BTEC Assessment activity 5.1 P1 M1

Light gates are often used to investigate velocity and acceleration. They use similar principles to some speed cameras.

1   Investigate how you can use light gates to measure velocity and acceleration in the laboratory. **P1**

2   The table below gives some measurements that were obtained from a trolley in an experiment using light gates. Calculate the average speed of the trolley for each set of readings, explaining your working. Calculate the overall average speed. **M1**

Quantity measured	1st readings	2nd readings
Total time taken between A and B (s)	0.04	0.08
Total distance travelled between A and B (cm)	5.8	11.7

Grading tip

When performing experiments for **P1**, **M1** and **D1**, always think about the errors that may occur in the experiments. How can you minimise errors? Do you need to change the equipment used to make the experiment more accurate or do you need to improve the method of data collection? Whatever experiment you decide to do, remember it needs to relate to a real-life application and you will need to evaluate it.

5.2 Graphs of motion

In this section: P1 M1 D1

Paul is a data analyst working for a drag-racing team. He needs to understand motion for the team to get the best out of their car.

Key terms

Uniform motion – an object moving with velocity or acceleration that is not changing.

Non-uniform motion – an object moving with velocity or acceleration that is changing.

Uniform velocity

Paul is analysing the data as the cars reach their terminal velocity. This is the maximum constant, or **uniform**, velocity that they reach when the force from the engine is balanced by the air resistance. Paul's team are using the red car and their rival team are in the blue car.

The data logged at the end of their race is shown below.

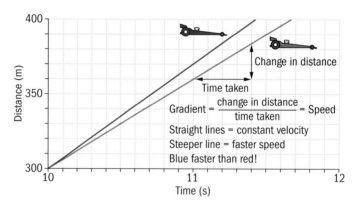

Distance–time graphs for Paul's team and their rival team.

Non–uniform velocity

Paul is now analysing the start of the drag race. This is when the cars accelerate, that is, speed up. The diagram in the margin shows a distance–time graph of the red and blue cars. Notice that the graphs are not straight lines but curves; they are non-linear. We say the cars are travelling with **non-uniform** velocity.

Uniform acceleration

Following some data analysis, Paul is now able to plot graphs of the velocity of the cars for the vital 5 seconds after the start of the race. This is shown in the velocity–time graph. You will notice that the velocity is increasing for both cars. Consider the red car: what do you notice about how fast its speed is increasing? You will see that it is increasing at a constant rate. For example, after 1 second it is at 10m/s and after

 Activity A

From the graph shown above, calculate the velocity of both cars when they have reached their terminal velocity at the end of the drag strip.

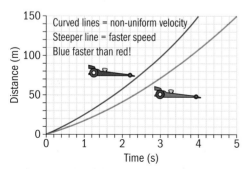

Distance–time graph of the red and blue cars.

2 seconds it is at 20 m/s. From the slope of the velocity–time graph we can get the acceleration. This is a very useful measurement for a race engineer to determine. We say that the acceleration is a constant and that the graph shows uniform motion.

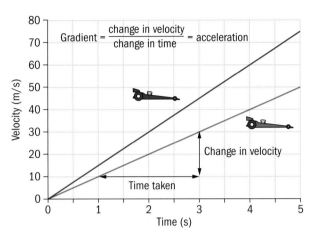

Velocity–time graph of the red and blue cars.

The area under the velocity–time graph also has real meaning. The area actually represents the distance travelled.

Worked example

Using the velocity–time graph, calculate the acceleration of the red car.

acceleration = slope = (change in y) ÷ (change in x)

acceleration = (change in velocity) ÷ (change in time)

acceleration = (50 − 0) ÷ (5 − 0)

acceleration = 10 m/s^2

BTEC Assessment activity 5.2   P1 M1 D1

1 Investigate how speed cameras that measure average speed work. **P1**

2 The following data were obtained on the A616 (single carriage way) in South Yorkshire at various data points. Note the national speed limit is 60 mph.

	Data point 1 (0 miles)	Data point 2 (1 mile)	Data point 3 (2 miles)
Speed (mph)	55	59	63

What is the average speed calculated from data points 1 and 3? **M1**

3 Explain the disadvantages of using the average speed trap method. **D1**

Activity B

From the velocity–time graph, which car has the highest acceleration? Explain your answer.

Science focus

The area under a straight line graph is simply the area of a triangle.

area of triangle = 0.5 × base × height

Activity C

Using the velocity–time graph, calculate:

(i) the acceleration of the blue car

(ii) the distance travelled by the blue car.

Functional skills

Mathematics

You can use your maths skills when using graphs to make a calculation. Remember to look at the scale on the axes carefully. Make sure that you are using the correct value for the grid intervals. They may be different on the y-axis and the x-axis.

Grading tip

Remember when planning your experiments for **P1** and **M1**, you need a wide variation of data points that cover small to large values of distance. You should aim for at least seven data points. Make sure when you are plotting graphs that the data points fit at least 70% of the graph paper.

For **D1**, you need to suggest improvements that you could make if the experiment was used in a real-life application.

5.3 Full of energy

In this section: P1 M1 D1

Key term

Gravitational potential energy – stored energy in a gravitational field.

Unit 2: You can find out more about different types of energy on page 38.

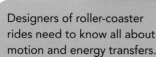

Designers of roller-coaster rides need to know all about motion and energy transfers.

Gravitational potential energy

If something has potential energy it has the potential to do work. One important example is **gravitational potential energy**: this is the potential energy due to the force of gravity. The higher something is the more gravitational potential energy it has. The symbol of this energy is E_P. Engineers calculate this energy using the equation:

$$E_P = \text{mass} \times \text{acceleration due to gravity} \times \text{change in height}$$

Did you know?

The acceleration due to gravity, on Earth, is approximately $10\,\text{m/s}^2$. This means that if you fall your speed will go up by $10\,\text{m/s}$ (approximately 22 miles per hour) every second.

Worked example

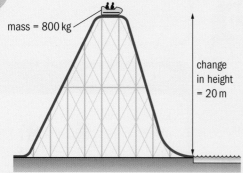

mass = 800 kg

change in height = 20 m

On a water-splash ride, a car with two people stops at a height of 20 m from the ground. The mass of the car, containing the people, is 800 kg. What is the gravitational potential energy of the car at this height?

$E_P = \text{mass} \times \text{acceleration due to gravity} \times \text{change in height}$

$E_P = 800 \times 10 \times (20 - 0) = 160\,000\,\text{J}$

The gravitational potential energy of the car (and passengers) is 160 000 J.

Activity A

What would the gravitational potential energy be if the water-splash car was at a height of 30 m?

Kinetic energy

Kinetic energy is movement energy and has the symbol E_K. Engineers calculate kinetic energy using the equation:

$$E_K = \tfrac{1}{2} \times \text{mass} \times (\text{speed})^2$$

Worked example

If the car carrying the two people is now travelling at a constant speed of 7 m/s, what is the kinetic energy of the car (and passengers)?

$E_K = \frac{1}{2} \times mass \times (speed)^2$

$E_K = \frac{1}{2} \times 800 \times (7)^2 = 19\,600\,J$

The kinetic energy of the car and passengers is 19 600 J.

Conserving energy

This diagram shows a car going over a roller-coaster ride. You can see how the potential energy at the top of the ride is transferred to kinetic, sound and heat energy as the car travels down the slope.

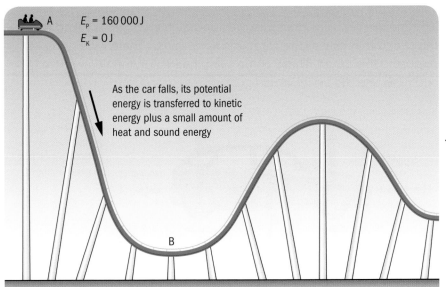

A $E_P = 160\,000\,J$
$E_K = 0\,J$

As the car falls, its potential energy is transferred to kinetic energy plus a small amount of heat and sound energy

B

Journey of a roller coaster.

Activity B

Refer to the diagram on the left.

(a) What is the potential energy at A and B?

(b) What is the kinetic energy at A and B, if we assume that no energy is transformed into sound and heat?

 BTEC ## Assessment activity 5.3 P1 M1 D1

Working in pairs, imagine you are the engineers trying to design a roller coaster.

1 Produce a plan for your dream roller coaster, marking your masses at a number of heights on the roller coaster. **P1**

2 Calculate the potential energy and kinetic energy you would expect at the heights indicated in Question 1. **M1**

3 What equipment would you need to measure both the potential and kinetic energy at the heights indicated in Questions 1 and 2? How might this be used as a safety device in a real-life roller coaster? **D1**

Grading tip

Make sure when you are using values of mass that these are in the SI unit of mass, which is the kilogram. The heights should also be in the SI unit, which is the metre.

5.4 Investigating forces

In this section: **P2** **M2** **D2**

Balanced and unbalanced forces

When you sit on a chair, you are exerting the force of gravity (your weight) on the chair. The chair also pushes upwards on you; this force is called a reaction force. These forces must be equal and opposite. We say they are balanced. When the forces are balanced we say that the net force is zero. The unit of force is the newton, with symbol N.

Activity A

What do you think would happen if your force due to gravity was bigger than the reaction force provided by the chair?

Crash test dummy.

Imagine driving a car, the force provided by the engine is called the thrust and that causes the car to move forwards. There will be forces opposing this thrust. These come from the friction between the road and tyres and the air resistance.

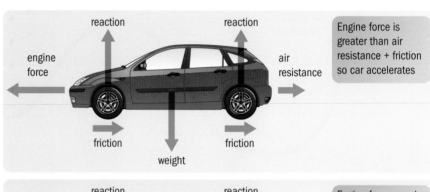

Engine force is greater than air resistance + friction so car accelerates

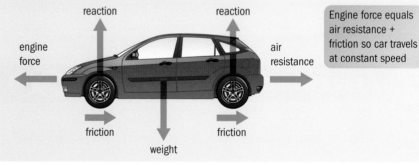

Engine force equals air resistance + friction so car travels at constant speed

The balanced and unbalanced forces on a car.

Anti-lock braking systems (ABS) are fitted to modern cars to prevent skidding when drivers brake hard.

Friction can also be useful. The friction force between the road and tyres is increased on a rough road, allowing the tyres to grip the road. This prevents skidding. This frictional force is reduced when there is ice or water on the road. The channels in the tread of a tyre allow water to escape from between the tyre and the road surface. A worn tyre has less tread. This means that the tyre will skid when there is only a small amount of water on the road.

Some useful quantities and their units	
Mass	kg
Acceleration	m/s² or m s⁻²
Force	kg m/s² = N (newton)

When a sky diver leaps from an aeroplane they are relying on the principle of balanced forces to limit their speed.

(a)

drag

weight

Weight (from force of gravity) is greater than drag (from air resistance) so sky diver accelerates

(b)

drag

weight

Weight is equal to drag so sky diver falls at constant velocity, called terminal velocity

The forces acting on a skydiver **(a)** immediately after leaving the aircraft and **(b)** when falling at terminal velocity.

It is a legal requirement for cars, light vans and light trailers to have a tyre tread depth of at least 1.6mm.

Engineers calculate the net force acting on an object using the equation:

$$\text{net force} = \text{mass} \times \text{acceleration}$$

 Activity B

If the net force acting on the 800 kg car decreases to 3000 N, what is the new acceleration?

 Worked example

What is the net force acting on an 800 kg car when it is accelerating at 5 m/s²?

net force = mass × acceleration

net force = 800 × 5 = 4000 N

The net force acting on the car is 4000 N.

 Did you know?

The force acting on anything on the Earth is called its weight. This can be calculated from
weight = mass × acceleration due to gravity

Acceleration due to gravity is approximately 10 m/s². The weight of something is expressed in newtons and not in kilograms.

When an engineer designs the safety systems for a car he needs to use the force due to acceleration equation. He can then work out how much force the airbags and crumple zones need to provide to keep the occupants safe as they decelerate rapidly during a crash.

 Assessment activity 5.4

1 Some skydivers wear wing suits that provide increased air resistance. How will their terminal velocity compare with that of a conventional skydiver? **P2**

2 Explain what happens when a skydiver opens his parachute. Draw diagrams that show the forces and how they affect his velocity. **M2**

3 A parachute designer is designing a parachute for use in a fast military jet. What forces would he need to consider? **D2**

Grading tip

The **P2**, **M2** and **D2** criteria expect you to carry out an experiment that investigates forces. Make sure that your experiment relates directly to a real-life application, for example friction and tyres, or terminal velocity and skydivers. When presenting your results, make sure that you have a table of results and that any graphs you draw are labelled correctly.

5.5 Forces in action

In this section: P2 M2 D2

Key term

Tensile force – a stretching force pulling at both ends of an object.

Air bags and seat belts provide forces that keep you safe in an accident.

Activity A

Why are crumple zones used in cars?

Science focus

The spring constant depends on what the spring is made from. It tells us the stiffness of a spring. The bigger the spring constant, the stiffer is the spring.

Type of spring	Spring constant (N/m)
Lorry suspension	105
Garage door	500

Impact forces

When an object is stopped during a collision, the impact time of the collision is very important. The shorter the impact time the greater the force needed to stop the object. In many collisions it is desirable to increase this impact time. Imagine kicking a brick; it's a painful experience so don't do it. This is because the time between your foot first touching the brick and your foot stopping is very short. Now try kicking a football. The experience is not too painful because the ball is designed to squash. This increases the impact time.

Engineers have developed crumple zones for cars. These are designed to squash, which increases the impact time during a collision. This means that the occupants of the car experience lower deceleration forces. Crumple zones are fitted to the front and back of modern cars which is why they are often write-offs after a collision.

Worked example

1 A car (with just the driver in) is travelling at 15 m/s (20 mph) when it collides with a stationary car and is brought to rest in 0.1 seconds.

(a) **Calculate the deceleration of the car.**

(b) **If the mass of the driver is 80 kg, what force would he experience?**

(a) acceleration = change in velocity ÷ change in time

acceleration = $(0 - 15) \div 0.1 = -15 \div 0.1 = -150 \, \text{m/s}^2$

(b) net force = mass × acceleration

net force = $80 \times 150 = 12\,000 \, \text{N}$

Forces acting on springs

When an object has its length increased because a force pulls it, there is a **tensile force** on it. An example of this is seatbelts. These stretch due to the force applied to them as they restrain a passenger in a collision. Another example of a tensile force is the stretching of a garage door spring, which slows down the closing of the door.

When something reduces its length because a force squashes it, we say there is a **compressive** force on it. The suspension systems of cars and aeroplane undercarriages are designed to cope with compressive forces.

Activity B

What is the difference between a compressive force and a tensile force?

Engineers can calculate the extension that results from a tensile force on a spring by using Hooke's law. This tells us that the amount a spring stretches is proportional to how much force is applied to it. If you double the force on the spring, you will double its extension.

> force = spring constant × extension

Worked example

2 **A lorry has a 400 kg load sitting directly above one of its suspension springs. If the spring constant is 6.5 × 10⁴ N/m, calculate how much the spring compresses.**

force = spring constant × extension

force = weight of load = mass × acceleration due to gravity

force = 400 × 10 = 4000 N

extension = force ÷ spring constant

= 4000 ÷ (6.5 × 10⁴) = 0.062 m

The spring compresses 6.2 cm.

Hooke's law is true only below a special point called the elastic limit. Materials that are not stretched beyond their elastic limit are said to behave elastically. Once you remove the force they will return to their original length. If you go past the elastic limit then the spring is deformed and is no longer elastic. Seatbelts that have been involved in a crash have to be replaced because they would have exceeded their limit.

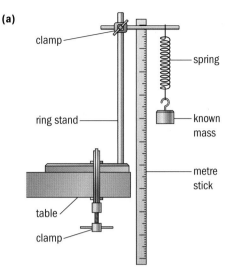

(a)

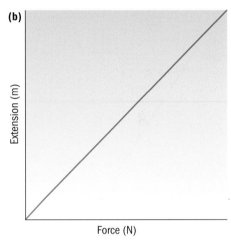

(b)

(a) An experiment with a spring to demonstrate Hooke's law. (b) The straight-line graph shows that the extension is proportional to the force, as predicted by Robert Hooke.

 ## Assessment activity 5.5

In a spring manufacturing company the springs are tested in the quality control department to ensure that each one obeys Hooke's law. Some of the data for one spring are shown in the table:

Force (N)	Extension (cm)
1.1	0.16
2.2	0.53
3.1	0.65

1 Looking at the data, does the spring obey Hooke's law? Explain your answer. **M2**

2 How would you improve the accuracy of the measurements made? **D2**

Grading tip

Before carrying out your experiment, make sure that you plan carefully. Think about what apparatus you will need and what results you are likely to get from your experiment. For **D2**, you will need to suggest improvements if your measurement technique was used in a real-life application.

5.6 Applications of refraction of light waves

In this section: P3 M3 D3

Key term

Normal – imaginary line used in light experiments, which is drawn at 90° to the surface of a mirror or to the boundary between two materials.

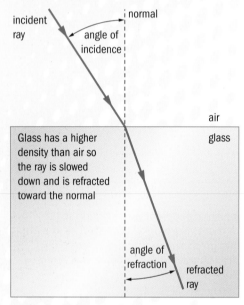

Refraction of light.

Activity A

What is the difference between a convex lens and a concave lens?

Activity B

What type of lens is in a healthy human eye?

When light passes from one material to another (such as from water to air) it changes direction. This is called **refraction**.

Convex and concave lenses

The human eye uses refraction to focus light so that we can see an image. The lens of our eye is a special lens called a convex lens. Convex lenses focus light to a point. There is also another type of lens called a concave lens which makes light diverge.

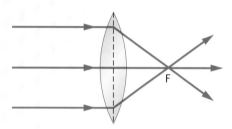

Convex lens focuses light parallel to axis at focal point F

Concave lens diverges light – parallel light appears to come from focal point F

Refraction by convex and concave lenses.

Ray diagrams can be used to show the effect of a lens when viewing an object.

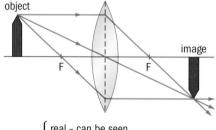

Image is { real – can be seen on a screen / inverted – upside down

Image is { virtual – where light appears to come from / upright – same as object

Ray diagrams showing real and virtual images.

Correcting eyesight

Concave and convex lens are used to correct for short sight and long sight. The diagrams on the right show how.

Activity C

In class, Jenny can't read the whiteboard but can read the text book. She realises that she needs to go to the optician. What type of lens should the optician prescribe?

Light from a distance converges to a point in front of the retina. The image is blurred

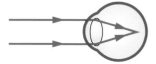

(a) Short sight – cannot focus on distant objects.

Concave lens makes light diverge. Distant objects can now be seen clearly

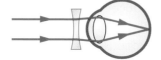

(b) Short sight correction.

Light from a close object converges to a point behind the retina. The image is blurred

(c) Long sight – cannot focus on near objects.

Convex lens makes light converge. Close objects can now be seen clearly

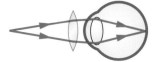

(d) Long sight correction.

Ray diagrams for eyesight correction.

Total internal reflection

If light is travelling from a denser material, such as glass, into air it will be partially refracted and partially reflected. If light reaches the boundary at the critical angle (or greater) it will all be reflected back. This is called total internal reflection. The diagram shows what happens. You see that the glass now acts as a mirror. This is how TV and Internet signals are transmitted in optical fibres.

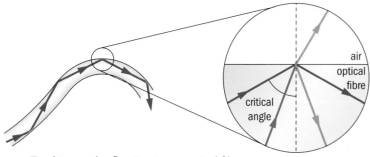

air
optical fibre
critical angle

If the angle of incidence is greater than the critical angle the light is reflected back into the optical fibre

Total internal reflection in an optical fibre.

One of the fascinating ways of using laser light and the property of total internal reflection is in keyhole surgery. Surgeons need to make only a small incision 5–10 mm across, through which a bundle of optical fibres can be inserted. These carry light into the body which is then reflected back to be viewed by the surgeon. They can then perform the operation through another small incision whilst observing what they are doing on a TV monitor.

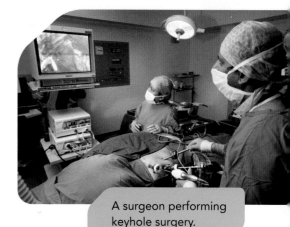

A surgeon performing keyhole surgery.

BTEC Assessment activity 5.6 P3 M3 D3

1 There is much software available to use for free online or for download that simulates the properties of lenses. Investigate the properties of lenses further using these software packages. **P3**

2 Draw a ray diagram for a convex lens being used as a magnifying glass. **M3**

3 What equipment would you need to set up an optical system in a laboratory? How would you set up your equipment to perform experiments on convex and concave lenses? What are the strengths and weaknesses of your set-up? **D3**

Grading tip

For **P3** and **M3**, make sure that all your ray diagrams indicate the object, image and focal point. For **D3**, remember that your investigation needs to be related to a real-life application, for example a telescope or a medical imaging device.

5.7 Applications of sound waves

In this section: P3 M3 D3

Key term

Ultrasound – sound waves with frequencies higher than the human ear can hear.

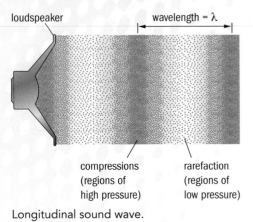

Ultrasound is used to view a baby in the womb.

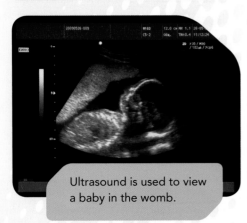

loudspeaker · · · · wavelength = λ

compressions (regions of high pressure) · rarefaction (regions of low pressure)

Longitudinal sound wave.

Unit 2: See page 48 to learn more about waves.

Sound waves are produced by something vibrating, like a loudspeaker cone or someone's vocal chords. The air is pushed together to form compressions and expanded to form rarefactions. The distance between two successive compressions is the wavelength. The regions of high and low pressure travel away from the loudspeaker in a **longitudinal wave**.

These vibrations travel faster when the particles are closer together such as in liquids and solids. For example, in air the speed of sound is about 330 m/s; in water it is about 1500 m/s; and in steel it is about 5000 m/s.

Activity A

Imagine you were in space. Would anyone be able to hear you scream? Explain your answer.

When engineers and scientists make calculations to predict how waves behave they need to know the frequency of the wave. The frequency of a sound wave is the number of compressions that pass a fixed point in one second. The human ear can hear frequencies from 20 Hz to about 20 kHz.

The speed v, frequency f and wavelength λ are related by the equation:

$$v = f \times \lambda$$

Each person's voice has a particular frequency spectrum (range of frequencies). This is called a speech signature. Computer software has been developed that shows a person's speech signature to be as unique as their fingerprints. This means that forensic analysts are able to identify criminals by recording and analysing their telephone conversations.

Ultrasound

Humans don't hear sounds that are above about 20 kHz. These are called **ultrasounds**.

Just as light is reflected, sound waves are also reflected. For example, echoes are caused by sound waves reflecting from surfaces.

Case study: Seeing with sound

Kate is a midwife who uses an ultrasound machine to obtain images of her patients' unborn babies. She uses ultrasound because it is does not harm the baby or mother like X-rays would.

The reflected ultrasound allows the doctor and midwife to determine the size of the baby and to look out for certain defects that the baby may have.

Before starting an examination Kate covers the ultrasonic probe with a jelly-like substance. Why does she do this?

How fast do you think the sound waves are travelling during the scan? Explain your answer.

Worked example

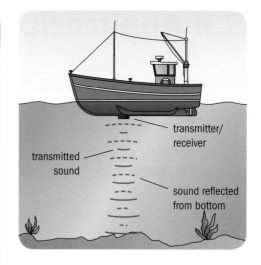

Ultrasound waves are used in SONAR (Sound Navigation and Ranging) to determine the depth of the seabed and to locate underwater objects. The ship in the picture sends an ultrasound signal and receives the echo of the ultrasound in 0.5 seconds.

Calculate the depth of the sea bed if the speed of the ultrasound in water is 1500 m/s.

Remember that speed = distance ÷ time

Make the distance the subject of the equation:

distance = speed × time = 1500 × 0.5 = 750 m

This is the total distance that the waves travel (going down and coming up) so the depth will be half this.

The depth of the sea is 375 m.

transmitter/receiver

transmitted sound

sound reflected from bottom

BTEC Assessment activity 5.7 **P3** **M3** **D3**

1 In the laboratory, how would you investigate how sound travels in a range of different materials? **P3**

2 What equipment would you use to make your measurements and how would you present your results? **M3**

3 How is ultrasound used in the food industry? What are the strengths and weaknesses of this application? **D3**

Grading tip

Remember that any investigations you perform need to relate to experiments that you will carry out in the laboratory. For **D3**, your experiments need to relate to a real-life application such as SONAR and you also need to evaluate your experiments.

WorkSpace
Robert Palmer
Smith & Son Optometrists

I work as an optometrist for a high-street optician, carrying out eye examinations.

Each appointment lasts about 20 minutes. In that time I perform a range of tests to check how well the eyes are functioning. By looking at the inside of the eye I can also check for diseases such as cataracts, glaucoma or even some cancers.

I test how well the lens in the eye can focus small objects like letters. If they seem blurred to the patient, I put an extra lens in front of their eyes. Different-shaped lenses refract the light in different ways, so by trying lenses with different thicknesses and shapes I can find a combination that produces a sharp image on the patient's retina. From this I can prescribe glasses or contact lenses to correct the patient's sight. I try to encourage all my patients to have eye checks at least once every two years.

Think about it!

1 Why should patients have regular eye checks?

2 If someone is very short-sighted, I often prescribe glasses with very powerful lenses to refract the light by a larger angle. Will these glasses be thick or thin? Why would contact lenses not be a good idea in this case?

5.8 Applications of electricity

In this section: P4 M4 D4

Key terms

Thermistor – a special type of resistor whose resistance changes with temperature.

Light dependent resistor (LDR) – a special type of resistor, made out of semiconductors, whose resistance decreases with increasing light.

Light emitting diode (LED) – a semiconducting diode that lights up when a current passes through it.

Electrocardiograph (ECG) – a machine that measures the electrical activity of the heart.

Defibrillator – a device that delivers electrical energy into the body to restore a normal heart beat.

Series and parallel circuits

Circuits that make up electrical equipment can be arranged either in series or in parallel. The lighting in your home is likely to be connected in parallel. If one light goes out none of the others will be affected. The diagram below shows two bulbs in series and two bulbs in parallel. Note the way the current and voltage behave in both circuits. The bulbs in the series circuit are dimmer because they are dissipating less energy.

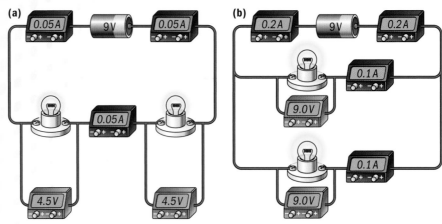

Supply voltage is split across the bulbs. Equal current flows through each bulb

Supply voltage is the same across each bulb. Supply current is shared through each bulb

Voltage and current in **(a)** a series circuit and **(b)** a parallel circuit.

Activity A

How does the voltage across the resistors behave in a series circuit?

Thermistors

A **thermistor** is a special resistor that can be used to detect changing temperature. For example, it can be used as the sensor in a device to monitor the temperature in a greenhouse. Some thermistors are manufactured so that they have a negative temperature coefficient. This means that their resistance goes down as their temperature increases. It is this change of resistance that can be used as an input signal to an electronic control device.

Activity B

How does a thermistor work?

Christmas lights used to be connected in series – if one bulb failed the rest of them went out.

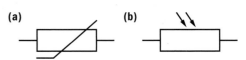

Electrical symbol for **(a)** a thermistor and **(b)** a light dependent resistor.

Light dependent resistor

A **light dependent resistor** (LDR) is another special type of resistor. Its resistance reduces when light strikes its surface. For example, in the dark the resistance of an LDR can be as high as 1 MΩ. When illuminated with light, its resistance can drop to a few hundred ohms. LDRs are useful as light sensors, for example as part of a burglar alarm. In this application, an LDR is placed in a lit place. When a burglar enters and casts a shadow on the LDR, its resistance increases. This change in resistance is used as the input signal to the circuit that sounds the alarm.

Light emitting diode

A **light emitting diode** (LED) emits light when a current passes through it. LEDs are used in many applications, such as the standby light in a TV, or the light that goes on and off if there is a problem with a washing machine.

Activity C

State one advantage and one disadvantage of using electric cars.

ECG and defibrillators

The human body is an amazing electrical conductor. This allows doctors to monitor the electrical activity of the heart by measuring small voltages within the body. This is done with an **electrocardiograph** (ECG).

It is now very common to see **defibrillators** in public places. These devices are used to restart someone's heart if they are having a heart attack. A computer in the defibrillator monitors the abnormal electrical activity of the heart. It then calculates what electrical charge is needed to shock the heart back into its correct rhythm.

BTEC Assessment activity 5.8 (P4) (M4) (D4)

1 Plan an experiment that you can carry out in the laboratory with bulbs, LEDs, LDRs or thermistors. Think about the equipment that is available to you. (P4)

2 What results do you expect to get from your experiments? (M4)

3 How can you use the results of your experiments to design a real-life application? (D4)

Grading tip

For (P4), remember that when you perform electrical experiments you connect the ammeters in series and the voltmeters in parallel.

Did you know?

Light dependent resistors are used in street lights so that they turn on as it gets dark.

Clock display made with LEDs.

  Electrical symbol for an LED.

Science focus

Electric cars are at the forefront in the drive to reduce the world's reliance on oil. Their electric motors do not produce any direct pollution and the stored energy they use can come from renewable power sources.

Advantages of electric cars	Disadvantages of electric cars
Lower overall pollution	Lower top speed
Quiet	Limited range
Cheap to run	Long charging time
Can recover and store energy that would normally be lost as heat during braking	

Just checking

1 Explain the difference between sound and light waves.
2 What information can you get from a velocity–time graph?
3 Write down the difference between kinetic energy and potential energy.
4 Explain how a simple speed camera works.
5 Describe two applications of Ultrasound
6 What is total internal reflection? Give one use of this.
7 Describe how a spring that obeys Hook's law behaves. Give one use of springs.
8 What is meant by an ECG signal? Where is it useful?
9 Sketch the symbols for the following electrical components:
   (a) thermistor, (b) an LED and (c) an LDR
10 Describe the voltage and current relationships for bulbs in series and in parallel.

Assignment tips

In this unit you will need to perform many experiments. This means that planning your experiment and presenting the results clearly is very important.

• Make sure that the experiments you plan relate to real life applications; this is necessary so that you can get the higher grades.

• When you plan an experiment, think about what equipment is available to you.

• Think about the results you expect to get from your experiments; Will these results be useful? They will need to help you to improve an application, if you want to get a distinction.

• Make sure that your results are presented in a table, with the correct symbols and units.

• When you plot a graph make sure that you use a sensible scale – your data should take up about 70% of the space on the graph. Make sure you label the axes and include a title.

For information on...	Visit...
motion	Motorsport mathematics
lenses and correcting eye defects	Freezeray: Interactive lenses
ultrasound and its applications	Ultrasound

6 Health applications of life science

Being healthy depends on lots of things such as what you eat and drink and how much exercise you take. Sometimes no matter how much you look after your body it can go wrong due to a disease or illness.

In this unit you will learn about the amazing way the human body helps to keep us healthy and how it fights disease. This can involve the body's natural defence system but sometimes we may need help from different areas of medicine to stop us becoming ill. You will use the understanding you have gained from Unit 3 'Biology and our environment' to develop your knowledge and skills on different things that can affect our health.

Your assignments will enable you to gain an insight into the many jobs that are available within the application of life science. You may take the role of a practitioner at a health centre advising people of different ages on the positive and negative sides of diet and exercise. Alternatively, you may practise the skills of a science journalist when presenting the facts for new medical treatments such as gene therapy. You will be able to understand the role of genetic counsellors when you prepare information that helps people make balanced decisions about gene therapy.

Learning outcomes

After completing this unit you should:

1 be able to investigate factors which contribute to healthy living
2 know how preventative measures can be used to support healthy living
3 be able to investigate how some treatments are used when illness occurs.

Assessment and grading criteria

This table shows you what you must do in order to achieve a **pass**, **merit** or **distinction** grade, and where you can find activities in this book to help you.

To achieve a **pass** grade, the evidence must show that the learner is able to:	To achieve a **merit** grade, the evidence must show that, in addition to the pass criteria, the learner is able to:	To achieve a **distinction** grade, the evidence must show that, in addition to the pass and merit criteria, the learner is able to:
P1 Assess the possible effects of diet on the functioning of the human body **See Assessment activities 6.1 and 6.2**	**M1** Explain how the diet and exercise plan will affect the functioning of the human body **See Assessment activities 6.3 and 6.4**	**D1** Evaluate your exercise plans and justify the menus and activities chosen **See Assessment activity 6.4**
P2 Design diet and exercise plans to promote healthy living **See Assessment activities 6.3 and 6.4**		
P3 Outline how the immune system defends the human body **See Assessment activity 6.5**	**M2** Describe the action of each component of the immune system **See Assessment activity 6.5**	**D2** Evaluate the effectiveness of vaccination and screening programmes **See Assessment activities 6.6 and 6.7**
P4 Identify the role of specific health screening programmes **See Assessment activity 6.7**	**M3** Describe the changes in the human body following a vaccination **See Assessment activity 6.6**	
P5 Carry out an investigation into the effects of antibiotics **See Assessment activity 6.8**	**M4** Using secondary data, carry out an investigation into the effectiveness of different kinds of medical treatment in the control of health. **See Assessment activities 6.8 and 6.9**	**D3** Evaluate the use of different kinds of medical treatments, justifying your opinions. **See Assessment activities 6.8 and 6.9**
P6 Describe what gene therapy is, giving examples of diseases and conditions associated with it. **See Assessment activities 6.10 and 6.11**		

How you will be assessed

Your assessment could be in the form of:

- an illustrated magazine article e.g. on healthy living
- a visual or audio presentation e.g. on preventative measures in healthcare
- a portfolio of information e.g. on different medical treatments.

Arsalan, 17 years old

I think this unit is quite interesting as it relates to our everyday lives such as the importance of diet and exercise and also how gene therapy is used to treat diseases.

This unit allows me to act out the role of some of the people who have jobs in health. It will help me to decide what I want to do and also help me to find the job I want to do following this course.

Catalyst

Healthy spider!

What does a healthy lifestyle mean to you and why do you need to maintain it?

Working in pairs, create a spider diagram to record your thoughts.

6.1 Health and disease

Nurses, doctors and other people working in healthcare need to know about illnesses and **diseases** and what causes them. This allows them to give patients the correct treatment and restore them to good **health**.

Physical illness

If you have ever had chicken pox then you will know that one of the signs is spots on your body. With a cold the signs may be a runny or blocked nose and a high temperature. These are **symptoms** that can be seen when you are unwell.

Key terms

Disease – a malfunction of the mind or body caused by departure from good health.

Health – a state of mental, physical and social wellbeing, not just the absence of disease.

Symptom – a sensation or change in bodily function associated with a particular disease.

Pathogen – an agent that causes disease, especially a living microorganism.

Activity A

What physical illnesses have you had in your lifetime? Have a discussion with a partner about the symptoms you had and the treatment you received, if any. Make a list of symptoms and treatments.

Mental illness

The way your brain works can be affected by mental illness. This can make you behave in unusual ways. This can last for hours, days or even years. It can be treated using therapy and medication. Many patients make a full recovery.

There are two types of mental illness.

- **Neurotic illness** – usually linked to emotional or social factors like losing a relative or stress in the workplace. An example of a neurotic illness is depression.

A nurse gives a patient drugs which have been prescribed by a doctor for an illness.

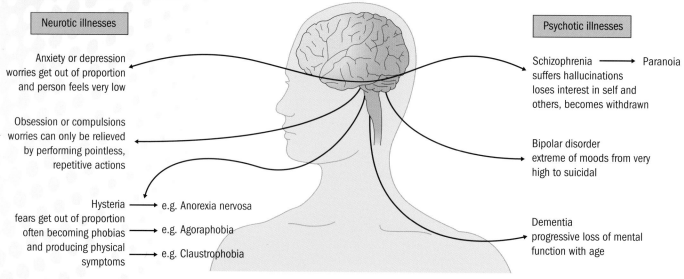

Neurotic illnesses

Anxiety or depression worries get out of proportion and person feels very low

Obsession or compulsions worries can only be relieved by performing pointless, repetitive actions

Hysteria fears get out of proportion often becoming phobias and producing physical symptoms

→ e.g. Anorexia nervosa
→ e.g. Agoraphobia
→ e.g. Claustrophobia

Psychotic illnesses

Schizophrenia → Paranoia suffers hallucinations loses interest in self and others, becomes withdrawn

Bipolar disorder extreme of moods from very high to suicidal

Dementia progressive loss of mental function with age

Mental illness can be grouped into two types: neurotic and psychotic.

- **Psychotic illness** – can result from emotional stress. Often it is caused by some form of damage or ageing to the brain. Someone who suffers from a psychotic illness may lose control of their behaviour and actions. Schizophrenia is an example of a psychotic illness.

Infectious diseases

If you catch a cold from someone you have caught an infection. Infectious diseases are caused by **pathogens** and are spread from other people or animals. Some diseases spread by animals include swine and bird flu.

Unit 3: See page 86 for more about pathogens.

Pathogen spread by	Examples
Droplets in air	Coughing, sneezing or breathing, e.g. cold, influenza, diphtheria and anthrax
Infected food and water	Contamination of food caused by bad hygiene, e.g. food poisoning Contamination of water by pollution and sewage, e.g. cholera and dysentery
Direct contact	Many sexual diseases, e.g. genital herpes, AIDS, *Chlamydia*
Objects	Sharing things such as hairbrushes or bed sheets, e.g. impetigo Sharing footwear or towels, e.g. athlete's foot
Insects and other animals	Organisms that spread disease from one person to another are called vectors, e.g. mosquitoes spread malaria, fleas spread the bubonic plague of 1665

Pathogens can be spread in five ways.

Non-infectious diseases

A non-infectious disease can't spread from one person to another. Examples include cancer, diabetes and inherited conditions such as cystic fibrosis or sickle cell anaemia.

Activity B

Make some notes about the main points of infectious diseases that you could teach to a class of younger children.

PLTS

Independent enquirers

You can practise this skill as you research information for your presentation to deliver to the rest of the class.

BTEC Assessment activity 6.1 **P1**

1 In groups, discuss why it is important that we lead a healthy lifestyle and avoid getting ill. Make a list of your reasons. **P1**

2 Produce a 5 minute presentation on 'Health and disease' to present to the rest of the class. Include dietary factors in your presentation. **P1**

Grading tip

You need to show that you know about the body in health and disease. For example, the body shows symptoms when you have an illness. **P1**

6.2 Diet and the human body

Key terms

Diet – what we normally eat and drink.

Overnutrition – where more food and nutrients are supplied than are needed for normal growth and development.

Obesity – when a person has an abnormally high amount of body fat.

Undernutrition – caused by lack of food or being unable to absorb foods.

Anaphylactic shock – severe allergic reaction that can cause death.

Science focus

The diet is made up of the following components:

- carbohydrates for energy
- fats for energy, protection and insulation
- protein for growth and repair
- fibre to add bulk and help move food through the gut
- water – essential for cell function and bodily processes
- vitamins and minerals (see table).

Eating too much and taking too little exercise can lead to obesity.

Diet and health

To stay healthy you need to eat a balanced **diet**. Diet involves the intake of foods and is not just something that we do to lose weight. A balanced diet includes all the components shown in the margin box, which must be eaten in the correct proportions. Dietitians design diet plans for patients to make sure they eat enough essential nutrients and not too many calories or fat.

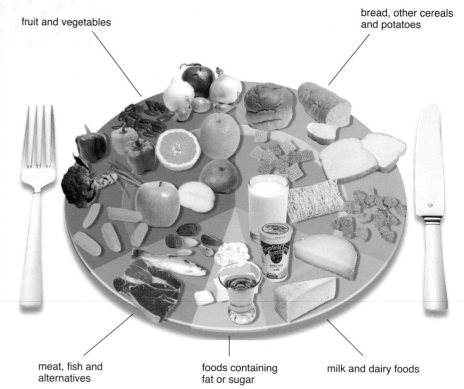

fruit and vegetables

bread, other cereals and potatoes

meat, fish and alternatives

foods containing fat or sugar

milk and dairy foods

This plate shows how much of each food group makes up a balanced diet. Does it match your diet?

Activity A

Discuss with a partner whether or not you eat a balanced diet.

Effects of eating too much or too little

Eating too much food with more calories than you need (**overnutrition**) can make you gain weight and sometimes cause **obesity**.

Undernutrition is caused by lack of food or being unable to absorb foods. If you eat fewer calories than you use each day you will lose weight. **Anorexia nervosa** is a psychological **eating disorder** that can lead to drastic weight loss. Sufferers become extremely ill and may die if not treated by professionals such as psychologists, nurses or psychiatrists.

	Nutrient	Source	Function	Result of deficiency
VITAMINS	A	Liver, yellow or orange fruits and vegetables	Helps you see in dim light	Night blindness
	B group	Dairy products, meat, fish and cereals	Helps in the release of energy from food in **respiration**	Affects skin, muscles and nerves
	C	Fresh citrus fruits and some vegetables	Helps keep skin, gums, teeth and bones healthy	**Scurvy** – swollen, bleeding gums and infections take longer to heal
	D	Dairy products, liver and exposure to sunlight	Helps absorption of calcium in bones and teeth	**Rickets** – dental problems, soft bones
MINERALS	Calcium	Cheese, yoghurt, milk and sardines	Required for strong bones and teeth	Rickets – dental problems, less effective blood clotting, muscle spasms
	Iron	Red meat, broccoli and liver	Helps form haemoglobin	**Anaemia**
	Iodine	Seawater, seafood, rocks and some soils	To help control metabolism	Mental retardation and stunted growth in children, and goitre

Examples of vitamins and minerals in a healthy diet and how a deficiency can lead to ill health.

Unit 10: See pages 198–201 for more about respiration.

Food allergies

Food allergies are when the body's **immune system** reacts as if a particular food is harmful. They can cause skin rashes, a runny nose and itchy eyes, headaches, diarrhoea, vomiting or stomach pain. The main foods which cause an allergic reaction are milk, eggs, nuts, shellfish, pork, coffee and chocolate.

Some people have a severe allergic reaction called an **anaphylactic shock** after eating peanuts. This is very serious and can lead to death. The symptoms are swelling of the throat, severe vomiting and low blood pressure.

Unit 6: See pages 148–149 for more about the immune system.

Did you know?

21% of women and 17% of men in England are obese and another 32% of women and 46% of men are overweight.

BTEC Assessment activity 6.2 **P1**

1 Make a diary of all the food you eat over a week to see how balanced your diet is. Write down everything, including the type of food, what components of the diet it contains, e.g. fat, protein, etc. and if it is high or low in fat and sugar. **P1**

2 Using the Internet and books, research bulimia nervosa and anorexia nervosa. In groups, discuss how these conditions affect health and how they are treated. Make a flowchart to show your results.

Grading tip

For **P1** you need to look at ways in which the human body and health are affected by how much or little you eat, how balanced the diet is and allergies to food. You need to show how these concepts relate to each other, for example by producing a flowchart.

PLTS

Reflective learners and Team workers

You will be able to work with others to form opinions based on the information you have gained about eating disorders.

6.3 Diet and lifestyle

In this section: **M1** **P2**

Key terms

Puberty – changes in teenagers when sex glands become functional and physical changes occur in the body.

Calorie – a unit used for measuring energy. The energy value of food is usually shown in kilocalories (= 1000 calories) although people often just say 'calories' when they mean kilocalories.

Person	Recommended average calories per day
Woman	2000
Man	2500
Teenager (15–18)	2755 (male) 2110 (female)
Baby (newborn)	690 (male) 515 (female)
Elderly person	1900 (male/female average)

An elderly person and a teenager require very different forms of food and exercise.

The amount and type of food you need to eat to stay healthy depends on your lifestyle and age. You will have very different energy requirements if you are training for a marathon or watching TV all day.

Age and energy requirements

Age changes our dietary requirements. A growing teenager will need more energy than an elderly person; a baby will need less than an adult. The table shows examples of the energy and nutritional requirements for people of different ages. You must understand that the number of calories required also depends upon how a person is built and their level of activity.

Activity A

Using the Internet or books, research the daily calorie intake recommended for someone of your build and level of activity.

To explain why these energy requirements differ so much you need to think about what is happening to the body at different ages.

- Teenagers are going through drastic changes including growth spurts and other changes associated with **puberty**. They will probably be quite active and participate in some kind of sport. This means that they will need more **calories** in the form of carbohydrates. As well as a balanced diet, they will need more protein to help with growth and repair.

- Elderly people are less active and do not require the same amount or types of food as a teenager. This is because they need less energy.

Activity B

Describe to a partner why the energy requirements of young and old people are so different.

Activity levels and energy requirements

Humans need energy from food just like a car needs fuel to move from one place to another. The further the car travels the more fuel it will need. This is the same for us but our energy requirements are much more complicated than those of a car.

Different activities burn off different amounts of energy as you can see in this diagram.

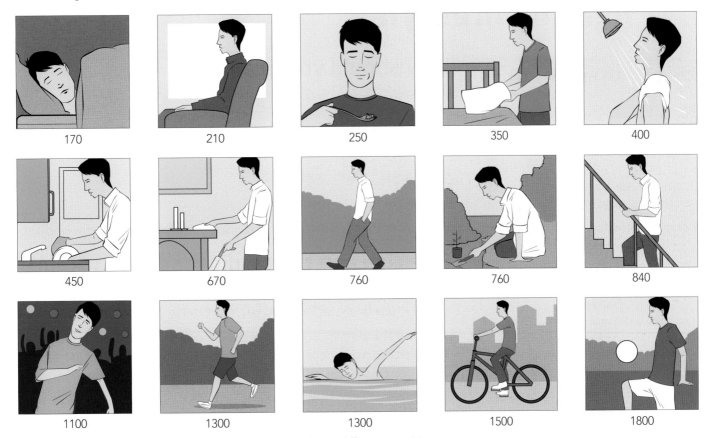

170	210	250	350	400
450	670	760	760	840
1100	1300	1300	1500	1800

Energy (kJ/hour) used up by a medium-sized teenager doing different activities.

BTEC Assessment activity 6.3 M1 P2

1 Use the information on these two pages to produce a chart explaining why energy requirements are different for a teenager and an elderly person. **P2**

2 **(a)** Produce two menus describing what diet you would recommend for a teenager and an elderly person. **P2**

   (b) Show this to a partner and explain why you have recommended each diet. **P2 M1**

Grading tip

For **P2** you need to use the information on these two pages to design a diet and exercise plan for a teenager and an adult which will help them maintain good health.

The **M1** criterion asks you to 'explain' so your work should be presented in the form of an extended piece of writing.

Did you know?

Energy is also measured in joules (J) as well as calories. 4.2 kilojoules (kJ) is equal to one kilocalorie (kcal).

You often see kJ and kcal side by side on many food labels.

Functional skills

ICT

You can use your ICT skills to produce and print the menus using a computer and printer.

6.4 The role of exercise in a healthy lifestyle

Key terms

Cardiovascular system – the heart, blood vessels and blood.

Respire – to use oxygen and glucose to release energy in cells.

Aerobic exercise – exercise which requires oxygen.

Exercise programme – a series of exercises to be carried out over a period of time, written by a trained person to help a particular person become fitter.

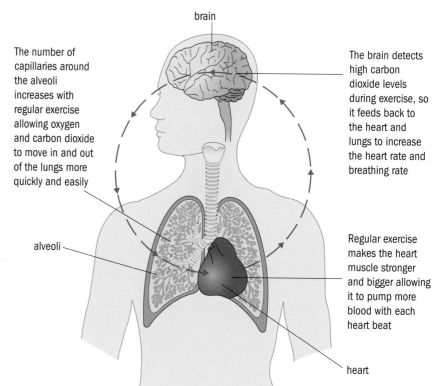

Exercise includes walking, swimming, weight training, martial arts, dancing and team sports like football and hockey.

Activity A

How can exercise improve the functioning of the heart and cardiovascular system?

Why should you exercise?

Regular exercise at least 3 or 4 times a week can help you:

- improve general health
- stop getting diseases
- have less fat and more muscle
- improve the way you feel
- improve the cardiovascular and respiratory functions of your body.

Exercise and the cardiovascular system

The main role of the **cardiovascular system** is to deliver oxygen to tissues in the body and remove waste products like carbon dioxide.

When we exercise, muscle cells **respire** more to release more energy. They also produce more carbon dioxide. There is an increase in heart rate and blood pressure. Blood flow is also diverted from other areas of the body to the working muscles.

Some of the positive changes to the heart from exercise include an increase in the size of the heart muscle. This makes it better at pumping blood around the body and lowers the resting heart rate.

brain

The number of capillaries around the alveoli increases with regular exercise allowing oxygen and carbon dioxide to move in and out of the lungs more quickly and easily

The brain detects high carbon dioxide levels during exercise, so it feeds back to the heart and lungs to increase the heart rate and breathing rate

alveoli

Regular exercise makes the heart muscle stronger and bigger allowing it to pump more blood with each heart beat

heart

The brain is involved in raising the heart rate and breathing rate during exercise.

Exercise and the respiratory system

When we exercise, our breathing rate increases. This increases the supply of oxygenated blood and glucose to the working muscles via the blood stream. This allows energy to be released in the muscles by the process of respiration. Regular **aerobic exercise** over time improves lung function and increases the blood flow to the alveoli in order to provide better gas exchange. This is of benefit to people when they undertake strenuous activity. They will not get out of breath as quickly and will be able to continue working physically hard for longer.

Activity B

How can regular exercise improve the functioning of the respiratory system?

What kind of exercise is best?

When a physiotherapist or personal trainer designs an **exercise programme** for someone, they need to decide what is suitable. You wouldn't expect someone who is extremely overweight and with a heart condition to run a marathon. The exercise needs to be right for each individual depending on their age, weight and fitness level. It also needs to consider factors such as the person's mobility, heart condition or other illnesses and conditions such as arthritis.

Sports injuries

If a person follows an exercise programme which is not suitable for their fitness level, it can cause physical stress and they may be injured. Injuries can include sprained tendons and ligaments. Tendons hold muscles to bones, and ligaments hold bones together at joints.

BTEC Assessment activity 6.4

1. In groups, discuss what factors need to be considered when planning an exercise programme for an elderly person and a teenager. **P2**
2. Describe the changes to the body functions following an exercise programme. Do some further research using books and the Internet. **M1**
3. List the questions you could ask yourself to help you form a conclusion about the positive and negative aspects of exercise programmes. **D1**

Did you know?

The average resting heart rate is 72 beats per minute. Athletes usually have a lower resting heart rate because exercise has improved their cardiovascular and respiratory systems. This means that they are better at delivering oxygen to working muscles. As a result their heart doesn't have to work as hard, so it can beat more slowly.

Science focus

Make sure that you do not confuse the term respiration, which involves energy release in cells, with the respiratory system, which is the lungs and associated airways.

Case study

A physiotherapist at work

Ayden is a physiotherapist who works in a health centre. He needs to prepare an exercise programme for an elderly woman who is quite healthy but is rather overweight.

What factors should Ayden consider when designing the exercise programme?

Grading tip

For **P2** when designing your exercise programme, think of the levels of activity and how often it should be done for both an elderly person and a teenager.

For **M1** research further how exercise will alter or improve how the body works.

For **D1** evaluate and justify means to look at the information you have and then form a conclusion based on your evidence.

6.5 The immune system

In this section:

Key terms

Immune system – the organs and cells that work together to defend the body against disease and illness.

Phagocytosis – the process that involves white blood cells surrounding and digesting bacteria or viruses.

Lymphocyte – a white blood cell in the immune system; can be B-type or T-type.

A macrophage surrounding a pathogen ready to digest it.

Pathogens are **microorganisms** that cause disease. Your body has several ways to stop pathogens entering the body so that you don't become ill. If pathogens get inside your body your immune system starts to kill them. It may kill them before or after you feel ill.

Unit 3: See page 86 for more about pathogens.

Non-specific response of the immune system

Non-specific immune responses are all the same no matter what type of pathogen there is. Non-specific responses include:

- barriers to pathogens entering and spreading in the body
- inflammatory response
- white blood cells (leucocytes) such as phagocytes.

First lines of defence

The ways pathogens are prevented from entering the body are listed in the margin box.

Activity A

Research the types of defence shown in the margin box. Make a table describing how each one stops pathogens entering the body.

Inflammatory response and phagocytosis

Pathogens can get into the body by a cut or damage to the skin. If this happens an **inflammatory response** is triggered. Damaged cells release a chemical called **histamine**. This sends signals to **macrophages** to come and kill the bacteria. Macrophages are white blood cells that contain digestive enzymes. To kill the bacteria, they surround and digest it. This is called **phagocytosis**.

Science focus

The body's defences
- Mucous membranes
- Skin
- Tears
- Nose hairs
- Cilia (small hairs lining the respiratory tract)
- Stomach acid

Activity B

Have you ever accidentally cut yourself? Explain how your immune system prevented you becoming ill from the cut.

Specific responses of the immune system

If any pathogens are not killed by the macrophages then other white blood cells called **lymphocytes** take over. This is a **specific immune response** because it depends on the specific antigens of the pathogen. There are two types of lymphocyte.

- **T lymphocytes** kill virus-infected cells and some cells that may have become cancerous.
- **B lymphocytes** produce antibodies to kill pathogens or destroy the toxins they produce.

How the specific immune system works

Antigens are found on the surface of pathogens. They act as an identity card. Our immune system uses these to identify that the pathogen is an invading organism that needs to be killed. Each particular pathogen has a different antigen. A B lymphocyte must produce specific **antibodies** for that specific pathogen to kill it and its antigens and stop it causing further damage.

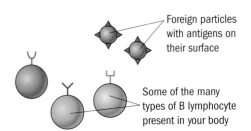

Foreign particles with antigens on their surface

Some of the many types of B lymphocyte present in your body

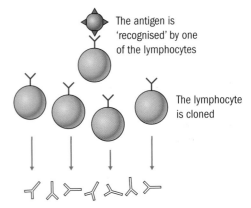

The antigen is 'recognised' by one of the lymphocytes

The lymphocyte is cloned

The clones produce antibodies which destroy the antigens

The specific immune response.

Case study: Analysing blood samples

Mahla is an immunologist who works in a hospital. She analyses blood samples to look for the different cells involved in the specific immune response. This helps her to identify what disease the patient is suffering from.

What types of white blood cells would Mahla be looking for?

Activity C

Working in groups, discuss how the specific immune system works to help us recover from chicken pox, which is caused by the herpes virus *Varicella zoster*.

 BTEC ## Assessment activity 6.5   **M2** **P3**

1 Produce a poster to describe how the body stops pathogens getting into it. **P3**

2 In groups, research further into the specific and non-specific immune responses and explain how they are different from each other. **M2**

Grading tip

For **P3** briefly outline each part of the immune system and how it protects us from becoming ill.

For **M2** provide a detailed description, including diagrams, of how specific and non-specific immune responses work.

Did you know?

There are thousands of different pathogens that can invade the body. For each pathogen's antigens, the immune system needs to produce an antibody to prevent us from becoming ill.

PLTS

Team workers

You can use your team working skills to explain how the specific immune system works.

6.6 Vaccinations and the immune system

In this section: **D2** **M3**

Key terms

Memory cells – immune system cells that remember antigens on the surface of pathogens and produce antibodies to kill them if they enter the body.

Vaccine – process which prepares the body to fight against certain infectious diseases should we come into contact with them in the future.

Immunity – having enough defences to avoid infection, disease or other unwanted invasion of the body.

Active immunity – immunity given by a vaccine to make the body produce antibodies.

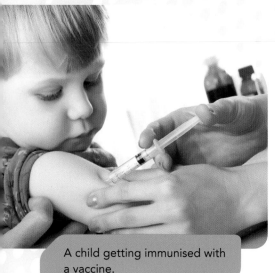

A child getting immunised with a vaccine.

Did you know?

Most vaccinations are given during childhood, usually by injection. They have dramatically improved the health of the people living in the UK.

Immunisation can protect people from certain diseases they may come into contact with in the future. Immunisation can be achieved by the use of vaccinations.

Changes in the body following vaccination

A **vaccination** takes advantage of the specific immune response. It contains antigens from pathogens. These cause your body to produce **memory cells**. These cells remember the antigen on the surface of a pathogen. If they meet the same antigen again they start to reproduce very quickly and produce lots of antibodies to help kill it. This is why you are unlikely to suffer from a disease if you have been vaccinated against it.

Activity A

What changes happen in the body following a vaccination?

Immunisation programmes

The NHS offers vaccination programmes for people of all ages. **Vaccines** are also available for people travelling abroad where they may come across pathogens for which they have no **immunity**.

Unit 3: See pages 90–91 for more about NHS vaccination programmes.

Activity B

Do you think humans should have vaccinations? Why? Why not?

How effective are immunisation programmes?

It is important to ensure that vaccines do more good than harm. In the past, there was a suspicion that the MMR (measles, mumps and rubella) vaccination given to children was linked to autism. There is no evidence to support this belief but it can worry parents deciding whether or not to have their child vaccinated. The graph shows evidence that the MMR vaccine has a very positive effect on reducing the spread of measles.

Unit 3: See pages 90–91 for more about measles, mumps and rubella.

How can the effectiveness of immunisation programmes be evaluated?

To decide if a vaccine or immunisation programme is effective, we can ask:

- Is the vaccine safe? Does it have any side effects?

- What is the evidence that a particular vaccine is good at preventing disease?

- At what age should the vaccine be given and how long does it last?

Booster injections

Once you have **active immunity** from a vaccine it can last a lifetime. Some diseases such as diphtheria need re-vaccination every few years to keep immunity. This is called a **booster injection** because it boosts the number of memory cells in your blood.

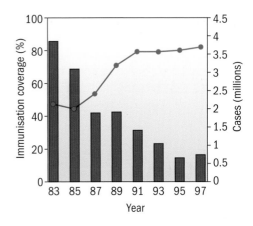

Number of cases of measles in the world

—●— % Immunisation coverage

Immunisation programmes have helped to reduce the spread of measles.

Case study: Why should my child be vaccinated?

Zara works as an assistant health visitor at a health centre. She has been asked to prepare a presentation on the MMR vaccination. Zara wants to stop people worrying that there might be side effects if they let their children have the vaccination. She is going to give her presentation in a series of workshops to the patients at her health centre.

What information should Zara include in her presentation?

Functional skills

English

You can use your reading and written English skills to find information on vaccinations and then answer the questions giving clear, grammatically correct descriptions and explanations that can be easily understood.

BTEC Assessment activity 6.6

1. Produce a flow diagram to show how vaccinations help to stop pathogens making us ill. Include information on how the immune system helps with this. **M3**

2. Use the information above and from Unit 3 on vaccinations to research further. Describe the advantages and disadvantages of one particular vaccination programme. **D2**

Grading tip

For **M3** use what you have learnt in Unit 3 on vaccinations and in this unit on the immune system to identify other vaccination programmes. Link these to how they help to maintain your health.

For part of **D2** think about the effectiveness of one vaccination programme and identify the advantages and disadvantages of it. Then explain how effective you think the programme is.

6.7 Screening for health

In this section: D2 P4

Key terms

Detection – to look for and find something.

Cervix – part of female reproductive system which forms the opening to the uterus.

Fetus – an unborn baby.

Diagnostic test – a test to confirm if a person has a disease or condition.

Ultrasound – a way of looking inside the body for problems, usually used in pregnancy.

Congenital – a condition that is there from birth.

What is screening?

Screening means looking for early signs of a particular disease in healthy people who do not show any symptoms of the illness. The earlier the disease is found the better the chance is of treating it.

Activity A

What is the overall purpose of screening and why is it important?

NHS screening programmes for cancer

Type of cancer	Purpose of screening	How it is done	How often
Breast	**Detection** of cancer at a very early stage and to check out any unusual lumps	X-ray (mammogram) of each breast	Every 3 years for women over 50 years old
Bowel	Detection of cancer at a very early stage	The test detects blood in the faeces. Bleeding may be harmless so further tests are needed, e.g. a colonoscopy to look inside the colon and get tissue samples	Every 2 years for men and women aged 60–69 years
Cervical	To identify abnormal cells Not a test for cancer itself	Cervical smear. A nurse inserts a device into the vagina to collect some cells from the surface of the **cervix**. The sample is sent to the laboratory for further testing	Every 3–5 years for women aged 25–64 years

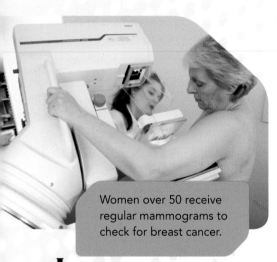

Women over 50 receive regular mammograms to check for breast cancer.

Science focus

Conditions for which antenatal diagnostic tests are offered

- Down's syndrome
- Fetal abnormality, including **congenital** abnormalities
- Sickle cell anaemia
- Thalassaemia
- Some infectious diseases

Antenatal screening

Antenatal screening looks at the risk that a **fetus** might have a disease or condition. If the risk is high the mother will be offered a **diagnostic test** to find out about the health of her unborn baby. A very common screening test uses **ultrasound** to obtain an image of the fetus. Some tests may cause miscarriage so they are offered only when the baby is at extreme risk of a particular disease.

Unit 5: See page 132 for a case study about ultrasound scans.

Health professionals can monitor the health of the fetus and prepare the care needed for when the baby is born. The mother may also consider ending the pregnancy.

Activity B

Using the Internet, research the different types of tests available that screen for the conditions listed in the margin.

Newborn screening

All newborn babies in England are screened for diseases including **phenylketonuria**, **congenital hypothyroidism**, **sickle cell disorder** and **cystic fibrosis**. Most babies will not have any of the conditions. For those that do, screening can allow early life-saving treatment or prevent severe disability.

Unit 3: See page 88 to find out about sickle cell anaemia and cystic fibrosis.

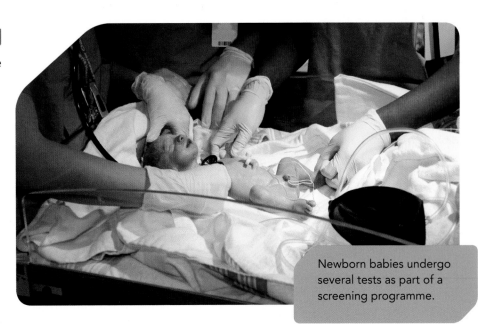

Newborn babies undergo several tests as part of a screening programme.

Vascular screening

Vascular screening involves looking at the cardiovascular system of healthy men and women. This screening allows people who may be at risk of strokes and heart attacks in the future to be aware of this. They are then advised how to change their lifestyle to minimise either of these happening.

Vascular screening can also be used as a diagnostic examination for people who may already have vascular problems.

Did you know?

The NHS breast screening programme alone saves 1400 lives a year in England.

PLTS

Reflective learners

Here you will form your own opinions using the information you have gathered.

 Assessment activity 6.7   D2 P4

1 Produce a fact sheet which describes how screening tests for cancer, antenatal care and newborn babies help to keep people healthy. **P4**

2 (a) Read through the information on these two pages. Using different websites, research further into one screening programme from Question 1. Make notes on the key points.

   (b) In groups, discuss how good this screening programme is, using the evidence you have researched. Present your findings to the class. **D2**

Grading tip

For part of **D2** think about the effectiveness of one screening programme by breaking it down into its advantages and disadvantages.

6.8 Antibiotics

In this section: D3 M4 P5

Key terms

Mutation – a change in the form of a gene.

Antibiotic resistant – a microorganism that can no longer be killed by a certain drug.

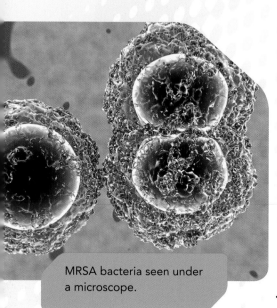

MRSA bacteria seen under a microscope.

Did you know?

The first antibiotic was discovered in 1928 by Alexander Fleming. He found penicillin growing as a mould on a culture of bacteria that he had accidentally left open to the air. Since then many other antibiotics have been developed.

Antibiotics are drugs that kill or prevent the growth of bacteria. Bacteria appear to mutate quickly because their **generation time** is so short. Resistance to an antibiotic might already be in a population of bacteria as a random **mutation**, possibly in just one bacterium. It is this one that is not killed, so it then reproduces and produces a whole population of resistant bacteria. This means that antibiotics used to treat some bacterial infections may no longer work.

Hospital-acquired infections

Pharmacologists work to produce new antibiotics to counter new types of bacteria. You may have heard of some bacteria often found in hospitals due to poor hygiene.

MRSA

Methicillin-resistant *Staphylococcus aureus* (MRSA) is resistant to most antibiotics (antibiotic resistant). It can cause blood poisoning, infections of the lungs, bones or heart valves and even death. After surgery or medical treatment we are more vulnerable to the infection it causes.

C diff

Clostridium difficile (C diff) is a bacterium that lives in the gut. It becomes a problem when antibiotics are taken which kill good bacteria in the gut as a side effect. This allows C diff to reproduce and cause diarrhoea and fever. It is often found in care homes and hospitals. Spread can be prevented by good hygiene.

Activity A

Why are these infections more common in hospitals?

Use and misuse of antibiotics

Antibiotics are used to treat other bacterial infections including:

- Tuberculosis (TB) – this infects the lungs and organs of the body.
- Cholera – this is transmitted in areas without clean water.
- Syphilis and gonorrhoea – these are sexually transmitted diseases.

Misuse of antibiotics can lead to the development of bacteria which are resistant to antibiotics. To make sure this doesn't happen:

- Never take antibiotics if you don't need them.

- Always take the full course of antibiotics. If you don't, some bacteria may survive and reproduce to cause another infection and develop resistance.

PLTS

Independent enquirers

You will be investigating the effects of different antibiotics on your own.

BTEC Assessment activity 6.8 · D3 · M4 · P5

Using the NHS website, research further the different kinds of bacterial infections found in hospitals. This will also provide you with some secondary data.

1 Explain in a paragraph why these infections are difficult to treat using antibiotics. **P5**

2 Produce and complete a table to show the advantages and disadvantages of using antibiotics to treat bacterial infections. **M4 D3**

3 In pairs, discuss why it is important not to take antibiotics if you don't really need them and to make sure you complete all antibiotic treatment. **M4 D3**

Grading tip

For **M4** and **D3** when you use secondary data, you need to compare the use of antibiotics with gene therapy (see pages 158–159), stem cells, blood grouping and transfusions (see pages 156–157) in controlling health.

Cheryl Piper
Pharmacy Technician, Charing Cross Hospital

I work in the dispensary, where I dispense medication to patients. Some are in the hospital to see a specialist, others are in-patients.

On a typical day I work with our pharmacy robot to issue medication to patients fast and safely. I have to liaise with doctors and nurses to ensure that the medication is appropriate for the patient. I also counsel patients on how to take their medication, and visit wards to ensure they have supplies of the medication they need.

Today I had to counsel a patient on how to take their course of metronidazole. I told them to take the tablets with food and with plenty of water and not to drink any alcohol, and that it's important that they take all the tablets even if they feel better before finishing the course. It's also helpful to the patient if I mention any side effects they may get.

Think about it!

1 Why is finishing a course of antibiotics important?

2 Why does it matter whether the patient takes the tablets with food or whether they drink alcohol?

155

6.9 Medical treatment in the control of health

In this section:

Key terms

Chemotherapy – drug treatment for disease, often used for cancer.

Embryonic tissue – tissue found in the very early stages of a new life.

Unspecialised – able to become any kind of cell at this stage.

Differentiated – specialised into one kind of cell able to perform a certain function.

A person receiving a blood transfusion.

Blood transfusions

A blood transfusion is the transfer of blood from one person into another person. It may be needed to replace blood lost during an operation, after an accident or during the birth of a child. Blood can also be given as treatment for conditions where you can't make your own healthy blood cells, such as sickle cell anaemia or cancer patients after **chemotherapy**.

Blood grouping and blood types

It is very important that when a person has a blood transfusion the correct type blood is given. But why is this?

Red blood cells have antigens on their surface and your blood plasma contains antibodies to those it doesn't recognise. The antibodies will trigger an attack to the antigens if they come into the body. A, B and rhesus are the most common and important antigens on red blood cells. Their effects must be understood so that a person is given the correct blood during a blood transfusion. The table shows different blood types and the antibodies they contain.

Unit 6: See page 149 for more about antigens.

ABO blood type	Antigen A	Antigen B	Antibody anti-A	Antibody anti-B
A	yes	no	no	yes
B	no	yes	yes	no
O	no	no	yes	yes
AB	yes	yes	no	no

A blood transfusion must be matched to your own blood type or the blood will clump together. This is called **agglutination** and can cause death. Before a blood transfusion, a **haematologist** mixes a sample of the patient's blood with the donor blood to check that it doesn't agglutinate. For example, A-type blood has antigen A but anti-B antibodies, so it cannot be mixed with B-type blood.

Did you know?

84% of the population is rhesus positive.

Activity A

Why is it so important to give patients the correct blood type in a blood transfusion? What is your blood type and what blood type(s) could you safely receive in a transfusion?

Rhesus factor

Many people have the antigen rhesus (Rh) factor on the surface of their red blood cells. Those who have it are called rhesus positive. If you don't have it you are called rhesus negative. Someone with AB blood type and the Rh factor is AB positive.

Stem cells

Stem cell researchers hope that one day **stem cells** will be used in medical treatments to replace missing, damaged or diseased tissues; for example, after strokes or burns, or in people with diabetes or Parkinson's disease.

What are stem cells?

Stem cells are found in **embryonic tissue**. You have them while you are a fetus and in some tissues after you are born, including your bone marrow.

Stem cells are **unspecialised** and can turn into any type of cell or tissue under certain conditions. Once they become a certain cell they have **differentiated** to perform a certain specialised function. Stem cells can be turned into specialised cells using different substances.

The future and stem cells

In the future, stem cells could be made into the following cells:

- insulin-producing cells so a diabetic person doesn't have to inject it
- heart muscle cells to repair a damaged heart after a heart attack
- nerve cells to treat spinal damage
- muscle cells to treat injuries
- red blood cells to be used in blood transfusions
- bone cells to repair weak or broken bones.

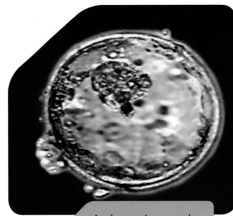

A photomicrograph of an embryonic stem cell.

Activity B

Where are stem cells found in the body?

Case study: Growing a new heart and liver

Dan works in stem cell research. He is trying to grow new hearts and livers from stem cells taken from different people. This has never been done before so he would like to be the first to do it.

What advantages are there of being able to grow hearts and livers using stem cells? Why might some people not want to use stem cells?

BTEC Assessment activity 6.9 · D3 · M4

1 Research into how antibiotics, blood transfusions and stem cells can be used to treat certain medical conditions. **M4** **D3**

2 Draw and complete a table to compare the different medical treatments you have researched in Question 1. **M4** **D3**

Grading tip

For **M4** and **D3** avoid concentrating on one kind of medical treatment. You must include an evaluation of the use of antibiotics, stem cells, gene therapy and blood transfusions to complete this assessment successfully.

6.10 Gene therapy

In this section: P6

Key terms

Gene therapy – the replacement of faulty genes by inserting healthy genes into body cells.

Genetic disorder – a disease or condition that can be transferred through the genes from parents to children, or is associated with the genes.

Vector – a virus or other agent that is used to transport DNA to a cell.

Gene therapy may give geneticists a way to treat or cure **genetic disorders** caused by a faulty gene. It could help sufferers of diseases like cystic fibrosis, muscular dystrophy, haemophilia and cancer. Gene therapy involves adding a healthy copy of the faulty **gene** into the cells of the body where they are needed.

How does gene therapy work?

Gene therapy has three key stages.

1 Find the gene involved in the genetic disorder.

2 Make lots of copies of a healthy version of the gene.

3 Transfer copies of the healthy gene into the patient.

The process must be repeated regularly as the cells die.

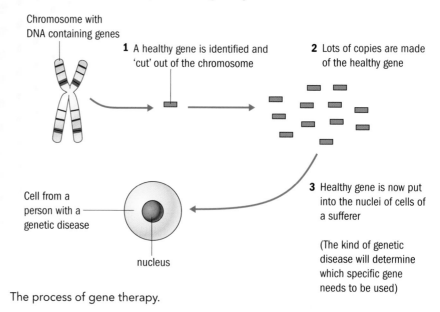

The process of gene therapy.

How are the healthy genes transferred?

The DNA can be directly injected into cells. The two most common ways genes are transferred into cells are by using **viruses** and **liposomes** as **vectors**.

- **Viruses** can enter cells of the body easily. Geneticists put a copy of the healthy gene into the virus so it will act as a vector to carry the gene into the cell. The common cold virus 'adenovirus' is one of the most common viruses used as a vector in gene therapy.

- **Liposomes** are fatty substances that can pass through the cell surface membrane. In gene therapy the genes are coated with liposomes to make it easier for them to enter the cell.

Case study

Genetic counselling

Sarah is a genetic counsellor working in a hospital. She works with families who have genetic diseases or conditions. They need specific information to help them understand what gene therapy is and how it can help them. Sarah wants to produce a leaflet to provide this important information.

What information could Sarah put in her leaflet?

Activity A

How can genes be transferred into the body to treat genetic conditions?

Problems with gene therapy

There are some problems that have kept gene therapy from becoming a good treatment for genetic diseases. These include the following.

- Patients need to undergo gene therapy many times.

- There is a risk of stimulating an immune response against a viral vector because the body sees it as a possibly harmful cell.

- Many vectors are viruses and could be harmful if the virus recovered its ability to cause disease.

- There is a risk of toxicity (poisoning), and immune and inflammatory responses.

- There is a need for better ways to deliver healthy genes to the body. The use of viruses and liposomes has the disadvantage of either severely damaging delicate cells or being very expensive and complicated.

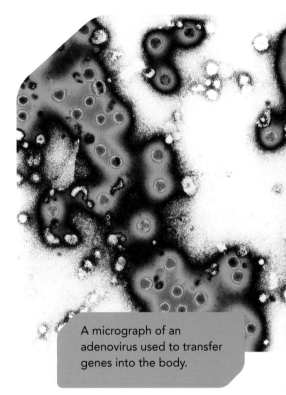

A micrograph of an adenovirus used to transfer genes into the body.

BTEC Assessment activity 6.10

1 Working in groups, using the information above, produce a leaflet describing what gene therapy is. **P6**

2 Research how vectors transport healthy genes into cells and draw a diagram to demonstrate this to your class. **P6**

3 Read the section on 'Problems with gene therapy'. Research this further and then write a brief magazine article describing your own feelings about the use of gene therapy to treat genetic diseases. **P6**

Grading tip

For **P6** you need to use your knowledge of genetics from Unit 3 and further research is essential to provide at least two detailed examples of gene therapy.

Did you know?

Gene therapy has provided the first cure for deafness. Geneticists have restored the hearing of a deaf guinea pig. This leads the way for a possible cure for human deafness using gene therapy.

Activity B

What problems are there with gene therapy?

Functional skills

ICT and English

There are lots of opportunities here to use your ICT and English skills to research for information, write articles and produce leaflets on gene therapy.

6.11 Using gene therapy to treat diseases

In this section: P6

Key terms

Inherited condition – a condition passed on from parents to children by genes.

Malnutrition – caused by not enough nutrients getting to the cells of the body.

Tumour – a lump made up of cells which can be cancerous.

Clotting factor – substances in the blood that act to stop bleeding by forming a clot.

Unit 6: See pages 158–159 for more about gene therapy using liposomes.

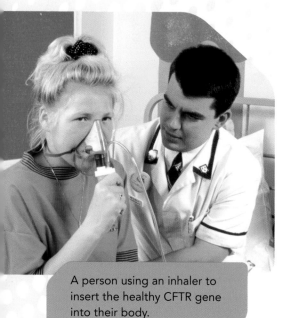

A person using an inhaler to insert the healthy CFTR gene into their body.

Gene therapy can help treat and may even cure certain diseases including inherited genetic conditions.

Unit 3: See page 88 for more about inherited genetic conditions like cystic fibrosis.

Cystic fibrosis

Cystic fibrosis is an **inherited condition** caused by a single damaged gene called the cystic fibrosis transmembrane regulator gene (CFTR). Lack of the CFTR gene causes the internal organs to accumulate thick, sticky mucus which causes inflammation and infections. This makes it hard to breathe and digest food.

Activity A

What is cystic fibrosis and how is it caused?

The main symptoms of cystic fibrosis include the following:

- a bad cough as the sufferer tries to clear their airways

- repeated chest infections due to a lot of thick sticky mucus containing bacteria which the sufferer can't get rid of

- prolonged diarrhoea because enzymes cannot get to the food to digest it well – this means that the food is not digested and fatty faeces are passed out

- poor weight gain as the thick sticky mucus in the intestine doesn't allow the nutrients to get into the blood stream – this can also lead to **malnutrition**.

Gene therapy involves using liposomes to insert the healthy CFTR gene directly into the lungs using an inhaler. This treatment is still on trial but will be used in the future.

Cancer

Normal body cells know when to divide, when to stop growing and when to die. Cancer cells continue to grow and divide out of control and they don't die.

A single cancer cell divides many times and forms a **tumour**. The tumour can kill the normal cells around it and damage the body's healthy tissues. Some cancer cells may break away and travel in the blood to other parts of the body and start dividing there to form another tumour – a secondary cancer. This makes a person suffering from cancer very ill.

Gene therapy for cancer is in the early stages. It will take time for it to be safe enough to treat the general public. Some successful trials include:

- targeting healthy cells to help them to fight cancer

- inserting functioning genes so that missing or mutated genes that could cause cancer do not have the chance to work and may eventually be turned off

- putting genes into cancer cells to make them more sensitive to chemotherapy or radiation therapy.

Haemophilia

If you ever cut yourself you will know that the blood flow quickly stops and a scab forms. People with haemophilia do not have one of the essential **clotting factors** and so they will bleed for a long time even from small cuts. If they bleed internally, this can cause serious blood loss. Haemophilia is an inherited condition. There are two types of haemophilia: haemophilia A in which factor VIII is not present and haemophilia B in which factor IX is not present.

Gene therapy for haemophilia sufferers could involve using healthy genes for blood clotting. These could then be joined with vectors and inserted into the sufferer to help the blood clot normally.

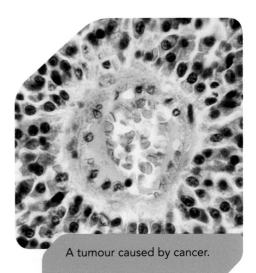
A tumour caused by cancer.

 Activity B

What are the benefits of using gene therapy in the future for the treatment of cancer?

 Did you know?

Haemophilia is usually found only in males.

 Case study: Geneticists at work – a true story

James is a geneticist working in a biotechnology laboratory. He is working on an experiment with dogs that have a genetic defect which gives them haemophilia. Healthy genes for the blood clotting factor IX were attached to a virus vector and then injected into the leg muscles of the dogs. They found the dogs began producing the blood clotting factor. After two months of treatment the blood started to clot as quickly as in a dog without the condition. The results lasted for more than a year and the dogs had no side effects from the treatment.

James found this method of direct injection of the healthy gene also lessens the risks associated with using viruses as vectors (see pages 158–159).

What possibilities could this experiment lead to?

 PLTS

Independent enquirers

You will work independently when researching into other genetic conditions treated using gene therapy.

 Assessment activity 6.11 **P6**

1 Read the information on these two pages. Then draw up a table. Put in the diseases, the symptoms and the treatment of each using gene therapy. **P6**

2 Using books or the Internet, research other diseases where gene therapy is used and make some notes, e.g. muscular dystrophy. **P6**

Grading tip

For **P6** give at least two examples of gene therapy used to treat different diseases or conditions stating clearly how gene therapy is carried out.

Just checking

1 Explain how diet and exercise affects our health at different stages of our lives.
2 List five things from this unit that can cause us to be ill.
3 Describe the different ways in which illness and disease can be treated or prevented.
4 In groups discuss what the immune system does. Produce a poster to show how it works.
5 Define what gene therapy is and make a bullet point list of how it can be useful.

Assignment tips

- Remember to use a wide range of different resources for each assignment in this unit. For the merit and distinction grades it is especially important that you research and select the correct information required for each of the grading criteria.

- Make sure that you answer each of the grading criteria using the correct command words such as explain, evaluate, describe etc. You will not achieve the grade for a task that is completed using the wrong command word.

- A full list of each command word and what it requires you to do is in your study guide. Make sure you keep checking the guide when preparing for and completing each task to be sure you are doing it correctly.

- Remember that a portfolio is an ordered file of evidence. For example, it could include a range of information to evidence that you understand what gene therapy is, how it works, how and when it is used and its advantages and disadvantages.

- Remember to use the knowledge and understanding you have gained from Unit 3 to help you to complete this unit.

- DNA and the genetic code (pages 72–75 and 88–89) will help you understand and successfully assess knowledge of gene therapy to complete **P6** in Unit 6.

- Information on vaccinations (pages 90–91) will help you successfully complete grading criteria on vaccinations and the immune system in Unit 6.

7 Practical scientific project

In this unit you will carry out a practical investigation to solve a problem. You will be able to choose a project that particularly interests you.

Before beginning to answer a scientific question it is very important to produce a plan of what you will do. The plan should show how you think you will solve the problem or answer the question. Deciding how you will tackle your project is one of the most important parts of this unit. You will need to be clear about what you are doing before you try to carry out any practical work. Your tutor can help in the planning process but they will not tell you exactly what to do.

Part of your planning should include carrying out any risk assessments that are needed to make sure that you work safely.

Once your plan has been agreed by your tutor you will be able to carry out the practical investigation part of your project and collect your results. You may also need to do more research on the topic of your project at this stage.

The final part of your project will be to consider your results carefully and decide what they tell you about the problem that you were investigating.

Learning outcomes

After completing this unit you should:

1 be able to plan a practical scientific project
2 be able to use appropriate practical skills
3 be able to analyse and present results.

Assessment and grading criteria

This table shows you what you must do in order to achieve a **pass**, **merit** or **distinction** grade, and where you can find activities in this book to help you.

To achieve a **pass** grade, the evidence must show that the learner is able to:	To achieve a **merit** grade, the evidence must show that, in addition to the pass criteria, the learner is able to:	To achieve a **distinction** grade, the evidence must show that, in addition to the pass and merit criteria, the learner is able to:
P1 Identify the health and safety risks associated with implementing this project **See Assessment activity 7.1**	**M1** Present a summary which clearly demonstrates how the information that was researched has contributed to the plan **See Assessment activity 7.1**	**D1** Evaluate the project, identifying any modification to the practical procedures and data collection methods used.
P2 Produce a project plan **See Assessment activity 7.1**		
P3 Assemble and use appropriate equipment safely to collect reliable scientific data	**M2** Discuss the importance of maintaining accurate logbooks.	
P4 Record scientific data		
P5 Analyse the scientific data obtained **See Assessment activity 7.2**		
P6 Produce a scientific report. **See Assessment activity 7.2**		

How you will be assessed

Your assessment could be in the form of:

- a written project plan
- a laboratory log
- observation of your practical work
- a written report on your findings.

Joanne, 15 years old

I carried out a practical investigation to decide how well plants grow in different conditions. I visited my local garden centre first of all to see the different types of plants and to talk to the staff. I always thought that all plants grew in the same conditions but the staff there told me that different types of plants like different amounts of water and sunlight. My tutor told me that I would need to pick the type of plants carefully in order to be able to measure the growth within a sensible time. He suggested that I grew some seeds so I could measure the seedlings as they grew.

It was great fun carrying out my own investigation and it was easy to do as I didn't need to set up any complicated equipment. One thing that I learnt from this work was always to write my results in my laboratory book. The first time I measured my plants I wrote the heights on a piece of paper but then I lost it. As it was the first measurements it didn't matter too much but I can see now that if it had been later measurements it would have left a gap in my results table.

Catalyst

Your project

What sort of investigation would you like to do? It is always a good idea to start with a question or a problem.

- How many bacteria are in milk? or What makes milk go off?
- How pure is the waste water put into a river by a chemical manufacturer? or How can the chemical manufacturer remove impurities from the water?

1 Write down between six and ten topics in science that interest you. Working in small groups, discuss these with others in your class to compare your ideas.

2 Identify two or three of your favourite ideas ready to discuss with your tutor.

7.1 Planning a scientific project

In this section:  P1 M1 P2 M2 P3 P4

Key terms

Hypothesis – a tentative statement or supposition, which may then be tested through research.

Manipulative skills – the way in which equipment or materials are handled.

Did you know?

Other people might already have carried out some investigations on the same topic as you. Finding examples of these might help you identify different ways in which things could be done.

Legislation	Purpose
Health and Safety at Work Act 1974	You have to take care of yourself and others when you do your project
Control of Substances Hazardous to Health (COSHH)	You need to identify potential risks in your work and what you need to do to keep things as safe as possible
Codes of practice	There are many different codes of practice which apply to specific types of work. You will need to make sure that your work meets these codes

Important legislation that might apply to your project.

Before you start any practical work it is important you are clear about what you are doing.

Project plan

You should start by thinking in general about what you might do. For example, you might think about investigating how the quality of milk or water is controlled. You should clearly identify the aim of your investigation and any further research that is necessary to complete the project. This is called the outline plan. This plan will show you which areas you need to research to help you draw up a detailed plan later in the project.

Information resources

Your tutor may give you some ideas about where to find information but you will need to identify some sources for yourself. Make sure that you only use information that is specific to your investigation.

The sources you use might include:

* previous investigations that you have carried out in class
* background reading from books or magazines
* research from the Internet.

Activity A

How can you record and present the sources of information you have used for your research?

How could you record in your plan how you have used this information?

Health and safety

Any practical investigation has risks involved. Laboratory workers (including you) have to make sure that they follow the laws and rules that cover the practical work they are doing. These laws and rules are called **legislation**. Your tutor will help you if necessary to make sure that you work safely.

Activity B

Why is it important that you carry out risk assessments for your project before you do any practical work?

Final plan

When you draw up your detailed plan you should think about the following:

- the **hypothesis** that your investigation intends to prove or disprove
- what equipment and materials you will need to carry out your plan
- what results you will record
- how often you will take measurements
- how you will record your results
- how you will reduce or remove any identified risks from your activities
- the dates for the milestones where you will review your progress.

Using appropriate practical skills

Experimental techniques

To carry out the practical techniques for your project, you need to be able to decide on and find the correct equipment and materials that you need. Your tutor will watch you doing this and will also make sure that you follow any health and safety rules.

You must practise the **manipulative skills** needed for the equipment you will use. You must also make sure that you take any measurements in the correct way. Remember to watch things carefully during your investigation.

Recording results

Use a laboratory log to record everything you do for your project and accurately record all of your results in it. You should do this even if some of the results are unexpected. (You can point these out and deal with them later.) In addition to the results of your measurements, remember to record how you are controlling other variables that might affect your investigation, e.g. the temperature, who took the measurement and the serial number of the equipment used. You will then be able to use this to help you write your report.

Carrying out a practical investigation.

BTEC Assessment activity 7.1 **P1 M1 P2**

You work at a horticultural research centre. You have been asked to investigate the growing conditions for a newly introduced plant. You are given 50 seeds of the plant.

1. Plan how you would use the seeds to find the best growing conditions. What conditions would you change and how would you keep others the same? **P2**
2. What are the health and safety issues of your practical and how would you carry out risk assessments? **P1**
3. Write a short summary to show how you have used researched information to help you plan the investigation. **M1**

7.2 Analysing and presenting results

In this section: **D1** **P5** **P6**

Once you have finished your practical work you need to analyse your results. Your analysis will help you decide what your results tell you and if you can prove or disprove the hypothesis your project plan identified.

Practical data

There are usually a number of different ways that you can present results. You will need to choose the ones that are most suitable for the data you have collected. It is important that you use the correct units and organise your data in the clearest way. Tables and graphs need to be labelled properly, so anyone can understand them. You need to decide whether your investigation was carried out with enough **precision** to give **accurate** results.

Validation of method and results

To test whether your method and results are valid you need to ask yourself:

- Did my chosen method for carrying out my investigation work well or did I have to make any changes?
- Did I have to repeat any parts of the investigation?
- How could errors happen in the readings that I took and how large could these errors have been?
- Did I successfully control the conditions?

Evaluation of your findings

In your evaluation you need to:

- decide how well your project met your aims
- decide whether your results were as you expected
- think about the errors in your investigation and what you did to minimise them.

What you found in your investigation should be linked to the scientific principles that you have learnt about or that you read about in the research you did.

Writing your report

Your report should tell another reader about your project. It must explain what your project is about and how you first planned it. It should state the hypothesis you were testing. Then it should describe what you actually did and the results you collected.

Your report must be well structured. Use different sections to make the information easy to follow. It must be written in the past tense and use phrases like 'the test tube was placed in the hot water' and not 'I placed the test tube in the hot water'. Don't forget to include references for the sources of information that you used in your project.

Key terms

Mean – the sum of quantities divided by the number of them.

Precision – the smallest change in a quantity which an instrument can measure *or* the degree to which measurements are repeatable.

Accurate – how close a measurement is to its true value.

Analysing the results.

Science focus

Data analysis

You may need to organise your data differently before it will give you the information that you need. For example, you may need to plot a graph from a table of results or work out the **mean** of your results before you can draw conclusions.

Functional skills

Mathematics and ICT

Choosing the best methods to organise and present your data will help you to draw conclusions and justify them.

Assessment activity 7.2 **P5 P6**

You are given the following data about the height of seedlings measured every 2 days.

Day	2	4	6	8	10	12	14	16	18	20	22	24	26	28	30
Height (cm)	1	1.5	2	2.5	4	6	7	9.5	11	13	14	14	14.5	15	15

1 Present this table in the form of:
  - a pie chart
  - a histogram
  - a line graph. **P5**

2 What conclusion can you draw about the growth of the seedlings? **P5**

3 Which of these graphs do you think is best for showing the information about growth of seedlings? **P5**

4 Share your findings with another member of your class. Do you both agree on the best format? If not, why do you think this may be? **P6**

Grading tip

Organising your data in different ways will help you to decide the best way to show what the results mean. You must make sure that all of your calculations are accurate and that you have used the correct units in order to satisfy the grading criteria. **P5**

 Activity A

Working in pairs or small groups, identify the different sections that you think a scientific report of a project should include. Make a list of these and discuss what information you think each section should contain.

 PLTS

Independent enquirers

Writing your report will help you develop skills in relating scientific theory to your work and using reasoned arguments to support your conclusions.

WorkSpace

Surjeet Singh

Formulation Group Leader, Glide Pharmaceutical Technologies Ltd

STEM AMBASSADORS ILLUMINATING FUTURES
Nationally coordinated by STEMNET

I am responsible for ensuring that the technical team uses standard operating procedures to formulate implants to a high quality. My team of pharmaceutical formulation scientists works with a wide variety of active drugs, proteins, peptide and vaccines to formulate them into a solid implant. The implant is designed with a tip and is robust enough to be pushed under the skin, where it delivers the active compound.

When we plan the manufacture of a new implant, the formulation team will first assess any health and safety issues (e.g. COSHH) of the active substance and the other ingredients used to stabilise it. Then a scientist will conduct feasibility investigations to decide how to manufacture the implant. The implants with the correct characteristics undergo several tests to show how well they penetrate the skin and how stable they are. We collect statistical data such as the mean and standard deviation on each batch to ensure that the implants are always very similar.

Think about it!

1 Suggest two or three methods of manufacturing solid dosage forms such as tablets.

2 Why are mean and standard deviation important in designing a manufacturing process?

Just checking

1 Identify seven areas to think about when planning your project.
2 What is GLP?
3 Outline why it is necessary to carry out risk assessments before carrying out your practical work.
4 Why is it important to record all of the results that you obtain during your project?
5 Identify one way in which you can validate the results you obtain from a practical investigation.

edexcel

Assignment tips

- Making your project too complicated will make it difficult for you to meet the assessment criteria. Equally, completing a project which is too simple will also make it difficult to achieve the criteria.

- When you make your plan it is important to clearly identify what you are investigating and what you think your results will show.

- In your report remember to include the risk assessments that you carried out before starting your practical.

- Try to identify different ways of tackling the project and clearly show these in your plan.

- Your plan should include identified points at which you will check how well you are doing. This may also mean that you will change your plan during these review points or because of the results that you obtain.

- Keeping a laboratory log of all of your activities will help you in completing your final report. It will also help you to produce evidence for **M2**.

- To achieve **M1** you need to produce a summary of your research to show how research from at least two different sources affected your plan. Remember that lots of the information that you gather may not help you with your plan and you should only use the information that you really need.

- Think about what went well with your project or where things could have been improved to help you evaluate the success of your project. This will help provide evidence for **D1**.

- Ask other people how well they thought you carried out your project. This could include people in your class (particularly if you have worked on a group project), the librarian and the laboratory technician.

- Your report could identify what you think you have learnt from carrying out the project or what parts you particularly enjoyed.

8 Science and the world of work

There are many organisations that use science in their everyday business. Some of these you can recognise immediately, for example the National Grid that brings electricity to our homes or the many pharmaceutical companies, such as GlaxoSmithKline (GSK) or Bayer, that make medicines. There are, however, many other organisations that don't obviously use science in their work.

For example, leading supermarkets will employ food scientists to investigate ingredients in their foods and the nutritional content of foods. Large bakeries will employ microbiologists to investigate different types of yeast in order to improve the process of making bread. Other organisations, such as hospitals, will have science laboratories that will carry out scientific tests on blood and urine, so as to help doctors find out why people are ill.

In this unit, you will investigate different types of companies that use science to make a product or provide a service. This may require you to go and visit a company or communicate with them via email, phone, by mail or by consulting their website. You will look at how the departments that make up the company work effectively, as well as go through the science and technology that is involved in the work they do.

Learning outcomes

After completing this unit you should:

1 be able to conduct research into how a science-based organisation operates

2 be able to investigate how scientific products or services are developed by a science-based organisation

3 know the health and safety legislation which refers to work within a science laboratory

4 know the key features of a working science laboratory.

Assessment and grading criteria

This table shows you what you must do in order to achieve a **pass**, **merit** or **distinction** grade, and where you can find activities in this book to help you.

To achieve a **pass** grade, the evidence must show that the learner is able to:	To achieve a **merit** grade, the evidence must show that, in addition to the pass criteria, the learner is able to:	To achieve a **distinction** grade, the evidence must show that, in addition to the pass and merit criteria, the learner is able to:
P1 Carry out research into how a science-based organisation operates **See Assessment activity 8.1**	**M1** Describe how the different departments of the organisation work with each other to meet the aims of the organisation **See Assessment activity 8.1**	**D1** Explain the advantages and disadvantages that the organisation has for people within the science and local community **See Assessment activity 8.1**
P2 Carry out an investigation into the process used to develop a scientific product or service **See Assessment activity 8.2**	**M2** Describe the scientific principles involved in developing the product or service supplied **See Assessment activity 8.2**	**D2** Compare the scientific product or service with one offered by a competing organisation **See Assessment activity 8.2**
P3 Identify the health and safety legislation which relates to working within a science laboratory **See Assessment activity 8.3**	**M3** Describe the role of health and safety legislation within a science laboratory **See Assessment activity 8.3**	**D3** Explain the consequences of not following health and safety legislation **See Assessment activity 8.3**
P4 Identify the key features of a working science laboratory. **See Assessment activity 8.4**	**M4** Describe the need for effective laboratory design. **See Assessment activity 8.4**	**D4** Evaluate the effectiveness of a laboratory design. **See Assessment activity 8.4**

How you will be assessed

Your assessment could be in the form of:

- an article e.g. about the structure of a company
- notes e.g. on health and safety legislation
- a design for a laboratory.

Georgia, 18 years old

This unit really helped me understand how a real company uses science to make money! Not only did I understand the science that they use but also how each department of the company works together. What made this unit really interesting is that it brought together biology, chemistry and physics in a real-world context. It also made me think about the type of organisation I would like to work for when I complete my studies.

Catalyst

What's the organisation

Imagine you are told you can work in any organisation that you want, as long as it uses science!

- What organisation would this be?
- What kind of science do you think is involved?
- How many people do you think work in this organisation?
- What would your role be in the organisation?
- Compare your thoughts with two other people in the class. Is there anything common between your answers?

8.1 How a science–based organisation operates

In this section: **P1** **M1** **D1**

You may think of science as something that only happens in schools, colleges and universities. Actually, lots of companies use science and scientists to test and create things – some examples are given in the table below. As someone with science qualifications you may be able to work in one of these companies.

Company/Organisation	Product/Service	Core business aim
National Health Service (NHS) 	For example, hospitals and clinics analysing samples of blood, urine, etc.	Provide healthcare services to the public
Forensic Science Service	Performing forensic tests on evidence	Supplying forensic science services to police forces in England and Wales
Nokia	Developing and manufacturing new mobile phone technology	Offering technology and communications solutions to customers

Science focus

Researching a science-based organisation

When researching organisations that use science you need to think about:

- what type of organisation it is (for example, is the company providing a service or making products?)
- what is its main aim – in other words, what is it trying to achieve?

Activity A

(i) Name one scientific or medical organisation that is different from those listed in the table above. Does it provide a product or a service? What is its core business aim?

(ii) Imagine your pet rabbit was sick and you wanted to take her to the vet. What kind of services could the vet provide?

Structure of an organisation

Organisations, whether providing a service or making products, will have an organised structure. The organisation may be a large company employing thousands of workers all over the world or it could be a small, family-run company with only three or four people employed.

Every company tends to have one person in overall charge. They may be called the Managing Director, or General Manager, or Chief Executive Officer. Below this person, most companies are organised into departments. Each department has a manager and, depending on the size of the department, there may be more managers of small sections within each department. This means that each member of staff will report to a manager. The manager is responsible for making sure that each member of staff knows what their job is, and that they carry it out to the best of their abilities.

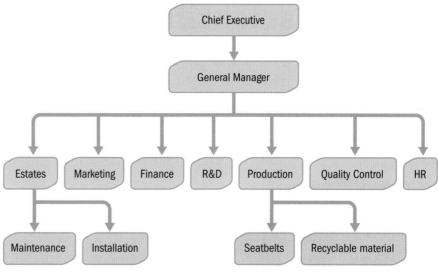

Organisational chart for a company that manufactures car seat belts.

Specialist departments

Table: Some important departments that exist in organisations and their main roles.

Department	Main role
R & D (Research and Development)	In the Research and Development department basic science and technology research is carried out. Inventions and products developed will be sold by the company in a few years' time. This is crucial to the future of the organisation
Finance	The Finance department is responsible for budgeting, company finance, ordering etc. This department helps make sure the company either makes money or stays within budget
ICT	The ICT or Information Communications Technology department is the section responsible for ensuring the company computer systems are maintained and are up to date. They also give technical assistance to all departments
HR	The Human Resources department is responsible for dealing with the staff who work for the organisation. This includes the recruitment and induction of new staff, salaries, pensions and promotions. They should also make sure that everyone in the company is treated well or at least within the law
Production	The Production department is responsible for making the product or products that the company sells
Quality Control	The Quality Control department ensures that the products are of good quality and that there has been no defect in the production process

BTEC **Assessment activity 8.1**

Investigate an organisation that you think you would like to work for. Make sure that the company is science based.

1 Find out how the company operates; does it provide a service or does it make a product? **P1**

2 Describe the structure of the company, showing how all the departments work together. **M1**

3 Explain the advantages and disadvantages of a company structured in this way. **D1**

Grading tip

When selecting an organisation, make sure that you have all the information that will enable you to attempt the **M1** and **D1** criteria. Try emailing or phoning the company to ask for more information, but make sure that you have first checked their website!

8.2 How scientific products and services are developed

In this section:

When a science-based organisation develops a product or provides a service, the organisation will depend on a number of departments all working together to make sure that the product or service is delivered at the correct time.

Research and Development (R & D)

This is where basic research takes place on a future product or technique. Sometimes it can take several years before a product is available to be used by members of the public. For example, the transistor is the main part inside the silicon chip which is used in all electronic products, such as TVs, computers, iPods and washing machines. It was researched in laboratories for nearly 10 years before it was manufactured using mass production.

Production

This is where the product or service is produced. For example, the assembly line for manufacturing silicon chips may involve producing thousands of electronic chips per minute. In a car manufacturing organisation it may involve making 700 cars per day. A production department could also be the laboratory where microorganisms are tested in a hospital.

Quality Control

In this department samples of the product are taken and tested to make sure they are working properly or are the correct weight, etc. For example, in a company producing breakfast cereals, there would be an instrument that would make sure each box contains the correct mass of cereal; if it doesn't the box is rejected. Pharmaceutical companies carry out extensive trials to ensure that their medicines are not dangerous to members of the public.

Technical Support

The support provided by this department could be in the form of technicians supporting engineers in the Research and Development department or servicing equipment on the production line. The technicians make sure that equipment is serviced and apparatus is correctly assembled. IT technicians will also support the computer network in all departments making sure that computers are functioning properly.

Marketing

This department is responsible for making sure that the product or service is advertised to customers. The department will also be responsible for carrying out research to find out how customers feel about the product or service.

A Research and Development laboratory.

A production line.

Activity A

(i) Which department is responsible for finding a new medicine?

(ii) If it was discovered that the medicine manufactured was making people sick rather than curing them, which department should have spotted the problem?

Industrial processes

A number of industrial processes are used to make products or carry out services in science-based companies. Some examples are listed in the table.

Feedback

Organisations always want to improve their products and services and so they place great importance in getting feedback, both good and bad. This feedback can be from within their organisation, from other companies or from members of the public. If the feedback is negative, the organisation may have to look again at how they produced their product or service, and possibly make modifications to the original design. They may even have to completely abandon the product and search for other things to make or service.

For example, in the car industry, some manufacturers had to recall cars that were sold because drivers complained of a problem with the car's braking system. Engineers later discovered that there was a defect in the design feature of the brakes.

Table: Some industrial processes

Technique	Use	Image
Chromatography	A method of separating compounds according to their solubility	
Distillation	A method of purifying liquids or separating mixtures of liquids having different boiling points	
Growth of microorganisms	Incubators are used for growing microorganisms in the production of medicines, e.g. antibiotics	
Filtration	Separating insoluble solids from a liquid. Normally a vacuum pump and filter paper are needed	
Use of centrifuges	Centrifuges separate liquids by weight. This is done by rapid spinning. Blood plasma and blood cells can be separated in this way	

BTEC Assessment activity 8.2

You are working in the marketing department of the organisation you researched earlier (page 175). Your job is to review how the company makes a product.

1 Select a product that the company makes and review the method used to make this product. (Hint: start with Research and Development.) **P2**

2 Now describe, briefly, the science and processes involved in making the product. **M2**

3 How does this product compare with a similar one that is produced in a competing organisation? **D2**

Grading tip

Before attempting the **P2** criterion, plan so that you are able to answer the tasks for **M2** and **D2**, as they are all related. Make sure that you can find out about the science involved in making the product you select and that you can find enough information on a second company that produces a similar product.

8.3 Working safely in the science workplace

In this section:

Key term

Risk assessment – an assessment that is made before an activity is carried out which identifies any risk to health or property that could occur because of the activity.

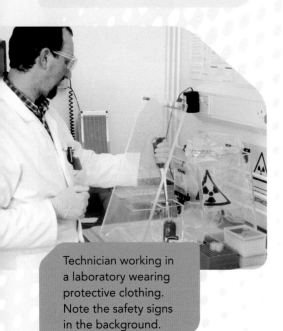

Technician working in a laboratory wearing protective clothing. Note the safety signs in the background.

Avoiding accidents and reducing the risk of causing an accident is everyone's responsibility. This includes employers and employees. All organisations, including laboratories and hospitals, have health and safety rules that must be followed by law.

Health and safety laws

The table below shows the laws (legislation) and regulations that you must know about in this unit. More information about these can be found from the **Health and Safety Executive** website.

Regulations	What it means:
Health and Safety at Work Act 1974 (HSAWA)	Employers must: • provide information and training on health and safety for employees • make sure equipment is maintained so it is safe, and protective clothing is used • provide a written safety policy and make sure that **risk assessments** are carried out **Protective clothing must be worn** Employees must: • take care of their own safety and that of others and cooperate with their employers • not interfere with any equipment or clothing that is for safety protection
Control of Substances Hazardous to Health 2002 (COSHH)	Employers must control substances that are hazardous to people's health. These substances are ones that carry a hazard sign on them and include chemicals, fumes, dust, gases and bacteria. Employers must provide protective clothing and guidance on safety limits, maximum exposure times, etc. Toxic Corrosive Radiation Biohazard
Reporting of Injuries, Diseases and Dangerous Occurrences Regulations 1995 (RIDDOR)	Employers, or those responsible for health and safety, must report workplace incidents. These could be actual accidents or near misses. The report can be made online, by phone, email or by post. RIDDOR advises that a report is made as soon as possible

Local Laboratory Rules (LLR)	The LLR will be shown on the wall in the laboratory. Those for a chemical or radioactive laboratory, would not allow anyone in the laboratory without specific protective clothing	 No smoking No drinking No eating
Codes of Practice	This is a document that each organisation will write. It explains how to comply with legislation. It is normally pinned to a wall in the laboratory	
Customs and Excise	Organisations have to follow strict rules when buying or selling chemicals and bacterial matter. These rules make sure that dangerous scientific materials will not fall into the wrong hands	
Radioactive Substances Act	This legislation ensures that radioactive substances are handled safely and that people follow safe procedures for disposing of radioactive waste	

Activity A

(i) Imagine you are carrying out an experiment with sulfur dioxide. What legislation do you need to follow?

(ii) If you accidentally get an electric shock because of faulty equipment, which organisation do you need to report to?

Risk assessment

To keep everyone safe, and to minimise damage to property, we need to plan for the possibility that an accident will happen. This is done in the form of a risk assessment.

A risk assessment looks at the likelihood of a hazard causing harm while an activity is being performed. To carry out a risk assessment properly the following three points should be considered:

- identify the hazard – this is the potential that something will do harm to you and other people while you are carrying out work

- identify what the harm will be to you or other people

- identify what you need to do and wear in order to reduce the risk of the harm taking place – this is called the 'control measure'.

When you work with hazardous substances, then the risk assessment needs to link with the COSHH guidelines. Sometimes there is a separate COSHH form as well as the risk assessment that includes other hazards.

BTEC Assessment activity 8.3 (P3) (M3) (D3)

You are an assistant technician preparing an experiment that measures current going through a bulb. You will be using a mains power supply, a bulb and an ammeter. You will be connecting the circuit using standard wires.

1 List the various legislations that apply to a science laboratory. (P3)

2 Describe the types of health and safety legislation that apply to this laboratory. (M3)

3 Explain what may happen if you don't apply the health and safety legislation when performing the experiment? (D3)

Grading tip

To meet (P3) ensure that you include the relevant legislation mentioned on these pages. To meet (M3), you need to make sure that the legislation you describe relates to a specific science laboratory and activity that is performed in the laboratory. To meet (D3), think about all the consequences to the laboratory and to the people who work in the laboratory of not applying the correct health and safety legislation.

8.4 Features of working science laboratories

In this section: P4 M4 D4

Whatever science-based organisation you research for this unit, their laboratories will be designed to fit the purposes of the company. For example, a microbiological laboratory will be designed differently from a space technology laboratory and from a forensic laboratory.

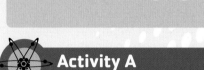

Activity A

Choose a science organisation. What kind of laboratories do you think they use?

Common laboratory features

In this section you will investigate different types of laboratories so that you are able to design your own dream science laboratory. Despite the differences that exist in laboratory designs there will be some common features. All laboratory features fall into the categories below.

Requirements	Examples
Health and safety	• Fire exit, fire extinguishers, fire blankets • Central gas/water taps • Circuit breakers
Service requirements	Each laboratory will need different services (gas, electricity, etc.), depending on what work is being done. For instance, a physics laboratory might need high-voltage electricity, while a chemistry laboratory will need gas taps.
Equipment	Depends on the laboratory, but most will have: • an Internet connection and a computer to enter results of experiments or to check on stock • testing equipment such as portable appliance testers used in PAT testing (electrical testing of equipment). In a microbiology laboratory there may be microscopes and in a physics or electronics laboratory there will be multimeters for electrical testing. In a chemistry laboratory there will probably be a spectrometer and some distillation equipment (for purifying solutions).
Furniture	All laboratories require furniture; this could include work benches and chairs.
Storage	Any laboratory will require ways of storing equipment and chemicals. The method of storage will depend on the item. For example, a fume cupboard is used to store chemicals that could give out toxic fumes. A fridge is required to store types of bacteria. Standard cupboards are used to store books, glassware and electrical equipment.
Glassware	Glassware for use in laboratories is normally heat-resistant (both for extreme cold and hot) and pure enough to be transparent and also unreactive to chemicals. Examples in a chemistry and a microbiological laboratory might include: beakers, pipettes, test tubes, flasks, Petri dishes and Florence flasks. They need to withstand cleaning at a high temperature to make sure that toxic chemicals and dangerous organisms are eradicated.

BTEC Assessment activity 8.4  P4 M4 D4

You are a science technician working for the organisation that you investigated for sections 8.1 and 8.2. The company have decided to build a new laboratory and you have been asked to advise on design and requirements.

1 List the equipment and furniture that you will need, including any health and safety requirements. Draw a rough plan of the laboratory and label important design considerations. **P4**

2 Describe why you will need the equipment and furniture you have listed. **M4**

3 Evaluate your design, including the furniture and equipment chosen, and make it clear why your laboratory will run smoothly and be a safe place to work. **D4**

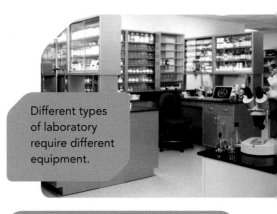

Different types of laboratory require different equipment.

Grading tip

For **P4**, make sure that when you identify the equipment and furniture they relate directly to a specific science laboratory. When attempting to meet **M4**, make sure you explain why you need the furniture, equipment and health and safety features for that laboratory. When looking at **D4**, think about how the features you identified in **P4** make your laboratory work well.

WorkSpace

Joanne Young
Healthcare Scientist in Radiation Protection, King's College Hospital

STEM AMBASSADORS
ILLUMINATING FUTURES
Nationally coordinated by STEMNET

My overall responsibility is to ensure that the hospital complies with all of the laws related to the use of radiation. My typical day would involve:

- testing X-ray machines to ensure they are working correctly
- checking that radiation doses for staff and patients are below set limits
- arranging for radioactive material to be safely delivered and disposed of
- writing reports and maintaining records.

There are several pieces of legislation on the use of ionising radiation:

- the Ionising Radiation Regulations 1999, policed by the Health and Safety Executive
- the Radioactive Substances Act 1993, policed by the Environment Agency
- the Ionising Radiation (Medical Exposures) Regulations 2000, policed by the Care Quality Commission.

If we do not comply with these regulations we could be prosecuted, leading to large fines and even imprisonment of the Chief Executive.

Think about it!

1 Why is it important to test X-ray machines?

2 Why is it important to ensure radioactive waste is disposed of correctly?

Just checking

1 Which department, in an organisation that uses science, involves discovering new materials?
2 Which department makes sure that the service carried out or the product produced is of good standard?
3 What are the common things that you would expect to find in any scientific laboratory?
4 Complete the table (below) about health and safety legislation.

Legislation	Details
	Relates to handling toxic materials
Risk assessment	
RIDDOR	

edexcel

Assignment tips

To get the grade you deserve in your assignments remember the following.

- When deciding on which company to choose for your investigation, make sure that you get all the information you need for all the criteria in this unit.

- Select an organisation that you would be interested in working for; this will make it interesting!

- Contact the organisation early with your enquiries. Remember, you can use normal mail, email, phone or make an appointment to see someone if possible. There may be lots of information on their website, so check this out first.

- Make your assignment interesting. Think about making the product in the laboratory, if possible.

- Refer to the chemistry, physics and biology core units when working out how the science product is produced, or the service is carried out.

- Remember health and safety is meant to make everyone safe; use the legislation when carrying out your experiments for the product.

9 Working in a science-based organisation

Whether you work as an assistant microbiologist in a hospital laboratory or as an engineering technician in a company making cars you will need to follow instructions correctly, organise your time effectively and be flexible in your working patterns. You will probably work as part of a team and use information technology in your day-to-day tasks.

This unit helps you to explore how science technicians organise their time effectively to be productive in their work and contribute to the success of their company. You will investigate the kind of technical skills that different types of science technicians have, including some of the common computing packages that they use.

Working safely is important in any work place, but in a science laboratory there are specific health and safety procedures that need to be followed; you will look at examples in this unit. You will also investigate how science technicians manage their laboratory in order to keep it running smoothly.

Learning outcomes

After completing this unit you should:

1 know the basic duties and responsibilities of a junior science technician/assistant practitioner

2 know the personal, communication and ICT skills needed to work in a science-based organisation

3 be able to describe key procedures and demonstrate safe working practices within a laboratory.

Assessment and grading criteria

This table shows you what you must do in order to achieve a **pass**, **merit** or **distinction** grade, and where you can find activities in this book to help you.

To achieve a **pass** grade, the evidence must show that the learner is able to:	To achieve a **merit** grade, the evidence must show that, in addition to the pass criteria, the learner is able to:	To achieve a **distinction** grade, the evidence must show that, in addition to the pass and merit criteria, the learner is able to:
P1 Identify the typical duties and responsibilities of a junior science technician/assistant practitioner **See Assessment activity 9.1**	**M1** Describe the typical duties and responsibilities of a junior science technician/assistant practitioner **See Assessment activity 9.1**	**D1** Explain how the typical duties and responsibilities of a junior science technician/assistant practitioner contribute to the effectiveness and efficiency of the laboratory workplace **See Assessment activity 9.1**
P2 Identify the personal, communication and ICT skills required by junior science technicians/assistant practitioners within an organisation **See Assessment activity 9.2**	**M2** Describe how the personal, communication and ICT skills of the junior science technician/assistant practitioner contribute to the work of a science-based organisation **See Assessment activity 9.2**	**D2** Evaluate how your own personal, communication and ICT skills could contribute to a science-based organisation **See Assessment activity 9.2**
P3 Describe the key procedures within a science laboratory **See Assessment activity 9.3**	**M3** Describe the need for effective laboratory practices and procedures within a science laboratory. **See Assessment activity 9.3**	**D3** Explain how following safe working practices and procedures maintains a safe environment within a science laboratory. **See Assessment activity 9.3**
P4 Demonstrate safe working practices in a laboratory. **See Assessment activity 9.3**		

How you will be assessed

Your assessment could be in the form of:

- induction materials for a new employee covering duties, skills and laboratory procedure
- a witness statement e.g. on your ability to demonstrate safe working practice.

Mohamed, 16 years old

From this unit I learned that whatever type of work you do you will need to make sure you get to work on time, organise your time and work well with people so that your organisation becomes successful.

Group work was encouraged during lessons, which I found helpful. I liked the investigations we did on health and safety, in particular a game involving working in various science laboratories. It was also fascinating when a practising science technician came to our lesson and told us what it was like to work in his organisation and the skills he gained working there.

- What areas of this unit might you find challenging?
- Which section of the unit are you most looking forward to?
- What preparation can you do to get ready for the unit assessments?

Catalyst

Where will you work?

Imagine you have a choice of places to work as a science technician. What kind of technician would you like to be? Now imagine that today you have just started work. With a partner, discuss what you would expect to do on your first day. Who would you expect to meet? What would your timings be?

Make your findings into a poster and present them to the class. Remember to have reasons for your responses.

9.1 Working as a junior science technician

In this section:  P1 M1 D1

Key terms

Good laboratory practice (GLP) – safe and efficient procedure that is followed by all people working in a science laboratory.

Employment law – law designed to ensure everyone working is treated fairly.

Activity A

Find out what type of technician would work

(i) in a brewery

(ii) in a plastics factory.

There are over 8000 science technicians and assistant practitioners working in the UK. The table below gives some places where they work.

Job title	Place of work
Veterinary nurse	Animal hospitals, vets
Dental technician	Dental hospitals, community dentists
Forensic science technician	National Forensic Science Agency, local police force, commercial forensic agencies
Medical laboratory assistant	Microbiological laboratories (for example in hospitals, environmental agencies, research and development departments)
Pharmacy assistant	Hospitals, high-street chemists, pharmaceutical industry
Engineering technician	Engineering companies and factories (for example, car manufacturing, steelworks, aerospace industry), university research and development departments
Science teaching technicians	Schools, colleges and universities

Junior science technicians and assistant practitioners work in a wide range of jobs, but they all have some common duties and responsibilities.

Case study: A biology technician's schedule

Mirvette is 18 and has been working as a junior biology technician in the microbiology laboratory at a local hospital for 18 months. She works 37 hours a week, and has 15-minute breaks during the day and a 45-minute lunch break. Mirvette follows **good laboratory practice**: for example, every day she makes sure that the laboratory is clean and tidy, the incubators are at 50°C and the vents are in good working condition, ready for the biomedical scientists to start their work.

On Mondays there is a departmental meeting from 10.30 to 11.30. Mirvette attends with the senior technician. The work for the week is discussed and data from last week is reviewed.

Every Wednesday Mirvette and the senior technician go to the A&E (Accident and Emergency) department to collect samples that may be contaminated with the MRSA bacterium. They also advise A&E on contamination and service some of their equipment. Once a year Mirvette carries

out PAT tests on the electrical equipment; she has undergone training so she is qualified to do this.

Mirvette has an appraisal once a year with the senior technician, in which her work is reviewed and targets for next year are agreed. They discuss training needs, as well as her salary.

Today, Mirvette's first task is to assemble a new autoclave for sterilising equipment, by following the instruction manual. Assembling and servicing the autoclave requires detailed understanding, which Mirvette learned when she attended a training course last month.

Next week she is going on holiday; she has 28 days paid leave every year.

Mirvette in her laboratory.

Employment law

When you are offered a job, your employer will give you a contract to sign. Even if you have no written contract, as long as you are working then it is assumed that you have accepted the conditions of the contract.

A contract of employment.

Activity B

Look at the case study on the previous page.

(i) List which tasks Mirvette does daily, weekly and yearly.

(ii) Which department does Mirvette service?

(iii) What is the name of the special meeting in which targets and training needs are discussed?

Science focus

As a young person you are protected from working more than 40 hours a week, and may not work between 10 pm and 6 am unless special arrangements are made. Your employer must also pay you at least the minimum wage, which will depend on your age. For young people who have left school (over 16 but under 18) the minimum wage is £3.57 (2009 rate).

- If you are pregnant, have given birth in the last 6 months or are breastfeeding, the law gives you and your baby special protection at work.

- Employment law also protects people from discrimination because of their gender, race, religious belief, disability or sexual orientation.

 BTEC ## Assessment activity 9.1 **P1 M1 D1**

Investigate the work of another technician from the table. Job descriptions from advertisements are a good place to look for information.

1 List the tasks carried out by your chosen technician on a daily, weekly and annual basis. **P1**

2 Make a flowchart describing the duties and career progression for this technician in their organisation. **M1**

3 Explain how the work of this technician helps their laboratory run smoothly and efficiently. **D1**

Grading tip

For **P1**, make sure that you relate duties and responsibilities to a particular science laboratory; don't be too general. For the **M1** criteria you need to describe the main points, such as the tasks, the support provided by the junior technician and career progression. For **D1**, think about what would happen to the laboratory if the technician did not do a good job, wasn't organised or could not read instruction manuals.

9.2 Personal skills in science

In this section: P2 M2 D2

Key term

Intranet – this is similar to the Internet but is available only to people inside an organisation; it includes email, websites and other useful online resources for workers.

Did you know?

You can use email software to remind you of important meetings and tasks.

Activity A

Imagine you are a forensic science technician, working in a forensic science service.

(i) List the type of people that you might work with.

(ii) Explain what you would do to make sure that your time is used effectively.

PLTS

Team workers

Activity A will make you aware of the importance of teamwork.

Time management

Science technicians and assistant practitioners have to carry out many tasks, so organising your time is important.

You can manage your time better by, for example, having a 'to do list' which lists the things you need to do each day, and by making a chart to prioritise the most important and the most urgent tasks.

Working with other people

- During your day-to-day work you will work closely with your supervisor. Your supervisor can give you advice about your work and support with training needs.

- As a junior science technician or assistant practitioner you will work with people in other departments too, for example human resources (HR). HR deals with your salary, holidays, sick leave and other entitlements.

- You will also deal with scientists and engineers in the same department and technicians from other departments. For example, Mirvette (see the case study on page 186) might work with technicians in other departments, such as histopathology, clinical chemistry, the blood bank and the hospital wards.

- As a technician you will work with external people as well, for example organisations who supply equipment and samples to your department and technicians who come to service equipment.

You need to be able to communicate quickly and clearly with all these people, using correct scientific language. The table below shows various ways of communicating in a science-based organisation.

Type of communication	What we use it for
Electronic	Emailing: technicians in other departments about equipment etc.; ordering enquires to companies; emailing supervisor about work issues
Verbal	Discussing work with supervisor, peers and members of the public
Written	Writing reports of experiments, progress of work, annual report on servicing of equipment, etc.
Video conferencing	Attending meetings via phone link with colleagues who are in different parts of the country
Phone/fax	Enquiring about orders, confirming timings for external servicing staff to come to the department, etc.

Computer skills

Using computers is an important skill in most organisations, but is particularly important in a science-based organisation.

- Spreadsheet programs are used to display and analyse experimental data.

- A database may be used to manage chemical stock and equipment.

- Management Information Systems (MISs) hold lots of different types of data for the organisation; for example, how many products are produced in a given time, or how much money is spent by each department. Science technicians contribute to the MIS, for example, by providing details about stock usage and the condition of equipment.

- The technician is responsible for scientific information provided on CDs, such as commercial catalogues and data sheets, or the CLEAPSS CD of information about chemicals, including hazards and how to dispose of waste.

Video conferencing in the workplace.

Internet and intranets

The internet is an important tool for technicians. They can use it for tasks as varied as updating their skills through online training and for ordering equipment. You need to know how to browse, use search engines and download and upload information.

An **intranet** is a 'local' Internet within your organisation. It allows an organisation to share information between departments quickly.

ICT skills are important for technicians.

BTEC Assessment activity 9.2 P2 M2 D2

Imagine you are a technician working in a car manufacturing company.

Write a diary about one day's work, and include the following:

(a) Which communication skills did you use, and with whom? Mention a little about what you would communicate in your work. **P2**

(b) List the ICT skills that you used in that day and how they related to your work. **P2**

(c) Describe how your skills in questions **(a)** and **(b)** contributed to the company. **M2**

(d) Now list your own skills and evaluate how these skills would contribute to an organisation that interests you. **D2**

Grading tip

When attempting the **M2** criterion, make sure you explain how the interpersonal skills you have allow you to do a good job as a technician in the organisation. The **D2** criterion needs to relate directly to your own skills. What are your skills and how will these skills help the work of a company?

Functional skills

ICT

The activity will allow you to evaluate your own ICT skills.

9.3 Laboratory procedures and safety

In this section: P3 M3 D3 P4

Key terms

Accident book – a record of any accidents that occur.

Incident book – a record of all near-misses and incidents that occur in the work place; this includes spillages and broken glassware.

Laboratory waste should be safely disposed of in clinical and chemical waste bins.

Technicians follow standard laboratory practices to ensure they work safely and to a high standard.

Case study: Key laboratory procedures

John, a junior pharmacy technician, works with a senior technician to manage three laboratories used by research pharmacists to develop medicines.

At the beginning of the financial year (in April), John and his supervisor received an annual budget for chemical stock, equipment, furniture, stationery and servicing of the laboratories.

John monitors the stock of chemicals and orders more before they run out so that the work of the pharmacists is not interrupted; this is called stock control. He checks catalogues and websites of three or four companies to get the best price and availability. The order goes through his supervisor, who passes it to the finance department. John makes sure that the department gets copies of the invoice and a receipt from the company, which he keeps in a file in the laboratory office. When the chemicals arrive he stores each one in the appropriate cabinet in order to avoid contamination.

At the start of the day, John checks that the temperature is right for each laboratory and that the vents are functioning. At the end of the day, he makes sure that the laboratories are clean and equipment is sterilised ready to be used again. This is part of good laboratory practice (GLP).

Sometimes, John needs to prepare samples of chemicals. Before he does so he writes a risk assessment of the procedure, identifying possible hazards and planning how the risk can be minimised by control measures. The control measures will include wearing the appropriate clothing, such as a lab coat, gloves to avoid touching corrosive chemicals and safety goggles to prevent splashes to the eyes.

John is aware that any accidents need to be recorded in the **accident book**. If an incident occurs, for example a chemical is spilled, but no one is harmed, then he records this incident in an **incident book**.

Activity A

Imagine you are working in a veterinary surgery. What kind of practices and procedures would you be involved in?

BTEC ## Assessment activity 9.3 P3 M3 D3 P4

You work in a laboratory which monitors water pollution. Your job is to help look after the laboratory and test samples that water engineers bring from local rivers.

1 Write in poster format what you do on a typical day. P3

2 Using a practical example, explain what you do to ensure that you are working safely P4 and how you make sure that your laboratory was working effectively. M3

3 Explain how your answers to questions 1 and 2 ensure that working in your laboratory is both safe and efficient. D3

Grading tip

To meet P3 and P4, you must write down specific procedures and work practices for work in a laboratory. For the M3 criterion you need to make the connection between what you do in your work and how that work makes your laboratory effective. For D3 you should explain how to work safely and why.

WorkSpace

Dr Stella Morafa

Principal Scientist, GlaxoSmithKline

STEM AMBASSADORS ILLUMINATING FUTURES
Nationally coordinated by STEMNET

I work for the largest pharmaceutical company in Europe. I am responsible for developing medicines to treat patients who have respiratory ailments such as asthma and chronic obstructive pulmonary disease.

When I was 16 I was not sure what I wanted to study in science and so decided the A-level route would not be right for me. Instead I studied the BTEC First in Applied Science followed by BTEC National Diploma in Science, as they encompassed all the sciences. My studies included a really interesting course called pharmaceutical manufacture, in which I learnt how to make toothpaste, creams, gels, soaps and tablets. At the end of my BTEC course I did a four-year Pharmaceutical Science degree. I liked it so much I went on to do a PhD. From the BTEC courses I learnt how to manage my time more effectively, how to work safely, how to follow good laboratory and good manufacturing practices and how to work with others.

In a typical day, I start off by checking emails and calendar reminders. I design experiments to understand the effects of drug molecules, which must be carried out safely and according to good manufacturing practice (GMP) regulation, and I supervise and train junior scientists. I also read scientific literature and implement what I learn.

Think about it!

1 What would happen if experiments to discover new drug molecules were reported incorrectly?

2 Why is it important to work safely in the laboratory?

Just checking

1 List two tasks that a science technician would perform:
- daily
- weekly
- annually.
2 Explain how you would support your colleagues whilst working as a science technician.
3 Describe the skills you would expect to possess as a science technician.
4 Describe three procedures and three practices that you would expect to be involved in if you were managing a science laboratory.
5 Write down the three key points you need to consider in order to work safely in a science laboratory.

Assignment tips

- When identifying the duties and responsibilities of a junior science technician, remember to cover the content in the specification. It's very easy to miss some of the duties and not meet the **P1** criteria. If the assignment brief is not clear, ask your tutor for clarification.

- It might help you to meet the **M1** criteria by inventing a diary of a junior technician, or thinking about a work experience that you have had.

- For **D1** it is important to mention why the duties and responsibilities of a junior science technician make the laboratory effective and efficient.

- For **P2** and **M2**, there are skills that all people working in science will have: for example, using a computer to order stock or presenting experimental data to other scientists and engineers. These skills need to be made clear in any discussion.

- **D2** requires you to evaluate your own technical skills. Think about the experiments that you have performed and the skills they required. You have other skills too: perhaps you are able to communicate well with people and can use a computer to write reports.

- For **P3** and **M3** you will need to give some detail about the type of work that science technicians do. Some tasks will be common to all jobs and others will vary; you need to refer to both in your assignments. For example, a pharmacy technician will carry out some tasks that will be different from those done by an engineering technician.

- For **D3** you have to be able to refer health and safety to working in a science laboratory. This will link to practice and procedure, which you covered in **P3** and **M3**.

- Remember that to meet **P4** you need to demonstrate safe working practice; it's not enough to talk about it. Normally you can demonstrate that you are working safely in the laboratory by carrying out a number of experiments safely.

10 The living body

Your body is an amazing machine, with lots of different systems working together to allow you to do things like eating, breathing, growing, moving and reproducing. All these human functions are able to happen because of the control we have from our nervous system and hormones. These are mainly coordinated by the brain, which controls the day-to-day working of our bodies.

This unit relates to you as a human. You will look at the structure and function of the digestive, respiratory, circulatory, nervous, endocrine and reproductive systems. You will gain knowledge of how these systems keep the human body working. This is essential if you wish to further your knowledge and understanding of human anatomy and physiology.

You will also be able to find out about jobs relating directly to the study of these human body system structures and functions. This could be by producing different pieces of evidence for your assessments acting as, for example, an enzymologist, a biologist or a neurologist. Other jobs where you will need this knowledge include working as a beautician or a fitness instructor.

Learning outcomes

After completing this unit you should:

1 know the role of enzymes as catalysts

2 be able to investigate individual body systems, relating their structure and functions to their role in maintaining health

3 know how the nervous and endocrine systems work to coordinate the body systems

4 know the structure and functions of the human reproductive system.

Assessment and grading criteria

This table shows you what you must do in order to achieve a **pass**, **merit** or **distinction** grade, and where you can find activities in this book to help you.

To achieve a **pass** grade, the evidence must show that the learner is able to:	To achieve a **merit** grade, the evidence must show that, in addition to the pass criteria, the learner is able to:	To achieve a **distinction** grade, the evidence must show that, in addition to the pass and merit criteria, the learner is able to:
P1 Outline the role of enzymes as catalysts **See Assessment activity 10.1**	**M1** Explain the factors affecting the functions of enzymes **See Assessment activity 10.1**	**D1** Analyse data to identify the optimal conditions of at least two parameters for the function of an enzyme **See Assessment activity 10.1**
P2 Carry out investigations into the structure and functions associated with the digestive, respiratory, circulatory and renal systems **See Assessment activity 10.2**	**M2** Explain the way the respiratory and circulatory systems interact to maintain cellular and body function **See Assessment activity 10.3**	**D2** Explain the consequences for the human body when one of these systems fails **See Assessment activity 10.3**
P3 Identify the components of a simple reflex arc **See Assessment activity 10.4**	**M3** Describe the difference between the somatic and autonomic nervous system **See Assessment activity 10.4**	**D3** Give possible causes of failure of the nervous system and explain the consequences **See Assessment activity 10.4**
P4 Identify the function of the main endocrine glands **See Assessment activity 10.5**	**M4** Describe the way hormones coordinate body functions **See Assessment activity 10.5**	**D4** Assess the difference between the way hormones coordinate body functions and the way the nervous system coordinates body functions **See Assessment activity 10.5**
P5 Identify the structure and functions of the male and female human reproductive system. **See Assessment activity 10.6**	**M5** Explain the process of hormonal control of the female reproductive cycles. **See Assessment activity 10.6**	**D5** Explain the way conception is controlled using replacement hormones. **See Assessment activity 10.6**

How you will be assessed

Your assessment could be in the form of:

- practical work e.g. on enzyme activity
- a report e.g. on the different body systems
- a presentation or report e.g. on the nervous system including a labelled diagram
- notes e.g. about the endocrine system including a labelled diagram
- a report e.g. on the reproductive systems including labelled diagrams.

Scott, 17 years old

I think this unit is very interesting because it relates to both animals and humans and how they are affected in health and disease. I like the way it discusses how people can be treated as this is a possible career path I would like to go into in the future. The case studies have given me an idea of different jobs and careers that I could do and I am interested in, and I am particularly interested in research – possibly as a molecular biologist. It also covers how different organisms adapt to survive, how we are all different because of our genes, how evolution happens and the importance of conservation. In this unit you also learn new skills to get you ready for the workplace, which is good. The practical work is also great because it shows us what really goes on in our body. I really think this is the best thing about this unit.

This unit will help me to prepare to pass this unit and help me to be organised to get the best possible grade for each assignment. The unit also shows the grading tips clearly and how to meet the (P) (M) and (D) criteria.

Catalyst

Body systems

1 Name the main body systems.

2 In groups, discuss what you think the function of each system is.

10.1 Enzymes

Key terms

Catalyst – a chemical that speeds up a chemical reaction without being changed itself.

Active site – a place on the enzyme where the substrate molecule must bind for the enzyme to work.

Substrate – the substance that is acted on by an enzyme.

Activity A

Working in pairs, use the table to test each other on the function of different enzymes.

What are enzymes?

Enzymes are biological **catalysts**. They are found in all living cells where they help to speed up chemical reactions. Enzymes are essential for life.

What do enzymes do?

Enzymes can break down large molecules into smaller ones, build larger molecules from smaller ones, or turn one molecule into another one.

Information about some enzymes is shown in the table.

Substrate	Enzyme	Found in	Function
Lipids	**Lipase**	Small intestine	In digestion, breaks down fats into fatty acids and glycerol
Proteins	**Protease**	Stomach Small intestine	In digestion, breaks down proteins into amino acids
Carbohydrates	**Amylase**	Salivary glands Small intestine	In digestion, breaks down starch into sugars
Hydrogen peroxide	**Catalase**	All cells	Breaks down hydrogen peroxide, a toxic waste product produced by **metabolism** in the body

The structure of enzymes

Enzymes are proteins found both inside cells (**intracellular**) and outside cells (**extracellular**). Each enzyme has a unique 3D shape with an area called an **active site** where the chemical reactions take place (see **(a)** in the diagram). A **substrate** needs to fit into the active site like a key into a lock **(b)** to enable the reaction to happen and **product** to be formed. In ideal conditions the enzyme is free to accept more substrate into its active site **(c)**. It continues to do this, making more product in the process.

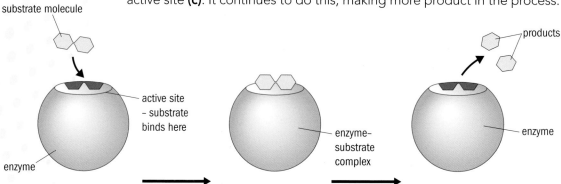

(a) The active site of the enzyme has a fixed shape into which only one particular sustrate will fit

(b) Once the substrate has slotted into the actice site, it can be activated

(c) The formation of products results in release from the enzyme, leaving the active site free for more substrate molecules to bind

A typical enzyme reaction.

Activity B

Explain how enzymes work. Use the diagram to help you.

Factors that affect the functioning of enzymes

Enzymes work best when they have the correct environment. Each one works best under different conditions; these are called the **optimum conditions**.

The following factors affect the optimum condition for the enzyme to function.

- pH – if the conditions are too acid or too alkaline this will affect the activity of the enzyme.

- Temperature – enzymes in the body work well at 37°C as this is human body temperature. Actually, our body enzymes would also work at 50°C but it would require a lot of energy for our bodies to be at this temperature and other complex molecules in the body would be **denatured**. If the temperature is too low then the enzymes will move around too slowly and the reaction is slowed.

- Substrate concentration – if there is not enough substrate or too much product is made in a reaction, then this will slow down the reaction. To avoid this happening, the product must be removed.

Did you know?

There are over 3000 types of enzyme in the human body.

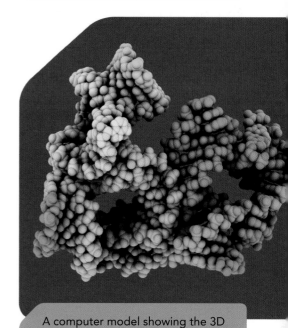

A computer model showing the 3D structure of an enzyme.

BTEC Assessment activity 10.1   P1 M1 D1

1 Using a number of labelled diagrams, describe the process of how enzymes speed up reactions. **P1**

2 Research in more detail the three factors listed above that affect enzyme activity. Make notes to help you explain these factors. **M1**

3 Use data to explain why it is important that the optimum conditions are present for enzymes to function correctly. **D1**

Grading tip

For **P1** you need to use the information on these pages and research further using books and the Internet to briefly state how enzymes speed up reactions.

For **M1** you should research factors affecting enzyme function in more detail, using graphs and a detailed explanation for each factor.

For **D1** you will be given some data that you need to analyse to identify at least two factors that affect how different enzymes work.

Functional skills

ICT and English

You can use your ICT skills to research for information on enzymes. You can use your English skills to give explanations about the ways enzymes work in different conditions in language that is clear and easy to understand.

10.2 The structure and functions of body systems

In this section: **P2**

Key terms

Absorption – the movement of molecules across a surface from one place to another, e.g. nutrients moving from the small intestine into the blood.

Peristalsis – a wave-like movement that moves food along the digestive system.

Alveoli – small sacs at the end of the bronchioles in the respiratory system where oxygen can move into the blood, and carbon dioxide can move out of the blood.

Renal system – made up of the kidneys to regulate body fluids and the ureters, bladder and urethra.

Filtration – the process in which fluids pass through a filter and some substances are removed.

Unit 6: See page 142 for more about the fate of nutrients.

Unit 3: See page 95 for more about the storage of excess nutrients.

Science focus

Digestion involves two processes

- Mechanical – chewing, swallowing and **peristalsis**.
- Chemical – enzymes breaking down large molecules, e.g. carbohydrates, fats and proteins.

Activity A

List the structures found in the respiratory system.

The body systems have structures that allow them to perform specific functions. This enables the body to function normally.

The digestive system

The main function of the digestive system is to break down food into smaller molecules. These can then be **absorbed** into the blood stream and delivered to body cells. Here these molecules are used to maintain cell and body functions. For example, glucose is used for respiration and protein is used for growth and repair. Any glucose that the body doesn't use can be stored in the liver and muscles as glycogen. It can then be used from the liver when glucose levels in the body are low.

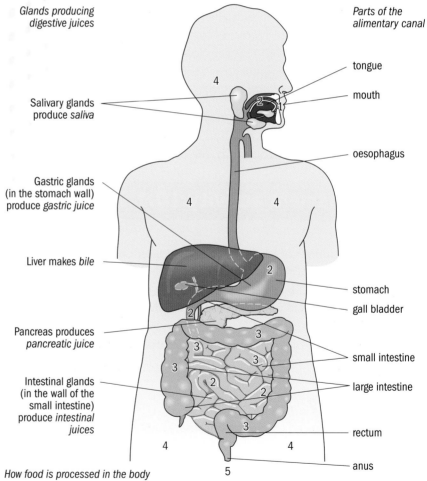

Glands producing digestive juices

Parts of the alimentary canal

Salivary glands produce *saliva*

Gastric glands (in the stomach wall) produce *gastric juice*

Liver makes *bile*

Pancreas produces *pancreatic juice*

Intestinal glands (in the wall of the small intestine) produce *intestinal juices*

tongue
mouth
oesophagus
stomach
gall bladder
small intestine
large intestine
rectum
anus

How food is processed in the body

1. Ingestion – food is taken into the mouth where digestion starts

2. Digestion – the large lumps of food are physically broken down into small pieces. Enzymes break the large complex molecules into smaller soluble molecules which can be absorbed

3. Absorption – in the small intestine digested food passes through the gut wall into the blood and lymph. In the large intestine, the remaining water, vitamins and minerals are absorbed

4. Assimilation – food is removed from the blood and used by the cells

5. Egestion – undigested waste is removed

The structure of the digestive system.

The respiratory system

The main function of the respiratory system is to supply the oxygen needed by living cells for respiration and to get rid of carbon dioxide, a waste product of respiration.

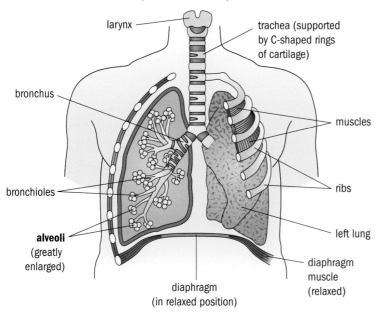

The structure of the respiratory system.

The circulatory system

The circulatory system is made up of the heart, blood vessels and blood. Blood is pumped around the body to deliver oxygen and nutrients to the cells for respiration. It also removes waste products such as carbon dioxide and water.

The renal system

The main functions of the **renal system** are:

- **filtration** to prevent larger molecules from being excreted
- **reabsorption** of amino acids, glucose, salts and water
- **excretion** of waste products including excess water and urea.

Fluid, salt and pH levels are all kept in balance by the kidneys in the renal system. Fluid is balanced with the help of anti-diuretic hormone (ADH). This is released from the pituitary gland in the brain when fluid levels are low. It helps the kidneys to reabsorb water back into the body.

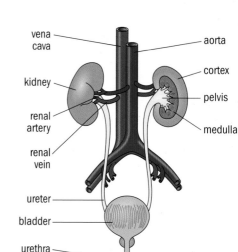

Oxygenated and deoxygenated blood in the circulatory system.

The renal system.

BTEC Assessment activity 10.2 **P2**

1. Using books and the Internet, research the function of each individual label for the body systems shown. **P2**
2. Produce a table and write about the specific functions for each body system. **P2**

Grading tip

For **P2**, avoid adding too much information for each function. Make sure your list of functions is concise and contains only the most important information for each label.

10.3 Body systems working together

In this section: M2 D2

Key terms

Plasma – the pale yellow watery part of the blood.

Platelets – disc-shaped cell fragments found in the blood which play an important role in blood clotting.

Haemoglobin – an iron-containing protein found in red blood cells that can bind to and transport oxygen.

The respiratory and circulatory systems work together

Next time you do some exercise think about how your breathing and heart rate are changing. You will find that they both increase. Why do they do this?

The respiratory and circulatory systems are very closely linked. Below is a diagram of the **alveoli**, which are the parts of the respiratory system where oxygen and carbon dioxide are exchanged. They are surrounded by blood vessels.

The alveoli

There are many alveoli in the lungs. They provide a large surface area for gas exchange. The walls of the alveoli are very thin so that oxygen moves into and carbon dioxide moves out of the blood stream easily.

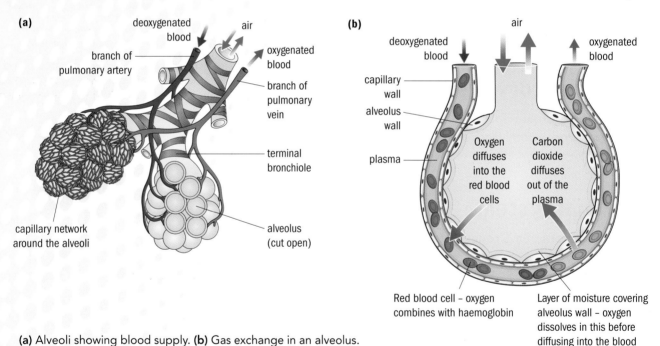

(a) Alveoli showing blood supply. (b) Gas exchange in an alveolus.

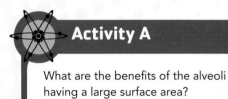
Activity A

What are the benefits of the alveoli having a large surface area?

Blood and the circulatory system

Blood is made up of **plasma**, **red blood cells**, a variety of **white blood cells** and **platelets**. The red blood cells contain **haemoglobin** which transports oxygen to every cell in the body. Diagram **(b)** above shows how oxygen moves out of the alveoli into the red blood cells. It is then transported round the body.

How is the oxygenated blood transported to the cells in the body?

The circulatory system consists of the heart, which pumps blood around the body, and three types of blood vessels.

- **Arteries** carry oxygenated blood under high pressure away from the heart to the body tissues and cells (apart from the pulmonary artery, which carries deoxygenated blood from the heart to the lungs).

- **Veins** return deoxygenated blood back to the heart (apart from the pulmonary vein, which carries oxygenated blood from the lungs to the heart). The blood then goes to the lungs where it gives up carbon dioxide and collects more oxygen.

- **Capillaries** are where oxygen, carbon dioxide and other products are exchanged between the blood and the body. Capillaries have very thin walls and form an extensive network inside all major organs and tissues.

Why do these systems need to work together?

Aerobic respiration involves the release of energy using oxygen and glucose. It happens in all living cells 24 hours a day and without it we would die.

The respiratory and circulatory systems work together in order to provide cells with enough oxygen and glucose for aerobic respiration. Cells also produce waste products, including carbon dioxide and water. These need to be removed from the body. When we exercise, the respiratory and circulatory systems need to work harder to do this, so they both speed up.

What would happen if the circulatory or respiratory system failed?

Failure of the circulatory or respiratory systems will result in serious medical conditions and will probably lead to death.

Respiratory system failure can result from choking, drowning or **cardiac arrest**. This stops oxygen reaching the body tissues.

Cardiovascular system failure such as a heart attack will stop the transport of oxygen to the cells and tissues of the body.

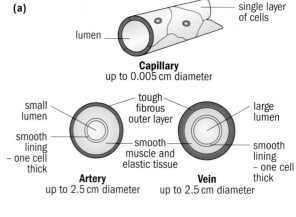

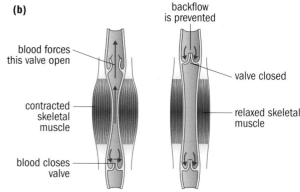

(a) The structure of the main blood vessels. (b) Valves in the veins prevent blood flowing back in the wrong direction, away from the heart.

BTEC **Assessment activity 10.3**

1 Using the information above and by further research, explain how the circulatory and respiratory systems work together to allow the body and cells to function normally. **M2**

2 Research further into the effects of respiratory and cardiovascular failure and explain, giving examples, how this affects the body. **D2**

Grading tip

For **M2** you need to provide a detailed explanation using diagrams of how these systems work together to enable normal body and cell functions.

For **D2** you need to do further research and explain using scientific terms what would happen if either system failed, giving three examples.

10.4 The nervous system

In this section: P3 M3 D3

Key terms

Neuron – a nerve cell found in the nervous system that processes electrical impulses and transmits them around the body to bring about a change to the body.

Synapse – the gap between two neurons across which a nerve impulse can travel.

Effector – a muscle or gland that brings about a response to a stimulus, e.g. the biceps muscle shortening to bend the arm.

Myelin sheath – the fatty covering surrounding a neuron which increases the speed at which nerve impulses travel along the neuron.

Activity A

What is the difference between the CNS and PNS?

Did you know?

The longest nerve in the human body is the sciatic nerve. This starts in the bottom of the spine and runs down the leg to the foot.

Activity B

Why is it important that we have reflex actions?

Why do we need a nervous system?

The nervous system gets information from the body, coordinates this information and then brings about a change to the body. It does this by coordinating muscles to allow you to do things like running, walking, talking and blinking.

What makes up the nervous system?

Nerve cells (or **neurons**) carry electrical information around the body. These electrical messages are called **nerve impulses**. There are three kinds of neuron: **sensory**, **motor** and **relay**. Messages pass from one neuron to another via **synapses**.

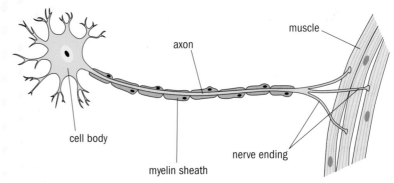

A motor neuron.

The organisation of the nervous system

The nervous system is split into two main parts.

- The **central nervous system** (CNS) is made up of the brain and spinal cord. Sensory receptors receive and coordinate messages sent to the brain and spinal cord. This brings about a change in the body.

- The **peripheral nervous system** (PNS) is made up of the nerves. It takes messages from sensory receptors to the CNS. It also carries messages away from the CNS to the **effectors**.

The peripheral nervous system is further divided into two parts.

- The **autonomic nervous system** (ANS) automatically controls the things that we need to be able to function normally as humans. These include increasing breathing and heart rate when exercising. You do not have to think about doing this; it is an automatic response.

- The **somatic nervous system** consists of peripheral nerve fibres that send information to the CNS and motor nerve fibres that go to skeletal muscle. It controls voluntary movement.

Reflex arcs

Your body uses **reflex actions** when you need to act quickly. This is a protective response and usually involves the spinal cord. The brain is also involved in complex reflexes such as learning to walk. The diagram below shows how a **reflex arc** produced by a reflex action enables the person to remove their hand from the drawing pin.

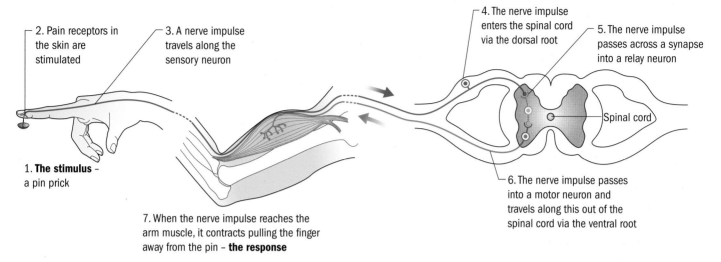

2. Pain receptors in the skin are stimulated

3. A nerve impulse travels along the sensory neuron

4. The nerve impulse enters the spinal cord via the dorsal root

5. The nerve impulse passes across a synapse into a relay neuron

Spinal cord

1. **The stimulus** – a pin prick

6. The nerve impulse passes into a motor neuron and travels along this out of the spinal cord via the ventral root

7. When the nerve impulse reaches the arm muscle, it contracts pulling the finger away from the pin – **the response**

A simple reflex arc showing the link between the receptor (in the skin) and the effector (the muscle in the arm).

What happens when the neurons go wrong?

Multiple sclerosis is a disease in which the messages sent by the neurons in the CNS are interrupted. This happens because the **myelin sheath** that surrounds and protects the neurons is lost. This causes a range of symptoms, e.g. muscle spasms and difficulty urinating, that can get worse over time.

PLTS

Independent enquirers

You will develop this skill when you research a different reflex arc.

Case study: Dealing with spinal injuries patients

Amy is an assistant to the neurologist at a large teaching hospital. Part of her job is to discuss with patients the reasons why spinal injury can cause different levels of disability. One of her patients is Katrina, a 30-year-old woman who was in a car accident. She broke her spine in the crash and is now in a wheelchair and unable to walk.

What information could Amy use to explain to Katrina why her spinal injury has caused her to be unable to walk?

 **Assessment activity 10.4** **P3** **M3** **D3**

1 Think of another stimulus that will initiate a response using a reflex arc. Draw and fully label this reflex arc. **P3**

2 Research the somatic and autonomic nervous systems in more detail and produce a table describing your findings. **M3**

3 Research how multiple sclerosis and spinal injury affect the functioning of the nervous system and explain the effect each has on the person involved. **D3**

Grading tip

For **P3** you should draw your own reflex arc and label it using all the correct terms.

For **M3** you should compare the somatic and autonomic nervous systems. A table is a good way to do this.

For **D3** avoid getting confused between the changes in the nervous system caused by multiple sclerosis and those caused by spinal injury.

10.5 The endocrine system

Endocrine glands and hormones

The endocrine system is made up of **glands** which produce chemical messengers called **hormones**. These are released into the blood stream and transported to the part of the body where they are needed.

Key terms

Gland – an organ in the body that makes substances, e.g. hormones, and releases them.

Secrete – make and release a useful substance.

Activity A

Using the table below, test your partner on the names of the different endocrine glands and the hormones they **secrete**.

PLTS

Self-manager

You will be fully responsible for working towards goals, showing initiative, commitment and perseverance when completing these tasks.

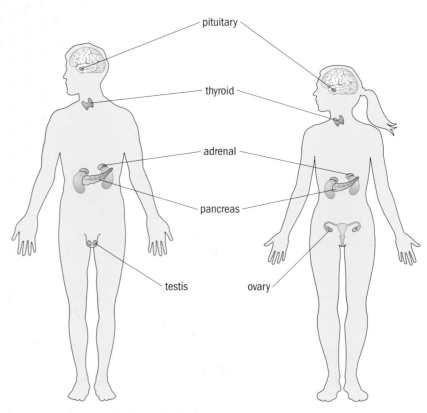

The main endocrine glands in the body.

Gland	Hormones and function
Pituitary gland	Releases eight hormones including **anti-diuretic hormone** (ADH) to control water balance and **growth hormones**. It also secretes **luteinising hormone** (LH) and **follicle-stimulating hormone** (FSH). In women, both of these hormones are involved in the formation and release of a follicle during the **menstrual cycle**. The pituitary gland also controls the functioning of almost all the other endocrine glands in the body
Thyroid gland	Produces **thyroxine** that controls metabolism, i.e. how fast the body burns oxygen and glucose to release energy from food
Pancreas	Secretes **insulin** and **glucagon** to regulate blood sugar and also secretes digestive juices
Adrenal gland	Secretes **adrenaline** and **noradrenaline** when the body is stressed. Adrenaline prepares the body for action
Testes (in males)	These secrete **testosterone** to control the development of male secondary sexual characteristics
Ovaries (in females)	These produce **oestrogen**, which controls development of female secondary sexual characteristics, and **progesterone**. Both are involved in the regulation of the menstrual cycle and pregnancy

The function of each endocrine gland and the hormones it produces.

How do hormones coordinate body functions?

When you are frightened, angry or excited your adrenal glands secrete adrenaline. This acts on a number of target organs and tissues and prepares the body for action. This happens in animals as well as humans; for example, when animals prepare to attack.

Adrenaline affects the body of an animal in many ways.

BTEC Assessment activity 10.5 P4 M4 D4

1 List all the endocrine glands and state the function of each. P4

2 Using books and the Internet, research and make notes on how the pituitary, thyroid and adrenal glands coordinate body functions. M4 D4

Grading tips

For P4 you need to identify the functions of the endocrine glands.

For M4 you need to show how other hormones coordinate body functions using diagrams and labels, or a flow diagram.

For D4 use the information you have learnt on the endocrine system and compare this with how the nervous system functions.

WorkSpace

Alisha Brookes
Invasive Cardiology Technologist in hospital cardiology department

I am responsible for assisting physicians in diagnosing and treating problems with the heart.

I schedule appointments, and I prepare each patient for surgery by shaving, cleaning and administering anaesthetic to the top of the leg. I assist the surgeons by inserting a catheter into the artery in the groin; the catheter goes through the artery to the heart to identify blockages. If a patient has an irregular heartbeat I carry out an electrophysiology test to find which area of the heart is causing it.

Some patients require surgery to insert a stent. This is a small tube that we slide into the artery using a catheter and place in the arteries to the heart to open up blockages. We monitor the blood pressure and heart rate of patients with special equipment throughout the procedure and tell the physician if there are any problems.

Following these procedures the patient can go on to live a normal life again.

Think about it!

1 Why is it important that patients are monitored throughout surgery?

2 What is satisfying about this job?

10.6 The reproductive system

In this section: P5 M5 D5

Humans reproduce sexually. Males and females produce sex cells called **gametes** (sperm and eggs). **Fertilisation** occurs when the nuclei of these gametes fuse together following sexual intercourse or assisted conception (IVF).

Key terms

Gametes – sex cells, the sperm and egg.

Fertilisation – this happens when the nuclei of the sperm and egg join together.

Oviduct – also called the fallopian tube. This is the tube an egg travels down to the uterus after being released from one of the ovaries every month.

The female reproductive system

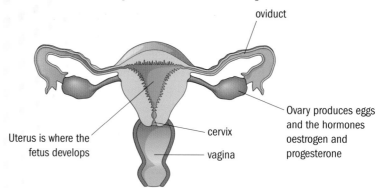

The female reproductive system.

oviduct

Ovary produces eggs and the hormones oestrogen and progesterone

cervix

vagina

Uterus is where the fetus develops

Science focus

Menstruation

Every month an egg matures and is released from an ovary. This is called ovulation. The egg travels down the **oviduct** to the uterus. If the egg is not fertilised by a sperm then the thick lining of the uterus breaks down and **menstruation** occurs. These events are all controlled by hormones: oestrogen, progesterone, luteinising hormone (LH) and follicle-stimulating hormone (FSH). LH and FSH are secreted by the pituitary gland.

The combined pill

This contains a combination of oestrogen and progesterone which prevents ovulation, and so prevents pregnancy.

Activity A

How does menstruation occur?

The male reproductive system

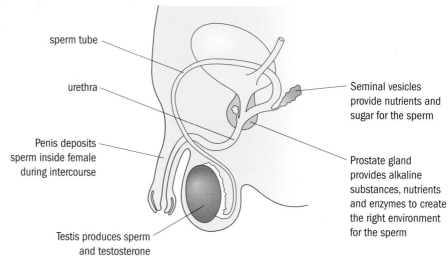

sperm tube

urethra

Penis deposits sperm inside female during intercourse

Seminal vesicles provide nutrients and sugar for the sperm

Prostate gland provides alkaline substances, nutrients and enzymes to create the right environment for the sperm

Testis produces sperm and testosterone

The male reproductive system.

Activity B

List the main structures in the male reproductive system and describe what their function is.

Formation of the embryo

Fertilisation leads to a zygote forming inside the oviduct. The zygote keeps dividing for a few days until it is a ball of cells and continues to move down the oviduct into the uterus. After about 5–6 days this implants into the wall of the uterus and becomes an embryo. At about 8 weeks we call the developed embryo a fetus.

Development of the fetus

The fetus continues to grow for the next 9 months, getting food and oxygen from the mother via her blood stream, the placenta and the umbilical cord. Waste is taken from the fetus via the umbilical cord, placenta and mother's blood stream.

Infertility

Sometimes it is not possible for a couple to conceive naturally. This can be due to factors including blocked oviducts, low sperm count or illness.

In-vitro fertilisation (IVF) allows a couple the chance to have a baby. IVF involves the following stages.

- The woman is given hormones, including oestrogen, FSH and LH, so that she produces several eggs.
- Once the eggs have grown to a certain size, they are removed.
- The eggs are mixed with sperm in a Petri dish.
- Progesterone is used to maintain the lining of the uterus ready for the fertilised egg(s) to be implanted.
- After a few days, one or two fertilised eggs are put back into the uterus.

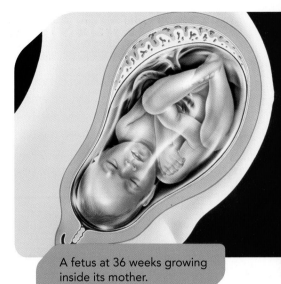

A fetus at 36 weeks growing inside its mother.

Science focus

Birth and the three stages of labour

At the end of the pregnancy the baby normally lies with its head facing towards the vagina ready to be born.

1 The mother's waters break and she gets contractions from the muscles in the uterus wall about every 20 minutes. The cervix also starts to widen.

2 The baby's head is pushed out of the cervix into the vagina and stronger contractions of the uterus occur.

3 The baby is delivered usually with the help of a midwife. The afterbirth is also delivered during this stage.

BTEC Assessment activity 10.6 P5 M5 D5

1 Research all the structures of both male and female reproductive systems. List the function of each structure. **P5**

2 Explain the function of each hormone involved in the menstrual cycle and how each one helps to regulate the cycle. **M5**

3 Research how IVF works and explain the importance of hormones in this process. **D5**

Grading tips

The process of how the hormones (on page 204) control the menstrual cycle needs to be explained using scientific facts and terms. **M5**

Explain in detail how hormones are used in IVF clinics to control the process of IVF, including how and when. **D5**

Just checking

1 List five different enzymes found in the body and explain what they do.
2 List the main body systems and state what each one does.
3 List five of the hormones found in the body.
4 What is the function of each of these hormones?
5 How are hormones used in IVF treatment?
6 Draw a labelled diagram of a neuron.

Assignment tips

To get the grade you deserve in your assignments remember the following.

- To award a pass, merit or distinction for this unit the people assessing your work will check that you have answered each task in the grading criteria correctly.

- You can apply the knowledge and understanding you have gained earlier on homeostasis (pages 92–95) to help you to complete P4, M4 and D4 in this unit.

- When completing practical reports and presentations for assessment for this unit, you will find useful help in your study guide.

Some of the key information you'll need to remember for this unit includes the following.

- Enzymes help to speed up chemical reactions in living things but their actions can be slowed down by environmental changes including pH and temperature.

- The different body systems are individually adapted to allow the body to perform its different functions effectively. Each body system is linked to the circulatory system.

- The endocrine and nervous system both send messages to bring about a change in bodily functions but differ in the way they deliver these messages in the body.

- Certain hormones are naturally involved in the regulation of the menstrual cycle but they can also be produced artificially to help childless couples have a baby with IVF treatment.

This unit will require you to do extra reading and researching from a variety of resources. Some excellent sources of information to help you include:

For information on...	Visit...
topics included in this unit	Wright D – *Human Physiology and Health* (Heinemann, 2007) ISBN 9780435633097
topics in this unit with animations	**BBC Science and Nature** website

Credit value: 10

11 Monitoring the environment

You will probably be aware from the television or newspapers of the environmental issues, such as climate change, that confront us today. Has the area where you live been affected by localised flooding due to torrential rain that is becoming more common?

Government bodies and charities such as Greenpeace carry out work to protect the environment. Without this protection we could lose many of the animals and plants that exist today.

In this unit you will learn about the structure of ecosystems and the balance between the different organisms in them. You will develop the skills used by real environmental workers to monitor ecosystems, like the environmental scientist in the photo whose job involves monitoring the quality of rivers. You will also learn how the actions of humans can affect ecosystems.

You will be assessed on your ability to take on the role of different types of environmental worker. You could be asked to write an article for an environmental magazine, or to do the job of an ecologist assessing a local area.

Learning outcomes

After completing this unit you should:

1 know the structure and operation of ecosystems
2 know how human activities influence ecosystems
3 be able to employ techniques involved in the monitoring of ecosystems
4 know how environmental protection is regulated.

Assessment and grading criteria

This table shows you what you must do in order to achieve a **pass**, **merit** or **distinction** grade, and where you can find activities in this book to help you.

To achieve a **pass** grade, the evidence must show that the learner is able to:	To achieve a **merit** grade, the evidence must show that, in addition to the pass criteria, the learner is able to:	To achieve a **distinction** grade, the evidence must show that, in addition to the pass and merit criteria, the learner is able to:
P1 Identify the structure and operation of ecosystems **See Assessment activity 11.1**	**M1** Describe the interrelationship between different components in the structure and operation of an ecosystem **See Assessment activity 11.1**	**D1** Analyse the roles of different components in maintaining the balance of an ecosystem **See Assessment activity 11.1**
P2 Identify human activities that influence ecosystems **See Assessment activity 11.2**	**M2** Describe how human activities influence ecosystems **See Assessment activity 11.2**	**D2** Discuss the long-term consequences of human influence on ecosystems **See Assessment activity 11.2**
P3 Use techniques involved in monitoring ecosystems **See Assessment activities 11.3 and 11.4**	**M3** Present results obtained from the techniques carried out to monitor ecosystems **See Assessment activities 11.3 and 11.4**	**D3** Evaluate the results and draw conclusions from the monitoring of ecosystems **See Assessment activities 11.3 and 11.4**
P4 Identify the role and rationale of agencies in environmental protection **See Assessment activity 11.5**	**M4** Describe the role and rationale of agencies in environmental protection **See Assessment activity 11.5**	**D4** Analyse the specific contribution of agencies to environmental protection **See Assessment activity 11.5**
P5 Identify reasons for protecting the environment. **See Assessment activity 11.6**	**M5** Describe reasons for protecting the environment. **See Assessment activity 11.6**	**D5** Evaluate reasons for protecting the environment. **See Assessment activity 11.6**

How you will be assessed

Your assessment could be in the form of:

- an annotated flow diagram of an ecosystem
- a magazine article e.g. on how human activity affects ecosystems
- a report e.g. on an ecological assessment of a piece of land
- a leaflet e.g. on the role of environmental protection agencies and organisations.

Ihsan, 19 years old

I found this unit very interesting. It relates to the natural world, the environment and to us as human beings. It has a lot of practical work in it, which I liked, and it tells you about how scientists work in different areas of environmental science.

The units are clearly laid out and it is easy to understand what you have to do for each assignment. They give you clear guidance and tips to help you gain a pass, merit or distinction.

I particularly enjoyed the field work in this unit. We went out and counted beetles in an area of woodland close to our school. I got quite good at using the pooter. We did the capture–recapture experiment and it worked better than I thought it would. I also set some pitfall traps near my home and came back later in the day to see what I had caught.

One day we went to a nature reserve. Kate, the wildlife warden, explained what she did to improve the habitat for breeding birds.

Catalyst

Working in groups, brainstorm the following questions:

1 What do you think an ecosystem is?

2 What do you already know about food chains? Give an example.

3 Can you think of any examples of humans affecting the environment?

11.1 The structure and operation of an ecosystem

In this section:

Key term

Ecosystem – a community of organisms living in a particular area interacting as one unit.

Woodland is just one example of a habitat. All of the organisms living in it make up an ecosystem.

Unit 3: The information on page 80 will help you to understand this topic.

Did you know?

All living organisms, both plants and animals, need to make energy 24 hours a day using the process of **respiration**. This is the opposite of photosynthesis. Respiration uses the products of photosynthesis, oxygen and glucose, to release energy. The process also releases carbon dioxide and water as waste products.

The structure of an ecosystem

The many different factors and components of an **ecosystem** can be described using the following terms.

Biosphere: the part of the Earth that is habitable. It is made up of many different kinds of natural habitat. Each habitat contains populations of their own plants and animal communities. Most of these are self-sufficient and have their own food and nutrient cycles.

Habitat: a place where an organism lives. It chooses this place for many reasons including the fact that there is food, shelter or it is hidden from predators. Some examples of habitats are woodland, seashore, grassland, rainforest, river or ocean.

Community: the organisms that live within one of these habitats form a community.

Population: the organisms of a certain species within the community, e.g. the frogs that live in a rainforest are called the rainforest frog population.

Ecological niche: within each community each species either feeds on and/or is eaten by other species in the food web. Each species occupies an ecological niche.

Biotic factors: the living components of an ecosystem, e.g. plants, animals, microorganisms and fungi.

Abiotic factors: the non-living components of an ecosystem, e.g. water, oxygen, nutrients and ambient temperature.

Activity A

In a fish tank you will find water, food and a variety of fish and plants. Write a paragraph describing each of these components, using the terms listed above.

Interrelationships within ecosystems

The main interaction between organisms in an ecosystem is eating and being eaten. Relationships also exist between biotic and abiotic factors.

Organisms and food

Green plants make their own food (glucose) by **photosynthesis**. They take in carbon dioxide from the air, and water via their roots. Using the energy from sunlight, the **chlorophyll** that is present in their leaves causes the carbon dioxide and water to combine together to produce glucose and oxygen.

Unit 12: To learn more about photosynthesis see page 228.

Green plants are at the bottom of the food chain. They supply all the other organisms within the ecosystem. They are called **producers**. Animals can't make their own food so they need to eat other living organisms, including producers. Animals are called **consumers**. A mixture of producers and consumers ensures that the ecosystem is balanced.

Activity B

Create a food web for a woodland habitat, naming the animals and plants at each level.

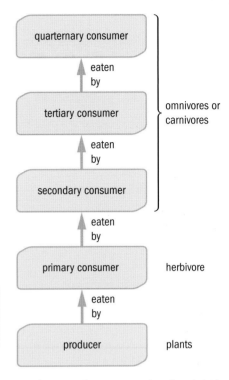

Producer and consumers in a food chain.

Unit 3: For more on food chains see page 80.

Relationships between biotic and abiotic factors

Here are some examples of interactions between biotic and abiotic factors.

- Plants need light for photosynthesis.

- Animals excrete waste providing the soil with nutrients.

- Plants take nutrients and water out of the soil.

- Microorganisms, including bacteria and fungi, feed on dead plant and animal matter to decompose it.

 **Assessment activity 11.1**

1 Discuss with a partner how you could put the information about a particular ecosystem into a flowchart. For example, if you choose the ocean you could start with a producer, i.e. plankton, and eventually end up with a top predator, i.e. a shark. **P1**

2 Imagine you are a local conservation worker. Look for interrelationships in your back garden or local park. Analyse how these interrelationships occur and what organisms are involved. **M1 D1**

Grading tip

For **P1**, make sure you include all the information in your flowchart.

For **M1**, use scientific facts to describe how different organisms interact with each other.

For **D1**, 'analyse' means to identify several relevant factors, show how they are linked and explain the importance of each one.

 PLTS

Self-managers

You will be good at organising time and resources and prioritising actions when producing a flowchart of the structure of an ecosystem.

11.2 Human activities and the ecosystem

In this section:  P2 M2 D2

Key terms

Biodiversity – the diversity of plant and animal life in a particular habitat.

Climate change – any long-term variation in the statistics of weather over a duration ranging from decades to millions of years.

Ozone layer – a region of the upper atmosphere, between about 15 and 30 kilometres above the Earth's surface.

Activity A

Give four reasons why the population of humans is increasing.

Safety and hazards

Hospital workers including doctors, surgeons, nurses and laboratory workers all need to dispose of body tissues, fluids and contaminated medical equipment safely to prevent the spread of infection.

Human population

People today have better health care and diets than ever before. This means that more babies survive birth, fewer people die from disease and people are living for longer. This has led to the human population doubling in the last 50 years. So what effect does this have on the environment? More humans on the planet means:

- more living space is required
- more food is needed
- more energy resources are required to power vehicles, workplaces and homes
- more pollution is produced
- diseases can be spread more easily.

All these factors can have a catastrophic effect on the environment as it struggles to keep up with demand.

Environmental issues

In order to provide more homes, food and energy for the growing population, more resources are required. This has led to the following environmental effects:

- depletion of the Earth's resources, e.g. coal, oil and gas
- damage to **biodiversity** with some species being wiped out within different ecosystems depending on the effect of humans in that particular area
- **climate change** largely caused by the release of methane, carbon dioxide and nitrogen oxide gases
- damage to the **ozone layer** causing an increase in the harmful ultraviolet radiation that reaches the Earth's surface.

Waste disposal

As humans we produce a lot of waste. Sewage is sent to treatment plants and household waste is taken to landfill sites or **incinerated**. However, the cost of this and the environmental implications of it need to be considered as we produce ever increasing amounts of waste. We can reduce the amount of waste by reusing and recycling as much as possible.

Unit 1: You can learn more about climate change, acid rain and the ozone layer on pages 28–29.

Pollution

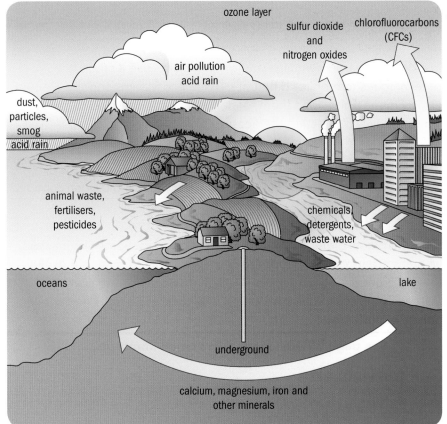

ozone layer

sulfur dioxide and nitrogen oxides

chlorofluorocarbons (CFCs)

air pollution acid rain

dust, particles, smog acid rain

animal waste, fertilisers, pesticides

chemicals, detergents, waste water

oceans

lake

underground

calcium, magnesium, iron and other minerals

Pollution caused by the human population.

Sustainable development

Sustainable development allows everyone in the world to have basic resources, such as food and housing, without causing harm to the planet for future generations. Recycling and saving energy are two important components of this. If they are managed effectively, progress can be made towards a sustainable future.

Activity B

Describe which types of pollution will become most serious as the human population increases. What measures could be taken to reduce them?

Functional skills

English and ICT

You could use your reading skills when reading sources of information and your ICT skills to produce images or tables to show the impact that humans have on the environment.

PLTS

Effective participators

You will discuss issues associated with human activities and their influences on the environment.

 Assessment activity 11.2 P2 M2 D2

1 List three human activities that influence the environment. **P2**

2 In pairs, discuss these three activities and how they influence the environment. **P2 M2 D2**

3 Find out more about these activities using books and the Internet. Write a short magazine article explaining the problem. **P2 M2 D2**

Grading tip

For **P2**, a list of how humans may act to cause change to ecosystems is sufficient but you could use pictures to make it more attractive.

For **M2**, you could use your list above to expand and describe how each one damages the environment.

For **D2**, discuss the advantages and disadvantages of human impact on the environment over a period of time.

11.3 Monitoring ecosystems: abiotic factors

In this section: 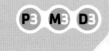 P3 M3 D3

Key term

Abiotic factor – physical, chemical and other non-living environmental factors.

pH meters give an accurate measurement of the alkalinity/acidity of soil or water.

 Activity A

Thinking about plants that like a wet environment, what size of soil particle would suit them best? Research what type of soil this would be.

Ecologists, environmentalists and climatologists use a range of techniques to monitor different parts of ecosystems and the environment. This allows them to identify any changes and to see if human activity is affecting the ecosystem.

Soil analysis

Soil can be analysed for pH, particle size, ion content and oxygen content.

pH of soil

Different plants prefer to grow in different pH soils. Some like it acidic whereas others prefer slightly alkaline soils. This information is useful to crop growers and gardeners. The pH of soil can be measured by dissolving the soil in deionised water and then using a **pH meter** or **universal indicator**.

Unit 17: See page 304 for more information about pH.

Soil particle size

Soil particles are surrounded by a thin film of water which plants can take up through their roots. To allow the water to be retained, the particles must be the right size.

Particle size too big	Water flows straight through the soil, deep into the ground taking nutrients with it, away from the roots of crops and plants
Particle size too small	After heavy rain, rainwater remains on the surface and the ground becomes waterlogged

You can measure soil particle size by seeing how quickly water drips through samples of soil using a funnel or by looking at different samples under a microscope.

Water analysis

Water can be analysed for pH and dissolved oxygen content.

pH of water

Again, the pH can be measured using a pH meter or universal indicator. If the pH is too acidic or alkaline this may indicate water pollution. If water is too acidic (the pH is lower than 4.5) then fish can die and plant growth is seriously affected.

Oxygen content of water

All living organisms require oxygen for respiration. Plants and animals that live in water use the oxygen that is dissolved in the water. Pollution often leads to low oxygen content, which means the organisms could die. The oxygen content is measured using a **dissolved oxygen meter**.

A dissolved oxygen meter.

 Case study: Testing the water

Ragini is an ecologist working for a city council. She has been called to a river downstream from a factory because there have been reports of dead fish.

What pieces of equipment should she take with her and why?

Air analysis

Air can be analysed for:

- oxygen content using a meter that detects the percentage of oxygen in the air
- particulates, which are tiny pieces of solid or liquid matter, such as soot, dust, fumes or mist.

Climate

Climatologists monitor the climate using various techniques. They monitor many factors including:

- rainfall
- temperature
- humidity
- wind speed.

Changes in these factors could provide evidence of climate change that can affect all life on Earth.

 Science focus

A microclimate is a small area which has a different climate from that of its surroundings. This could be a small patch of land which is sheltered and has an ideal temperature for growing certain plants. Nearby there could be areas which might be cooler and more open to the wind and rain where the same plants would not grow.

In the UK, vineyards are often planted in places with a warm microclimate.

Grading tip

To meet **P3**, you need to be able to use different methods in the field to monitor soil, water, air and the climate.

To achieve **M3**, you need to present your results clearly (e.g. in a table).

To achieve **D3**, you need to evaluate what your results tell you about the ecosystem.

 **Assessment activity 11.3** **P3** **M3** **D3**

1 Use the Internet and books to research further into how scientists monitor (a) air and (b) soil. Describe the techniques ecologists use to monitor these two things. **P3**

2 Explain why ecologists use the techniques you described in Question 1. **M3** **D3**

3 How do climatologists monitor the climate? Research this in groups and produce a poster. **P3**

 Functional skills

English and ICT

You will use your reading and Internet search skills to complete this assignment.

11.4 Monitoring ecosystems: biotic factors

In this section: P3 M3 D3

Key term

Sampling – the technique of selecting appropriate specimens for analysis.

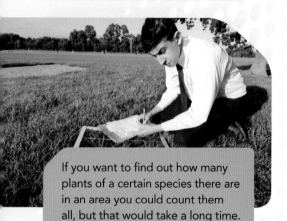

If you want to find out how many plants of a certain species there are in an area you could count them all, but that would take a long time. This pupil is using a quadrat to sample the plants in this field.

Grading tip

The quadrat you use may be a different size so make sure you allow for this in your calculations. Also, remember to adjust the number of samples you take to tie in with the area you are investigating.

Science focus

Microbial activity is when microorganisms break down dead plant and animal matter. Humus, which is the top layer of soil, is formed from this decomposing matter. It contains vital nutrients for plants to grow. If microbial activity slows down due to human impact on the environment, humus is reduced. As a result the whole ecosystem can be affected.

Sampling plants

A change in the numbers of certain species can often be the first visible sign that something is wrong in an ecosystem. Ecologists study the distribution of plants using different methods of **sampling**, for example **quadrats** or **line transects**.

Quadrats

A quadrat is a square wooden or metal frame that usually measures one square metre. To sample plants using a quadrat, follow these steps.

1 Throw the quadrat randomly onto the area of land you are investigating.

2 Count the number of plants of the species you are sampling within the quadrat.

3 Record the number.

4 Repeat steps 1–3 until you have sampled the area well enough – 10 times for an area about the size of a football pitch (6500 square metres).

5 To find the mean number of plants per square metre, add the number of plants in each sample and divide by the number of samples.

Activity A

How many times would you need to throw out the quadrat in an area measuring 26 000 square metres?

Line transects

To use a line transect, lay out a piece of nylon rope or tape straight across the area you want to study. Any species touching the line should be recorded along the whole length of the line. You can also record the species at regular intervals along the tape. Input your data into a table so that you can then analyse them.

Sampling animals

You need to use different methods to sample animals because they move around. You also need different methods for different habitats. To find out what types of species are present in an area of woodland, for example, you can use **pitfall traps** or **pooters**. If you want to estimate the size of a population you should use the capture–recapture method.

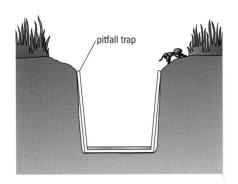

pitfall trap

A pitfall trap is set in the soil to trap insects, amphibians and reptiles that fall in as they move along the ground. It should be loosely covered to prevent rain getting in.

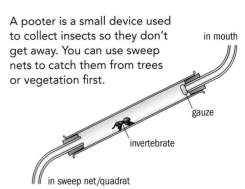

A pooter is a small device used to collect insects so they don't get away. You can use sweep nets to catch them from trees or vegetation first.

in mouth

gauze

invertebrate

in sweep net/quadrat

Point transects

Point transects are used in bird surveys. This involves standing at positions chosen randomly in the sample area and counting how many birds can be seen. Sometimes it is based on the number of birds that can be heard.

Safety and hazards

You must be careful not to get bitten by small insects such as midges when you are sampling. You must also protect the animals or insects from any harm wherever possible.

Case study: Sampling crabs with capture–recapture

Alice is a marine ecologist working for a wildlife trust. She needs to check the crab population to see if it has fallen due to pollution. She plans to use the capture–recapture method.

Alice collects an initial sample (n_1) and finds she has 150 crabs. She then marks these crabs with a non-toxic marker pen and puts them back. This allows the sample to mix with the rest of the population. Later she takes a second sample (n_2) and finds there are 110 crabs. Out of this sample, 60 are

marked ones. She uses this formula to calculate the total population.

$$population = \frac{\text{number in sample 1} (n_1) \times \text{number in sample 2} (n_2)}{\text{number of marked individuals in sample 2} (n_m)}$$

Calculate the population of crabs using Alice's data and the formula above.

Assessment activity 11.4   P3 M3 D3

1 Imagine you are an educational officer at a wildlife reserve. Write instructions, with diagrams, based on your practical work, telling children how to sample both plants and animals. **P3**

2 Using small pieces of paper to represent creatures, practise the capture–recapture method and carry out a calculation for your results. **P3 M3 D3**

Grading tip

For **P3**, **M3** and **D3**, you will need to be well prepared and understand what you need to do before you start your investigation. Also ensure you prepare your results table in advance to help you to record your data as you carry out your investigations.

Functional skills

Mathematics and ICT

You will use your mathematical skills to analyse data for various ecosystem monitoring techniques, your writing techniques to write a report and your ICT skills to create tables.

PLTS

Team workers and Independent enquirers

You will work together in teams to monitor ecosystems, and plan and carry out your investigation in a safe manner.

11.5 Agencies and the environment

Protecting the environment and local ecosystems is a big task and UK government bodies work alongside other agencies to do this.

Legislation

Some of the **legislation** that governs conservation in the UK and around the world involves the introduction of **acts**. These are laws passed by Parliament to protect certain sites of scientific interest, wildlife and the environment. In the UK these include the following Acts of Parliament.

Key terms

Legislation – the laws that are made by Parliament.

Act – a specific law passed by Parliament.

Regional – concerning a particular region or district.

- National Parks and Access to the Countryside Act 1949 – this helped to set up many of the National Parks that exist today and has allowed people to access the countryside.

- Wildlife and Countryside Act 1981 – among other things, this act makes it an offence to intentionally kill, injure, or take many wild birds, their eggs or nests.

- Environment Act 1995 – this created a number of new agencies and set new standards for environmental management.

Government bodies

In the UK the Department for Environment, Food and Rural Affairs (DEFRA) is the government department that is responsible for the environment, food and farming, and rural matters. DEFRA's aim is sustainable development. This should enable all people to have a good quality of life with all the basic needs such as food and water without causing any harm to future generations.

There are laws to protect the environment in the UK, for example the Clean Neighbourhoods and Environment Act 2005.

Other government agencies that have responsibility for the environment and wildlife include:

- **Natural England** – involved with the protection of landscapes, rural communities, nature and wildlife

- **Scottish Natural Heritage** – involved with protecting the landscape and the conservation of wildlife

- **Countryside Council for Wales** – involved with the protection of rural countryside and the protection of landscapes

- **Northern Ireland Environment Agency** – involved with the protection and conservation of the natural environment and built heritage.

There are also many non-governmental organisations, many of which are run as charities. These include the National Trust and the Royal Society for the Protection of Birds.

Activity A

In groups, research the work of one of the UK government agencies listed on the right. Then make a poster to show the importance of this organisation in environmental protection.

Regional and local environmental protection

Some areas in the UK have special protection.

- **National Nature Reserves (NNRs)** – these are the best examples in the UK of a particular habitat.

- **Sites of Special Scientific Interest (SSSI)** – these have the best wildlife and geological sites in the UK.

At a **regional** and **local** level there are many areas of conservation including nature reserves, listed buildings or green belts.

The New Forest National Park. Currently there are 14 National Parks in the UK. Nine of these are in England, making up 7% of the land area.

Activity B

Find out if there are any nature reserves, listed buildings or green belts near where you live.

Global protection of the environment

The loss of natural habitats and the welfare of endangered species is a worldwide problem. The cooperation of governments around the world is needed for conservation to be successful. One example is the Convention on International Trade in Endangered Species of Wild Fauna and Flora (CITES) that was signed in 1973. This agreement is between many countries. It prevents the trade in wild plants and animal species which occurs mainly by smuggling and threatens their survival.

Did you know?

National Parks are protected areas of the countryside. Each one is managed by an organisation that protects the area and makes it accessible to visitors. To protect them, many activities, such as building, are strictly controlled within their boundaries.

 BTEC Assessment activity 11.5

1 Working in pairs, select one aspect of legislation and research this, e.g. Wildlife and Countryside Act 1981 – the illegal collection of birds' eggs. Then act out a role play where one of you acts as the person enforcing the legislation and the other acts as the person breaking the law.

2 (a) Using the Internet and books, carry out further research into the agencies involved in the local, regional, national and global protection of the environment. (b) Produce a table and add information on what each agency does, why they do it and how they achieve it. **P4 M4 D4**

Grading tip

For **P4**, you need to give the basic facts of different organisations and what their aim is.

Science focus

These charities aim to provide protection to plants and animals around the world.

- World Wide Fund for Nature – WWF works with governments, businesses and communities to set up conservation programmes to preserve natural habitats.

- Greenpeace – campaigns for environmental change against governments and corporations and accepts funding only through individuals and foundation grants.

- Friends of the Earth – works to provide a sustainable and environmentally friendly society.

11.6 Reasons for protecting the environment

In this section:

Key terms

Aesthetic – concerning the appreciation of beauty or good taste.

Economic – relating to the production, development and management of material wealth.

Environmental – concerning the natural world and how human activity affects it.

Ethical – addressing moral feelings or duties.

Social – relating to society and its interrelationships.

You would probably agree that it is important to protect the environment. But why?

Science focus

Conservation can involve trying to encourage species to move into habitats or the removal of some species into zoos or wildlife parks to protect them from extinction. Some plant species which are found to be in danger of **extinction** are also grown in special greenhouses to conserve the species.

Coral reefs and tropical rainforests have a massive diversity of different species that must be protected. But there are also ecosystems in the UK that need to be conserved.

Unit 3: See page 82 for more information about human effects on the environment.

There are five main reasons why we need to protect the environment.

1 **Aesthetic:** most people think the natural world is beautiful and would like to see it protected for that reason alone. If you like spending time outdoors and want to see more of the natural world you could even choose a career in environmental science.

Activity A

Find out if there are any areas which are aesthetically pleasing near where you live. Make a list and describe what makes each area special.

2 **Economic:** many areas of conservation also contain resources that provide economic benefit. For example, rainforests contain many plants from which we can produce medicines. Also eco-tourists can bring in much needed money to the area. Rainforests regulate weather, recycle nutrients and detoxify water. As a result they are quite self-sufficient. In some places, rainforests have been cleared to produce land for housing and farming. Fishing from the ocean also provides people with food.

3 **Environmental:** a decline in the population of one species can affect the balance of all of the organisms within an ecosystem. For example, if too many whales and seals are hunted then the numbers of krill and crustaceans, on which they feed, will increase. However, the numbers of organisms lower down the food chain will start to decrease as they are eaten by the increased number of krill. If the hunting of a particular animal is prevented, then the balance of organisms within that ecosystem is restored.

4 **Ethical:** we all have a right to live on the planet. However, human action including deforestation, hunting, production of waste and pollution has helped to bring about the extinction of many species of animals and plants. Most people think it is wrong for humans to have an impact on the environment without taking some responsibility for it too.

5 **Social:** one example of the social benefit gained from protecting the environment is the refurbishment and conservation of historic city centres. This is a proven method by which communities and neighbourhoods can be improved.

Did you know?

Lonesome George is the only surviving tortoise of its species from Pinta Island in the Galapagos. Goats introduced to the island ate all the vegetation that the tortoises previously survived on.

Grading tip

Think about what the command words are asking you to do when you complete **P5**, **M5** and **D5**. For example, 'Evaluate' means bring together all the information and then make a judgement.

BTEC Assessment activity 11.6 **P5** **M5** **D5**

1 Describe why it is important that we protect the environment and the ecosystems within it. **P5** **M5**

2 Do some further research using books and the Internet and produce a spider diagram of the ecological, economic, aesthetic and ethical reasons for environmental protection by conservation. **P5** **M5**

3 For two of the reasons for protecting the environment given in Question 2, write a short paragraph explaining whether you think these are good reasons, and why. **D5**

Functional skills

ICT

You will need to select appropriate information to form your opinion or support your point of view.

James Collins
Air Quality Engineer, Carstairs Gunesekara Ltd.

I am the principal air quality engineer working for an environmental technology consultancy; I'm responsible for the management and development of the air quality team and for their projects.

My team of environmental scientists undertake air quality assessments of mineral and waste facilities and residential and business developments, and provide information for planning applications and support for local authority work.

The team is also responsible for calculating carbon footprints for proposed developments, and looking at climate change, including daylight and wind assessments.

On a typical day I meet my team to discuss a project on air quality monitoring in London for a local authority. I then talk to our clients, the local authority, to discuss what we plan to do. I also contact other local authorities and companies to develop new business. I produce a report of my team's findings on air quality from a previous contract to give to my manager before submitting this to the client.

Think about it!

1 What factors might affect the quality of the air?

2 How could monitoring air quality help to prevent climate change?

Just checking

1 Define an ecosystem, a habitat, a community and a population. Explain how they interact.
2 List three ways in which humans have an impact on the environment.
3 Describe two ways of sampling for a) plants and b) animals or insects.
4 List some of the agencies involved in protecting the environment and describe how they do this.
5 Give two reasons why it is important to protect the environment.

Assignment tips

To get the grade you deserve in your assignments remember the following.

- You need to demonstrate an independent approach to completing your work. This includes researching for information. As well as the Internet you could look at a range of other sources like books, journals, DVDs and newspaper articles.

- Diagrams and photographs are useful in capturing people's attention and making the information interesting.

- If your assignment asks you to prepare a scientific report you will need to write this up to include a plan, equipment, results, graphical evidence, a conclusion and an evaluation of your findings.

Some of the key information you'll need to remember includes the following:

- A successful ecosystem includes organisms that rely on each other or on environmental factors that enable all organisms to work together and stay in balance.

- Human behaviour (such as using transport) can lead to long-term changes in the environment which affect an ecosystem's ability to stay in balance.

- Sampling and monitoring animals and plants is an important part of an ecologist's work.

- Legislation can ensure that animals and wildlife are protected from harm.

You may find the following websites useful as you work through this unit.

For information on...	Visit...
legislation and conservation	Wikipedia: conservation in the UK DEFRA website

12 Growing plants for food

Food chains begin with plants. Plants trap sunlight energy and make food. We eat plants, foods made from plants and/or animals that have eaten plants. There are numerous jobs, here and abroad, connected to producing or distributing food. These include farming, plant breeding, quality control, sales, agrichemicals, biotechnology, nutrition and food hygiene.

The world population has grown a great deal over the last 200 years. Cities and towns have expanded leaving less land for farmers to grow food. Scientists can help by making better fertilisers and pesticides, carrying out breeding programmes to develop plants that give higher yields, and developing genetically modified crop plants.

In this unit you will learn about the structures and functions of the main parts of plants and how this relates to the food we eat. You will also learn about the different types of food plants, including where in the world they grow. When discussing growing plants for food, we also need to think about the economic and political aspects of growing enough food to feed the world's population. You will also learn about how we can use biotechnology and plant breeding to help feed the world, and the advantages and disadvantages of using fertilisers.

Learning outcomes

After completing this unit you should:

1 be able to investigate the structure and function of the main components of plants

2 know where the major food plants of the world are grown

3 know the economic relationship between food production and population size

4 be able to investigate the role of plant breeding and technology

5 know the effects of a range of fertilisers on food production.

Assessment and grading criteria

This table shows you what you must do in order to achieve a **pass**, **merit** or **distinction** grade, and where you can find activities in this book to help you.

To achieve a **pass** grade, the evidence must show that the learner is able to:	To achieve a **merit** grade, the evidence must show that, in addition to the pass criteria, the learner is able to:	To achieve a **distinction** grade, the evidence must show that, in addition to the pass and merit criteria, the learner is able to:
P1 Identify the structure and functions of plant components in relation to food production **See Assessment activity 12.1**	**M1** Describe the structure and functions of plant components in relation to food production **See Assessment activity 12.1**	**D1** Explain the structure and functions of plant components in relation to food production **See Assessment activity 12.1**
P2 Carry out practical investigations into the materials which are stored in plants **See Assessment activity 12.1**	**M2** Describe how the materials are stored in the plant **See Assessment activity 12.1**	**D2** Explain how these materials are used for food production **See Assessment activity 12.1**
P3 Outline where the major food plants of the world are grown, noting the climate and typical production figures **See Assessment activity 12.2**	**M3** Compare the major food crops across the world, indicating the relationship between climate, food production and population **See Assessment activities 12.2 and 12.3**	**D3** Compare the advantages and disadvantages of the major food crops, particularly in terms of nutritional value
P4 Identify issues relating to food supply in national and global terms **See Assessment activities 12.3 and 12.4**	**M4** Assess the influence of economic, political and environmental factors on food production **See Assessment activities 12.3 and 12.4**	**D4** Analyse the influence of food plants on the demography of the world **See Assessment activities 12.3 and 12.4**
P5 Carry out an investigation into plant-breeding technologies **See Assessment activity 12.5**	**M5** Explain the plant-breeding techniques that have led to 'improved' varieties of major food plants **See Assessment activity 12.5**	**D5** Evaluate the advantages and disadvantages of plant-breeding technologies **See Assessment activity 12.5**
P6 Describe the effects of particular fertilisers on food production. **See Assessment activity 12.6**	**M6** Compare the effect of organic and non-organic fertilisers on food production. **See Assessment activity 12.6**	**D6** Evaluate the problems associated with overuse of fertilisers. **See Assessment activity 12.6**

How you will be assessed

Your assessment could be in the form of:

- a practical investigation e.g. of the materials stored in plants
- a written report e.g. on countries who can and can't feed their population
- an investigation or presentation e.g. on plant-breeding technologies
- a newspaper article e.g. comparing different fertilisers.

Sean, 17 years old

I enjoyed the growing plants for food section. I used to think plants were boring. I didn't realise that everything we eat comes originally from plants. Even cola drinks come from kola nuts with sugar from sugar cane.

Our class went to a local plant science and microbiology company and found out about the latest research into genetically modified plants.

There has been a lot in the news recently about food security – whether we will be able to produce enough food in future to feed the population. In this unit we thought about that and looked at how politics and climate change might affect food production.

Catalyst

What plants did you eat yesterday?

Draw a table like the one below.

Complete it by writing down everything that you ate yesterday. An example has been filled in to help you. Include snacks as well as meals.

Compare your food diary with another person's in your group.

Is there any time during the day that you ate something that did not come from a plant or from an animal that ate plants?

Meal	Food	Is it a plant or part of a plant?	Is it made from a plant?	Is it from an animal that ate plants or parts of plants?
Breakfast	cornflakes		yes	
	sugar		yes	
	milk			yes
	boiled egg			yes
	toast		yes	
	butter			yes
	marmalade		yes	
	apple	yes		

12.1 The structures and functions of the main parts of plants

In this section: P1 M1 D1 P2 M2 D2

Key terms

Photosynthesis – process by which plants make food, using carbon dioxide, water and sunlight energy.

Organ – group of tissues that work together to perform a function.

Plants make and store food that we and other animals can eat. There are many science jobs involved with growing plants for food. For example, some scientists work as food science technicians. Others may work in horticulture, assisting in market gardens.

Leaves, stems and roots

Leaves are the plant organs that make food. They do this by a process called **photosynthesis** in chloroplasts of leaf cells. Chloroplasts are green because they contain chlorophyll (a green pigment). Chlorophyll uses carbon dioxide from the air, water from soil and energy from sunlight to make sugars and the waste product oxygen. Enzymes are used to do this.

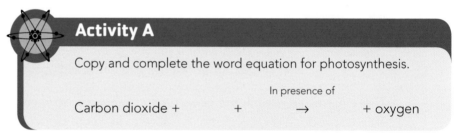

Activity A

Copy and complete the word equation for photosynthesis.

In presence of

Carbon dioxide +　　　　 +　　　 →　　　 + oxygen

Leaves are flat and thin and have a big surface area for catching sunlight. They have pores underneath their surface where carbon dioxide enters the leaves and moves to the cells that make food.

Roots keep plants anchored in soil and take up water and minerals from the soil.

The stem carries water and minerals up to the leaves and holds the plant up so the leaves face the Sun.

Why do plants have flowers?

Flowers are the reproductive **organs** of plants. Most flowers have both male and female parts. The male parts make pollen. Insects or wind take the pollen to the female parts.

Pollen fertilises the structures called ovules inside the ovary (a female part). The fertilised ovules become seeds. Inside each seed is an embryo (this can become a new plant) and its food. We can eat the food stored in seeds. Seeds develop in the ovary and the two together become a fruit. We can eat the food stored in fruits.

Water is lost from the leaves

Sugars are made in the leaves but are needed all over the plant

Water and minerals move up stem in xylem

Sugars move down stem in phloem

Water is absorbed from the soil, but is needed all over the plant

How roots, stems and leaves are involved in making food.

Unit 3: Page 70 tells you about plant cell structure.

Activity B

Think of some vegetables we eat that are the flowers of plants. Write a list and compare your list with that of others in the class.

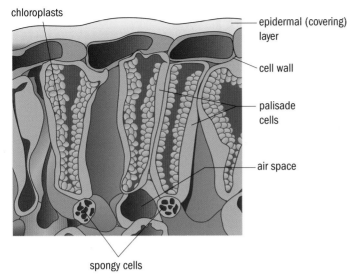

- chloroplasts
- epidermal (covering) layer
- cell wall
- palisade cells
- air space
- spongy cells

Section of a leaf showing cells and chloroplasts.

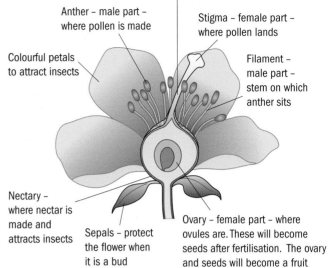

- Style – female part – pollen tube grows down here to the ovules
- Anther – male part – where pollen is made
- Stigma – female part – where pollen lands
- Colourful petals to attract insects
- Filament – male part – stem on which anther sits
- Nectary – where nectar is made and attracts insects
- Sepals – protect the flower when it is a bud
- Ovary – female part – where ovules are. These will become seeds after fertilisation. The ovary and seeds will become a fruit

Diagram showing male and female parts of a flower.

Energy stores in plants

Food	Plant energy store	Main type of stored material
rhubarb	stem	sugar
cabbage, lettuce	leaves	starch
parsnips, beetroot, yam	root	starch
sugar beet, carrots	root	sugar
oranges, grapes, tomatoes	fruits	sugar
wheat, rice, corn	seeds	starch
peanuts, walnuts	seeds	oils
peas, beans	seeds	protein

Unit 6: Page 142 tells you about why we need these vitamins and which foods have them.

Science focus

Humans and animals can get vitamins A, B, C and E from plants. For example, vitamin C can be found in oranges, kiwi fruits, green vegetables and many other fruits and vegetables. Vitamin E is found in apricots, peanuts, sunflower seeds, sweet potatoes and wheat, as well as many other plants.

BTEC Assessment activity 12.1 P1 M1 D1 P2 M2 D2

1 **(a)** Explain to your partner why leaves are flat and thin – how does this help the leaves to make food? How do roots and stems help plants make food? **P1 M1 D1**

   (b) If your teacher can supervise you, test some leaves, stems, roots, fruits and seeds for sugar, starch, oils and protein. These are the sort of tests that a food technician would carry out. **P2**

2 Look at the picture of a section of a leaf. Discuss with your partner all the ways that cells in leaves are good at doing their job (making food). Pool your ideas with the rest of the class and make a list. **M2 D2**

Grading tip

To obtain **P1 P2** you need to know which parts of the plant contain sugar, starch, oil and protein.

To reach **M1 M2** describe the parts of the plant and the cells that make and store each type of food.

For **D1 D2** explain how the cells that make the food are suited to doing this. Explain how each type of food can be used to make food that we eat – for example starch from wheat seeds is used to make bread, biscuits and cakes.

12.2 Food plants of the world

Key terms

Cereal plants – plants in the grass family, with narrow, parallel-veined leaves, grown for seeds.

Pseudocereals – broad-leaved plants, not grasses, that are grown for seeds.

Staple crops – crops that provide starch. They make up the bulk of the diet for most people.

Coeliac disease – genetic disease in which sufferers cannot digest gluten protein. This leads to abdominal bloating, diarrhoea, vomiting and muscle wasting.

Quinoa.

Did you know?

Buckwheat and quinoa do not contain the protein gluten. They can be eaten by people with **coeliac disease**. People who have this disease can't eat gluten, which is in wheat flour and other cereals. Quinoa also has all the essential amino acids that people need so it is better than wheat as a source of protein for vegetarians.

We eat many different types of foods. We all eat more foods that come from plants than foods that come from animals (meat and dairy products). Some people eat little or no animal foods. Plants can provide starches and sugars, oils and proteins (and fibre) so they can give us a balanced diet.

Cereals

Cereal plants are grasses that are grown for their seeds. You can eat them as whole grains and you will get starch, oils, protein, vitamins and minerals from them. If you eat foods made from these grains you will get mainly starch. Examples of these types of foods are bread, biscuits and cakes.

Cereals are also called **staple crops** because a large part of our diet needs to be starch as starch gives us energy. Maize (corn), wheat and rice are the crops most widely grown. They provide nearly half the energy needs for people of the world.

- maize and wheat grow in temperate regions like England and North America

- oats and barley grow in colder regions like Scotland

- rice grows in temperate and tropical regions such as China, Vietnam and India

- many African countries grow cassava, millet and sorghum.

Pseudocereals

These are plants that aren't grasses but their seeds contain a lot of starch and can be used like cereal seeds. Important **pseudocereals** are buckwheat, grown mainly in Russia, and quinoa, grown in the Andes of South America. Buckwheat flour is used to make noodles and pancakes. Quinoa can be used like rice or as flour to make things like bread.

Oil-producing plants

Some plants, like olives and palm trees, have seeds with a lot of oil in them. We use the oil for cooking, salad dressings and spreads. You need some oil in your diet to make cell membranes.

- Olives grow in Spain, Italy and Greece.

- Oil palms grow in Malaysia, Indonesia and Colombia.

We can get oil from other seeds such as sunflower, soya beans and peanuts.

Protein-producing plants

Legumes are plants that have seeds in pods, such as peas, beans, lentils and peanuts, and they all contain protein. Farmers grow other legumes, like clover, soya, alfalfa and lupins, to feed livestock. In societies where people don't eat much meat, grains (e.g. corn, wheat, millet and rice) and legumes (e.g. lentils) eaten together provide good protein. This is because the amino acids that are missing from cereal grains are found in legumes. Brazil, India, China and Germany grow a lot of legumes.

Activity A

(i) Explain why, in India, rice and dhal (made from lentils) are often eaten together.

(ii) Explain why peanut butter sandwiches are good for vegetarians.

Sugar-storing plants

Many fruits have a lot of sugar in them.

- Sugar beet plants store sugar in their roots. They are grown mainly in France, Germany and the USA.

- Sugar cane plants store sugar in their stems. They are grown in Brazil, India and China.

- Bees make honey (which contains sugar) from the sugary nectar of flowers.

Sugar is used in cooking and to make processed food, sweets and drinks.

The price of sugar is increasing because

- some sugar has been used to make biofuel and so less is available for eating

- climate change means some parts of the world cannot grow as much sugar cane as before

- in China it is more profitable to use land for industry than for growing sugar cane.

Case study

Extracting oil from olives

Eric is working in Spain on a small cooperative farm. He helps pick olives and he operates the press to extract oil from the olives. He is improving his Spanish as well as learning about the scientific processes involved in extracting oil. The olives are ground with stones to make a paste and then liquid is squeezed from the paste. The liquid is put in a centrifuge to remove water. There are two exits from the centrifuge, one for water and one for oil. The oil is then filtered to remove any solid particles.

Explain, in terms of density, what happens to water during the centrifuge process. Which exit is at the top – the one for water or the one for oil?

PLTS

Independent enquirers and Self-managers

In this activity you will need to direct your own research as an independent enquirer and manage your own work.

Assessment activity 12.2 P3 M3

1 You are writing an article for a magazine. In the article, show an outline map of the world. On the map show where the following major food plants are grown: potatoes, rice, corn, sugar, wheat, millet, sorghum and soya beans. **P3**

2 Describe the climate in each of the areas where the food plants are grown and show how much of each food is grown per year. **M3** Compare the advantages and disadvantages of the crops. **D3**

You can research the information on the Internet. Put the names of each of the food plants, individually, into a search engine (such as Google) or look them up in books on geography or agriculture.

Grading tip

To meet **P3** show where each crop is grown, what the climate is like and how many tonnes are grown each year.

For **M3** show how climate (rainfall, temperature and sunshine) affects food production.

To achieve **D3** compare the nutritional values (e.g. starch, fat, protein and vitamin content) of the different food crops.

12.3 Food production and population size

In this section:  M3 P4 M4 D4

Key terms

Agriculture – the science of growing and harvesting crops and of rearing livestock for food.

Irrigation – building channels to take water to land where crops are growing.

Irrigation system.

Did you know?

Borlaug received the Nobel Peace Prize in 1970. He died in September 2009.

Activity A

(i) How does the fall in water table levels affect crop growing?

(ii) Some insecticides kill useful insects. What would happen to crops if bees were killed because of the use of insecticides?

About 12 000 years ago humans began to settle in small groups and establish farms. This is known as settled **agriculture**. They grew plants and later they tamed wild animals. They then developed **irrigation** to make sure their plants were watered. With each advance in agriculture the world population increased. When the increase in food supplies levelled off, so did the increase in world population.

The Green Revolution

By the middle of the twentieth century people in African and Asian countries did not have enough to eat. A plant breeder from the USA, Norman Borlaug, introduced new varieties of wheat that contained larger amounts of edible grain known as high-yield. The short thick stems of the plants supported many seed heads and they did not become top-heavy and fall over. He also developed varieties of rice that could produce two crops per year.

By the 1970s, in Asian countries like India and China, wheat production had trebled and rice production had increased even more.

These new crops needed lots of water and they needed fertiliser to add minerals to the soil. They also needed pesticides to kill weeds and insects.

As a result, by the 1990s

- the increased yield per unit area of land had levelled off

- the water table (depth of water beneath the soil) had dropped because more water was being used for the new crops

- the use of fertilisers and pesticides had caused pollution and led to some health problems among communities in African and Asian countries.

The Green Revolution failed to reach most African countries. This was due to corruption and lack of infrastructure (roads and railways). The populations in these countries continued to grow but food production decreased leading to famine and the need for food aid.

The Malawi Miracle

The African country of Malawi was different from many other African countries. From 2006, small farmers were allowed to buy seeds of high-yield crops and fertiliser at one-third of the normal cost. There were bumper harvests with enough to feed the population and even to export some. This showed that investment in agriculture can make a difference.

Collapse in food production

The war between Eritrea and Ethiopia in the late 1990s meant that farmers lost land and homes and so were unable to plant crops, leading to a collapse in food production.

Nowadays, even though this part of Africa is fertile

- the agriculture is not organised

- there are few irrigation systems

- farmers are able to lease only a small amount of land

- farmers burn animal manure for fuel and do not spread it on the fields as fertiliser.

As a result, the crop yields are small and are inadequate to feed families, leading to nearly half the population's children being underweight and malnourished.

We need another revolution

For the past 10 years the world has eaten more food than it has produced and we have been using our saved grain. As the climate has changed, many areas are becoming hotter and drier, making it harder to grow crops.

Fossil fuels are used to make fertilisers and pesticides. As fossil fuels become more expensive so do fertilisers and pesticides, which in turn makes food more expensive.

The world population is growing and will be 9 billion by 2050, putting more strain on the food supply. As countries like China get richer, the demand for meat increases. China cannot grow enough grain to feed the increasing number of animals on its farms so has to import feedstuffs like grains and legumes. Brazil exports soya beans to China. Brazil increases its growing area for soya beans by cutting down rainforests.

We need another Green Revolution to increase food production across the globe by 70%–100% by the year 2030. Scientists are developing strains of genetically modified crops that are:

- drought-resistant

- pest-resistant

- able to fix nitrogen so don't need fertiliser

- higher in nutrients – for example sorghum that has more oil and protein.

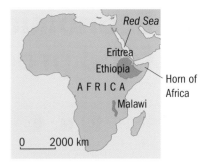

Map showing Ethiopia, Eritrea (horn of Africa) and Malawi.

See Unit 20 for more detailed information on genetically modified crops.

Functional skills

ICT, Mathematics and English

You will develop ICT skills in researching and selecting information and in preparing your written report. Use tables and graphs to illustrate your report. You will use some mathematical skills in preparing tables and graphs of data. You will use English skills in writing and presenting your report.

Grading tip

To reach **P4** identify the issues relating to food supply, nationally and globally. Think about transport costs; think about if poor countries grow crops to sell they may not grow enough food for themselves; what would happen if some of the countries that sell food to us refused to do so?

For **M3** and **M4** you must also compare the major food crops of different parts of the world. Show the relationship between climate, food production and population. Assess how political factors (government help, civil wars) and environmental factors affect food production.

Analyse how the distribution of different types of food plants in the world has affected the populations of those different parts of the world. **D4**

BTEC Assessment activity 12.3

You are an environmental advisor reporting on countries that cannot feed their populations. Research this topic: find out whether the UK can produce all the food it needs or has to import some. Where do we import it from? Do the same for China, India, the USA, Canada, Australia, France, Nigeria, Botswana, and some other countries of your choice, and prepare a written report. **M3 P4 M4 D4**

12.4 Economics, policies and food production

In this section:  P4 M4 D4

Key terms

Sustainable – to practise agriculture in such a way that it can be maintained – the soil does not become depleted or exhausted.

Biofuels – fuel made from plant matter by fermenting sugar or starch to alcohol.

Economics

An economically productive farm produces enough food to feed the farmer's family and some surplus to sell at markets. In many less developed countries there are small farms that may not grow enough food to feed the family.

Policies

Governments sometimes introduce policies to try and make farming more efficient. In 1958 The European Union introduced the Common Agricultural Policy. It aimed to:

- increase productivity – growing more food to eat
- give farmers a fair standard of living
- give consumers good-quality food at fair prices
- preserve the countryside for future generations.

There have been changes to the policy over the years. There are advantages and disadvantages to having such a policy:

Advantages	Disadvantages
Farming became more efficient	Jobs were lost as fewer people worked on farms
	More fertilisers were used and more pollution caused
	Hedgerows were removed and wildlife lost habitats (places to live)
	Excess food was stored in warehouses and freezers
Farmers were encouraged to grow oil-seed rape for vegetable oil	Left fewer fields available to grow other crops
Quotas were introduced so that farmers could not overproduce food and farmers were paid to set-aside some land, helping wildlife	
Farmers were paid subsidies to help keep prices stable for consumers	Bigger farms benefitted more than smaller farms
Farmers were paid to conserve certain areas, like marshland and moors for wildlife habitats	

Activity A

Suggest some ways in which food can be produced sustainably.

Sustainability of food production

You can see from pages 232–233 that the world needs to produce more food to feed our growing population. Many people say that this should be done in a more **sustainable** way and in a way better for the environment, in other words, in an eco-friendly way.

We need to:

- use less fertiliser and fewer pesticides which would also mean using less fossil fuel – fossil fuels are used to make fertilisers and pesticides

- grow more legumes (peas and beans) and less grain (corn and wheat) as legumes have more protein and can fix nitrogen in the soil so the soil doesn't need as much fertiliser

- use crop rotation – e.g. growing legumes one year and corn the next so that the minerals in the soil do not get used up

- put compost on soil or grow green manure as these hold water in the soil.

In addition, we may be able to genetically modify crops so that they need less water, fertilisers and pesticides and are able to fix nitrogen.

Organic and non-organic farming

Many people in rich countries can afford to make the choice to buy meat and vegetables grown organically. These products are more expensive to produce because less can be grown or reared in the same space than conventional crops and livestock as no chemicals, pesticides or fertilisers are used. Organically grown food is not proven to be better for your health but it does have advantages, as well as some disadvantages.

Advantages	Disadvantages
Animals are not intensively reared and have better lives	This method of farming cannot produce enough food to feed the world
Less machinery and fossil fuels are used	In areas where growing crops and rearing animals is difficult, there won't be enough manure or compost made
Plant and animal waste is recycled as compost and manure	Manure can also pollute waterways and add extra nitrogen to water
Organic farming is more sustainable as it does not damage the environment as much as non-organic farming	Biological control methods are not very efficient at removing pests

BTEC Assessment activity 12.4 (P4) (M4) (D4)

1 You are working in the information department for DEFRA, the government's Department for the Environment, Food and Rural Affairs. Make a poster that describes in simple terms the differences between organic and non-organic farming. **P4**

2 Show why food produced organically is more expensive. **M4**

3 Explain why, although many aspects of organic farming are desirable, it will not be able to feed the ever-growing world population. **D4**

Did you know?

Nitrogen fixation – some bacteria can use nitrogen from air to make ammonium compounds. Some of these bacteria live in the root nodules of legume plants and make amino acids from the ammonium compounds. Legume plants get amino acids and can make proteins for growth, without having extra fertiliser.

Science focus

Developed countries are trying to find new sources of fuel because fossil fuels will run out. Sugar from sugar cane and starch from corn can be fermented to make ethanol. Ethanol can be mixed with petrol to make gasohol, which can be used in some cars. It can also be used on its own as a fuel in some specially modified vehicles.

However, growing crops for **biofuels** reduces the amount of land for growing food.

An organic pig farm.

Grading tip

For **P4** identify differences between organic and non-organic farming.

To reach **M4** assess the reasons why organically grown food is more expensive.

Analyse which parts of the world have populations that cannot afford organically grown food to gain **D4**.

12.5 Plant breeding and biotechnology

In this section: P5 M5 D5

Key terms

Mutation – change to the genetic material, either genes/DNA or to whole chromosomes.

Diploid – having two sets of chromosomes.

Polyploid – having more than two sets of chromosomes.

Selective breeding – artificial selection by a breeding programme. Humans select crops (or livestock) with desirable characteristics and breed from them.

Activity A

Explain the main difference between sexual and asexual reproduction.

Governments of all countries are worried about food security – will we be able to grow enough food to feed our populations? Increasing population sizes and climate change will affect this ability. For example, for every 1°C rise in the temperature of the Earth there is a 5% reduction in crop yields. Plant breeding and biotechnology may be able to help increase food production.

Sexual and asexual plant reproduction

Plants can reproduce (breed) sexually. Flowers are the organs for sexual reproduction (see pages 228–229). After pollination there is fertilisation when the pollen (male nucleus) and the ovule (female nucleus) join and seeds develop inside fruits. Each seed can germinate and grow into a new plant of the same species. The male and female nuclei are both made by meiosis (a type of cell division) and the offspring may be genetically different from the parents.

Some plants can reproduce without making seeds. Potatoes are swollen parts of underground stems and each one can grow into a new plant. Strawberries put out runners on the end of which a new plantlet grows. Many plants can grow from cuttings. Small pieces of the parent plant, such as pieces of stem, are placed in clean soil. The pieces grow new roots and leaves and become new plants. The offspring are all genetically identical to the one parent.

Selective breeding

When humans first began farming 12 000 years ago, they domesticated plants and began **selective breeding**. They saved seed from the plants that grew best and gave the biggest yield. Some related plants were crossed. Tomatoes that are edible and disease-resistant came from crossing:

wild tomatoes that were disease-resistant but poisonous to humans   ×   wild tomatoes that were susceptible to disease but had edible fruit

There are centres of origin in the world for all cultivated plants that have been genetically altered by selective breeding. These centres are where the wild ancestors of plants grow. These plants may be the source for new genes for plant-breeding programmes.

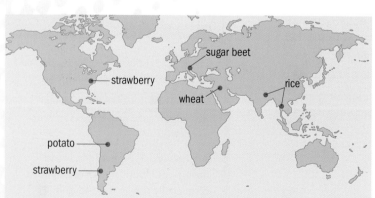

The centres of origin for wheat, sugar beet, rice, potatoes and strawberries.

Genetic modification of food crops

See Unit 20 for more information on biotechnology.

All the cultivated plants we eat are genetically modified by years of selective breeding. This has transferred genes from related plants by cross-breeding.

Polyploid crop plants	Number of sets of chromosomes
bread wheat, oats, kiwi fruit	6
bananas, apples	3
strawberries, sugar cane	8

Gene technology

Gene technology	Example
Switching off genes	In 'Flavr savr' tomatoes the gene that makes tomatoes go mushy when they are ripe is switched off so they ripen without going mushy on the vine. They keep fresh longer and taste better
Switch on genes in another part of the plant	Golden rice makes vitamin A in the rice grains and not just in leaves. This could help prevent poor children in developing countries who have very few vegetables to eat suffering from a lack of vitamin A. Vitamin A deficiency can lead to blindness
Biofortification	Sorghum grows in dry regions but is low in some essential nutrients. Through biofortification a version of sorghum has been developed that has more nutrients and is disease-resistant so pesticides aren't needed for it to grow successfully
Gene transfer	A type of maize has been developed that contains the same types of fatty acids that are in fish oils. These fatty acids are needed for nerve and brain development, making the plant useful as fish stocks run out

Science focus

Mutations

Many plants can become **polyploid** – their cells have more than the usual two (**diploid**) sets of chromosomes. This makes the nucleus bigger so each cell is bigger and the fruits and seeds are bigger. Bigger plants give a higher yield.

Plant breeders can cause **mutations** in crop plants using chemicals, such as colchicine, and physical methods, such as ionising radiation.

The mutations can be in the form of gene mutations (e.g. to make plants resistant to disease) or chromosome mutations (e.g. to make plants polyploid).

Case study

Florence Wambugu

Florence Wambugu is director of Africa Harvest, a company in Kenya that develops genetically engineered crops to improve food supplies to people in Africa. She has received honours and prizes for her work. She has developed disease-free bananas (a staple food) that have twice the normal yield.

Find out how bananas and plantains (a close relative of bananas) are eaten in Kenya.

  **Assessment activity 12.5** **P5** **M5** **D5**

You are a biotechnologist researching the various plant-breeding technologies. Find out about the technologies and give a short presentation to the rest of your class. Include information about selective breeding, GM crops, food security, the effect of climate change and the new green revolution. **P5** **M5** **D5**

Grading tip

For **P5** describe the plant-breeding technologies – selective breeding and genetic modification.

To gain **M5** give some explanations of the plant-breeding technologies that have led to improved food production.

Think about some of the advantages and disadvantages of the technologies to achieve **D5**.

12.6 How do fertilisers affect food production?

In this section: P6 M6 D5

Key terms

Eutrophication – poisoning of water courses by excess nitrate which causes algal bloom, death of plants, depletion of oxygen and death of fish.

Fertiliser burn – damage to leaves caused by too much nitrate fertiliser.

Table: Some of the different minerals required by plants

Mineral	Why plants need it
Nitrogen (as nitrates or ammonium compounds)	To make amino acids and proteins which are needed to make new cells and organelles
Potassium	To help make proteins and to open the leaf pores to let in carbon dioxide
Phosphorus	To make cell membranes and adenosine triphosphate (ATP)
Magnesium	To make chlorophyll

Activity A

What would a plant look like if it did not have enough magnesium?

Plants need minerals from soil for healthy growth. Soil that is intensively farmed often does not contain enough minerals for healthy plant growth, so they are added to the soil as fertilisers.

Organic versus non-organic fertilisers

On mixed farms, where there are animals and crop plants, animal manure is used as a fertiliser for crops. Growers also use the unusable parts of crop plants for compost as fertiliser. Human sewage, seaweed, animal blood and bone meal (made from crushed animal bones and blood from slaughterhouses) can also be used. These are all organic fertilisers.

Just using organic fertilisers isn't enough to give the yields of crop plants that we now need to feed our large population. In 1940 scientists discovered how to make ammonium-based (non-organic) fertilisers very cheaply. This helped farmers increase their yields. Making these fertilisers uses fuel, which used to be cheap and plentiful, but is now more expensive.

Problems with overusing fertiliser

Eutrophication

Eutrophication occurs if too much non-organic or organic fertiliser is used:

- when it rains, excess fertiliser runs off the land, flowing into rivers, lakes and streams
- the extra nitrogen and phosphorus content in water causes more algae to grow than usual
- the algae stops light reaching plants in the water
- water plants die and decompose (break down)
- the bacteria which cause the decomposition use oxygen from the water
- fish can't get enough oxygen and they die or leave the area.

Fertiliser burn

If too much nitrate fertiliser is used on plants they can suffer from **fertiliser burn**. The leaves wilt and look as if they have been scorched – they are curled at the edges and have yellowy brown patches. If the leaves drop off the plant then the plant cannot make as much food and will not grow.

Growers can treat these plants by adding water to the soil and making sure that the water drains off and takes the extra nitrogen with it. But the extra nitrogen could enter streams and cause eutrophication.

BTEC Assessment activity 12.6 P6 M6 D6

You are a newspaper reporter. Write an article showing the advantages and disadvantages of organic and non-organic fertilisers. **M6**

You will need to find out about the different fertilisers using the Internet. Research how they are applied and how they affect the growth of crop plants. **P6**

You will also need to find out what problems they may cause to the environment. **D6**

Grading tip

Describe the effects of organic and non-organic fertilisers on food production and on the environment to gain **P6**.

For **M6** compare the effects of organic and non-organic fertilisers on food production and on the environment.

To achieve **D6** evaluate the problems associated with overuse of fertilisers. 'Evaluating the problems' means thinking about all parts of the problems and coming to a conclusion.

People are buying more organic produce – maybe because they don't want to eat food that has been sprayed with fertilisers.

Activity B

(i) Explain why having fewer leaves will lead to a plant making fewer fruits or seeds.

(ii) Find out how too much nitrate in drinking water may harm humans.

WorkSpace

Jamil Luzon

Science Assistant, International Rice Research Institute

I am working for a year as a science assistant at the International Rice Research Institute in the Philippines. I work in a department that is developing new ways of controlling pests of rice, as these pests reduce the yield of rice.

I am responsible for collecting samples from research plots where we grow rice and try out different pesticides. Each day I collect 10 rice seedlings from each test plot and examine the leaves to see if they have been damaged. The pests I look for are army worms, leafhoppers, leafminers, midges and weevils. I record how many leaves from each sample are affected. I also look for damage done to the roots, as this indicates tadpole shrimps and crayfish. The less damage seen, the more effective the pesticide that is being tested.

I have to keep careful records of my findings on a database so that the research scientists can access these results.

Think about it!

1 Why is it important to reduce the number of pests of rice plants?

2 Why do you think we need to develop new pesticides for these pests?

Just checking

1 Draw a simple plant cell and show where the chlorophyll is for making food.
2 Draw a simple diagram of a plant and label stems, roots, leaves and flowers. Add some annotations to show examples of each part that may be eaten.
3 How do fertilisers and pesticides help to increase food production?
4 Explain how we would not have any food at all if there were no plants.
5 How can genetic modification of plants help to increase food production across the world?
6 How does making biofuels affect food production?

Assignment tips

To get the grade you deserve in your assignments remember to do the following.

- When writing a scientific report include a plan, equipment, risk assessment, results, and a conclusion of your findings.

- When writing a magazine article, remember your target audience and make sure you include only the necessary information. Be objective – don't be emotional or take sides but present both sides of the argument.

Some of the key information you'll need to remember includes the following.

- Food security refers to the ability of a country to feed its population.

- Staple crops are those that are good sources of carbohydrate and make up the bulk of the diet.

- Legumes are plants such as peas, beans and clover.

- Policies made by governments may help to make farming more efficient or more sustainable and better for the environment.

- Leaves are organs that are specialised for making food.

- Many parts of plants, such as roots, stems, leaves, flowers and seeds store food and can be eaten.

- Plants can reproduce sexually or asexually.

- Plants can be genetically altered by selective breeding or by using gene technology.

You may find the websites below useful.

For information on...	Visit...
nutrition	Nutrition data
food policy	The Society of Biology
Africa Harvest	Africa Harvest
organic farming	DEFRA
sustainable farming	Sustainable Harvest International

13 Investigating a crime scene

Investigating a crime scene is about answering questions – the five Ws – Who? What? Where? When? Why? In order to answer these questions you must gather evidence, analyse it and draw appropriate conclusions.

In this unit you will have the opportunity to role play being a scene-of-crime officer (SOCO) or a crime scene investigator (CSI). You will perhaps know some aspects of the role from watching a range of crime-related television programmes. How much of what you see on the television accurately reflects the role? In real life the activities that you see the television CSI carrying out are frequently completed by different people.

Much of the forensic science that is shown on television and that you will be familiar with is founded in 'Locard's principle'. Edmond Locard developed the theory that 'every contact leaves a trace'. This means that a criminal will leave something at the crime scene and/or take something away with them.

This principle does not just apply to criminals but to any person. This means that the processing of a crime scene has to be conducted in a way that will preserve the evidence. This will be one of your very important tasks during this unit as you investigate a scene-of-crime.

Learning outcomes

After completing this unit you should:

1 be able to investigate a scene-of-crime
2 be able to use appropriate scientific techniques to analyse evidence which has been collected from the scene-of-crime
3 understand the relationship of forensic science to the law, including the criminal justice system.

Assessment and grading criteria

This table shows you what you must do in order to achieve a **pass**, **merit** or **distinction** grade, and where you can find activities in this book to help you.

To achieve a **pass** grade, the evidence must show that the learner is able to:	To achieve a **merit** grade, the evidence must show that, in addition to the pass criteria, the learner is able to:	To achieve a **distinction** grade, the evidence must show that, in addition to the pass and merit criteria, the learner is able to:
P1 Carry out an investigation to collect evidence from a crime scene **See Assessment activites 13.1 and 13.2**	**M1** Describe the processing of a crime scene, explaining how the techniques used obtained valid forensic evidence **See Assessment activites 13.1 and 13.2**	**D1** Evaluate the processing of a crime scene, interpreting how the valid evidence collected could be used in a criminal investigation **See Assessment activites 13.1 and 13.2**
P2 Demonstrate the most appropriate methods to record and preserve evidence from the crime scene **See Assessment activities 13.1 and 13.2**		
P3 Produce a simple plan to analyse biological, chemical and physical evidence from the crime scene **See Assessment activities 13.3 and 13.4**	**M2** Produce a detailed plan to analyse biological, physical and chemical evidence from the crime scene **See Assessment activities 13.3 and 13.4**	**D2** Assess the potential risks associated with analysing biological, physical and chemical evidence from a crime scene **See Assessment activity 13.4**
P4 Carry out experiments to analyse biological, chemical and physical evidence from the crime scene	**M3** Describe the patterns found from the evidence and make connections **See Assessment activity 13.4**	**D3** Explain the patterns found from the evidence and make connections **See Assessment activity 13.4**
P5 Outline the results of the investigation as a statement to the court **See Assessment activity 13.5**	**M4** Explain the conclusions drawn from the investigation as a statement to the court **See Assessment activity 13.5**	**D4** Justify the conclusions drawn from the investigation as a statement to the court **See Assessment activity 13.5**
P6 Discuss the role of the Forensic Science Service in the criminal justice system. **See Assessment activity 13.6**	**M5** Identify the links between the Forensic Science Service and the criminal justice system. **See Assessment activity 13.6**	**D5** Explain the relationship between the Forensic Science Service and the criminal justice system. **See Assessment activity 13.6**

How you will be assessed

Your assessment could be in the form of:

- an observation of you processing a fictional crime-scene
- an evaluation e.g. of the process and problems you encountered when processing the crime-scene
- a plan e.g. of the analysis of evidence
- an observation of your analysis of evidence, including a portfolio with experimental details etc.
- a role-play e.g. of a mock trial.

Susan, 16 years old

This unit was really exciting and helped me to see how science is used to help solve crime. I really enjoyed the part where a group of us had to work together to find any evidence left at the scene of crime and the dressing up was great.

Learning about the importance of recording and documenting the scene and the evidence was interesting and my teacher used examples of real crimes to help us understand what could happen if it wasn't done properly.

Once we had collected the evidence, I had to plan what I would do with it to try and solve the crime. I had fingerprints, hair samples, footprints and a note written in different coloured pens which we had taken from the scene. I had to compare these with ones that belonged to various 'suspects'. I really enjoyed these practical tasks.

Our teacher told us about lots of other techniques which we weren't able to do and sometimes we could watch videos of them being carried out. Some of these were quite complicated, like the DNA fingerprinting and the blood splatter analysis, and I was glad that I didn't have to actually do them on our evidence.

One of the things we did in class was to pretend that we were giving evidence in court and so I had to do a written statement and then do a talk to my group on what I had found.

Catalyst

Crime scene

Look at the photograph on the opening page of this unit.

1 What sort of evidence is being collected?

2 How could this evidence be helpful in catching and convicting a criminal?

3 What other types of evidence do you know of that can be collected from a crime scene?

13.1 Investigating a crime scene

Key terms

Crime scene – the total area which needs to be examined or is involved in the criminal investigation.

Evidence – anything that gives information about a crime (or a potential crime).

Contamination – the presence of small amounts of one substance in another, e.g. if one piece of evidence comes into contact with another, small amounts can be transferred from one to the other.

PLTS

Team workers

Working with others in processing a mock crime scene will give you the opportunity to develop your team working skills. You will need to work closely with others to make sure that all the evidence is collected and you fulfil your responsibilities within the team.

The use of crime-scene tape. How do you think this protects the scene?

Whatever the **crime scene**, there are a number of actions that need to be carried out. These are: an initial assessment, securing the crime scene, recording the scene, searching for **evidence**, gathering the evidence and then packaging and labelling it.

This usually means that a team of people will work together to process the scene.

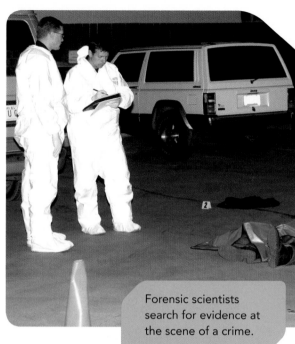

Forensic scientists search for evidence at the scene of a crime.

Initial assessment

The first action to be taken at the crime scene is to make the area safe, for example turning off the ignition if a vehicle is involved. The initial assessment must then be used to determine whether a crime has taken place. Any injured people or victims must be identified. Also, any witnesses or suspects in the area must be identified and kept separate from each other. Part of the initial assessment is to identify hazards and risks at the scene. The team will then take action to reduce these.

Securing the crime scene

Steps must be taken to preserve the scene in the state in which it was found.

The more people who enter the scene the more likelihood there is of **contamination**, so access must be restricted. The scene may also need to be protected from weather conditions or from view of the public or media. In real life, crime-scene tape and tents are used to do this but it is unlikely that you will use a tent for your mock crime scene.

Recording the scene

Sketches, notes, photographs and video recordings are all ways in which the original state of the crime scene can be recorded. You may have the opportunity to try out some of these methods as preparation for your assessment.

Searching the scene

A crime scene must be searched systematically to make sure that all of the evidence is found and recorded. There often isn't a second chance to find evidence. This is because searchers can disturb an area as they search it.

Activity A

Using available books and the Internet, research the various search patterns that can be used for a crime scene. Draw diagrams of four of these patterns and make some notes about how each is carried out.

The initial assessment of the scene will have included an assessment of the risk of any hazards present. This will then help the people carrying out the search to know what protective equipment they need to use. This personal protective equipment (PPE) includes disposable suits, masks and gloves. These protect the wearer from hazardous materials and chemicals including acids and poisons. Biohazards, for example body fluids, may also be present as well as physical hazards such as sharps: knives, razor blades or syringe needles.

The clothing worn by the searchers also prevents contamination of any evidence found during the search.

The location of evidence is carefully marked as it is found. Photographs can then be taken to show the evidence in its original state and position.

Safety and hazards

Risk assessment is an important process which is carried out to reduce the risk from the hazards present and to make sure that evidence is collected safely. Adherence to regulations such as the Control of Substances Hazardous to Health (COSHH) is just as important at the crime scene as it is in the laboratory.

Did you know?

Locard's principle applies not only to a criminal leaving a trace but also to a crime-scene investigator who could leave traces.

BTEC Assessment activity 13.1

1 What are the six sets of actions that are carried out in the processing of a crime scene? **P1** **P2**

2 Why is each of these actions important in obtaining evidence? **M1**

3 Identify three different types of evidence that might be collected at a crime scene and evaluate how each might be used in a criminal investigation. **D1**

Grading tip

Work carefully to make sure that you find, mark and document all of the available evidence in your simulated crime scene. This must then be collected and packaged using the right methods to achieve criteria **P1** and **P2**.

Knowing how these methods ensure you collect the evidence correctly will help you gain criterion **M1**.

To achieve **D1** you will need to be able to show that you know how the evidence collected can be used in an investigation.

Photographing evidence. Why do you think the board marked 3 is there?

13.2 Collecting evidence

In this section: P₁ M₁ D₁ P₂

Key term

Chain of continuity – a complete documentation of the progress of evidence from the crime scene to the court.

Did you know?

Sometimes footprints or fingerprints can be found more easily by using ultraviolet light or different fingerprint powders. This is because they make the prints show up more clearly.

Collecting forensic evidence in full personal protective equipment (PPE). Can you give two different reasons why full PPE is important?

Types of evidence

A methodical search of a crime scene will help an investigator to find and record the position of many types of evidence. This evidence can include hair, fingerprints, blood and other body fluids, footprints and other impressions, insects, and trace evidence such as fibres, glass or paint. The collection of these different types of evidence will need a range of equipment and techniques.

Unit 13: See pages 248–249 for more information on different types of evidence.

Gathering the evidence

A crucial part of the processing of a crime scene is that any evidence collected is not contaminated. For this reason people who enter a crime scene must wear suitable clothing.

It is important that any potential evidence is identified and collected using appropriate methods. A simple evidence collection kit can include paper envelopes, boxes, plastic bags, a torch, magnifying glass, disposable pipettes, sterile swabs, moulding compound, tweezers, sticky tape, fingerprint powder, brushes, labels and markers.

Activity A

What do you think each of the following could be used for when collecting evidence at a crime scene?

moulding compound   tweezers   pipette

Packaging the evidence

Different types of evidence need to be packaged in different types of containers. This is to preserve the evidence and ensure that it remains uncontaminated. For example, liquids will be packaged in glass vials, bottles or jars whereas items of dry clothing are more likely to be packaged in paper bags. Wet cloth is best packaged in polythene bags. Small items such as hairs or chips of paint may be wrapped in folded paper and then put into small bags.

An example of an evidence collection kit.

Labelling the evidence

The record of who has had responsibility for the evidence at every stage is very important. This is known as a **chain of continuity**, or the chain of custody, of evidence. It helps to ensure the validity of the evidence if it is later presented as an exhibit in court.

Summary of the processing of a crime scene.

Science focus

The correct labelling of evidence is very important.

The evidence label must have:

- an evidence number
- a description of evidence
- the place, time and date collected
- the name of person the evidence was taken from (if relevant)
- the name of person who collected it
- a record of who has handled the evidence at any time.

Grading tip

In order to meet the **P1** and **P2** criteria for this unit, you will need to show that you can collect, record and preserve evidence from a simulated crime scene using suitable methods.

To meet the higher criteria **M1** and **D1** you must show that you have a good understanding of the procedures that you have used. You could do this by providing a detailed description of how you processed the crime scene. Then you could state how you collected the evidence and interpret how the evidence could be used to decide whether or not a suspect is guilty.

BTEC Assessment activity 13.2 P1 M1 D1 P2

To help you prepare for your assessment on investigating a scene-of-crime you may find it helpful to consider the techniques used to gather and package evidence.

1. Copy the table on the right. Then, using the Internet and any books available to you, carry out some research to complete it. **P1**
2. Explain how each of the collection methods you have listed helps obtain valid forensic evidence. **M1**
3. Identify four important pieces of information which should be recorded on an evidence label. **P2**
4. Evaluate the importance of a chain of continuity. **D1**

Evidence	Collection method	Packaging
Hair or fibres		
Clothes		
Footprints		
Dried blood		
Liquids		

13.3 Types of evidence found at crime scenes

In this section: M2 P3

Key terms

DNA – deoxyribonucleic acid, contains the genetic instructions for the development of living things, the building blocks of life.

Toxicology – the study of substances harmful to life (poisons).

Entomology – the study of insects. Forensic scientists use insects to give information about a crime.

Trace evidence – small pieces of evidence which need to be collected at a crime scene.

Unit 13: See page 250 for a picture showing the different types of fingerprints.

Science focus

Toxicology tests are used to identify the presence of drugs, alcohol or poisons in the body. Because hair is fed by the blood stream, it can be analysed to find out if there are any drugs in the blood, and, if so, the amount.

Activity A

Working in a small group, carry out some research using the Internet, your work from other lessons and text books to help you design a poster on blood. You will need to include information about how blood is made up and the different blood groups.

Many types of evidence are left at crime scenes, including hair, fingerprints, blood and other body fluids, **DNA**, footprints and **trace evidence**. A scene-of-crime officer will need to know how to investigate all the different types of evidence that can be found at a crime scene. You should remember that the successful solving of a crime most often relies on the use of a number of different types of evidence.

Hair

It is very difficult to stop hair falling from your body. Hair found at a crime scene may belong to a victim or the person guilty of the crime. The structure of the hair can help to give information about where it came from. For example, an animal hair is tapered at the end. Similarly, there are differences between hair from different people and from different areas of the body. All these differences can be seen under a microscope.

Fingerprints

The skin on finger ends has distinctive patterns. These patterns can be used for identification because:

- the fingerprints of a person stay the same throughout their life
- no two people share the same fingerprints.

Entomology

Evidence from insects can help an investigator find the time of death and any change in location of a body after death. Tables of the expected life cycle development of different insects are available to investigators. These, together with details about the weather conditions, can be used to estimate the time of death.

Unit 3: You can read a case study about the use of insects in forensic science on page 79.

Trace evidence

Trace evidence is any evidence that is present in small quantities. This is usually things like fibres, glass or paint, or traces of saliva from partly eaten food, for example.

DNA

Deoxyribonucleic acid (DNA) can be found within almost every cell of the human body. The DNA contains a genetic blueprint that can be used as a means of identification. DNA is present in all body fluids and can also be found in hair. A national database of DNA profiles is used to help to identify suspects from DNA found at the scene of a crime.

Blood

A lot of very different evidence, in addition to DNA, can be associated with blood found at a scene-of-crime. Different blood groups are identified by looking at the substances found in the blood.

Blood patterns

The physical presence of drops of blood or the patterns made when blood splatters can also be very important in determining valuable information about the scene-of-crime. The shape and size of blood stains depend on the volume of blood, the speed at which the blood was travelling and the surface on which it landed.

Unit 6: You can learn more about blood groups on pages 156–157.

Case study: Combined evidence

In December 2002, three men were convicted of murdering Kevin Jackson in Halifax. Kevin had chased after three men who were trying to steal a car. When one of the suspects was arrested, a screwdriver was found in the boot of his car. Microscopic drops of blood were found between the blade and the handle. DNA analysis of this blood matched Kevin Jackson. Further analysis showed that the marks inside the lock of the car which the men were attempting to steal were made by the same screwdriver.

What other evidence might have been found at the crime scene and on the screwdriver which would have linked the suspects to the scene?

Marks and impressions

Footprints can provide very important evidence which links a suspect to a scene. If the print is sufficiently clear it can indicate the size, make and amount of wear on the shoe. This can be compared with shoes belonging to the suspect. It is also possible to get an indication of the height of the wearer from a print.

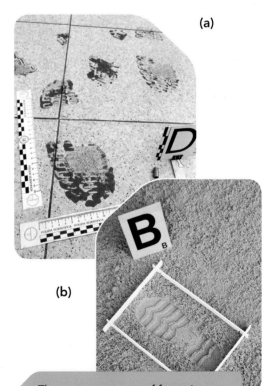
(a)
(b)

There are two types of footprint: **(a)** two dimensional, like a mark left in blood, and **(b)** three dimensional, like a mark left in soft sand. A casting is used to create an impression of a 3D print.

BTEC Assessment activity 13.3

1 How could fingerprint evidence be used in a criminal investigation? **P3**

2 A footprint in a pool of blood was found at a scene-of-crime. Give four different ways that this evidence could be used in a criminal investigation. **M2**

Grading tip

A simple plan of how you will analyse the evidence that you have gathered will be sufficient to gain the **P3** criterion.

A more detailed plan which takes into account any factors which affect what you can do to analyse your evidence is needed to gain **M2**.

For both of these criteria you will also have to consider the techniques covered on pages 250–251.

13.4 Using scientific techniques to analyse evidence

In this section: M2 D2 P3 M3 D3 P4

Key terms

Presumptive tests – quick tests which decide whether a sample is probably a specific substance or is definitely not the substance.

Conclusive tests – tests which are used to confirm that a sample actually is a specific substance.

DNA profiling – a technique used to help identify individuals based on the make-up of their DNA.

Anthropology – the study of humans. Most forensic anthropology deals with the study of human bones.

A variety of techniques are used to analyse forensic evidence. Many involve comparison or identification with similar samples from a suspect. Some of the more advanced techniques are outlined here together with those you are likely to carry out yourself.

Comparison microscopy

A comparison microscope has a single viewer through which two samples can be seen side by side. These can be taken from the scene-of-crime and a suspect. The technique is often used to compare hair or fibre samples, pieces of glass, tool marks or the marks on bullet casings.

Activity A

You can carry out a comparison of samples separately on a standard microscope. Why do you think the side-by-side comparison is more accurate?

Fingerprints

Fingerprints found at a crime scene can be 'visible' or 'latent'.

- **Visible prints** can be seen by just looking, often because the hand was covered in something like oil or blood. They are usually photographed.

- **Latent prints** cannot be seen without the use of a special powder or chemical reagent. The choice of powder/reagent depends on the surface. These prints are photographed and then 'lifted' onto card using sticky tape.

The pattern and characteristics of the fingerprints taken from the scene-of-crime can then be compared with those already held on a national database or with those taken from a suspect.

camera linked to microscope

two stages of the comparison microscope

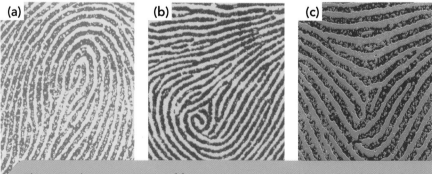

A comparison microscope.

camera transmits the images to a screen

(a)   (b)   (c)

There are three main types of fingerprint patterns: (a) loop, (b) whorl and (c) arch. They are all made up of many individual ridges. It is these ridges that give individual fingerprints there characteristics.

Activity B

Carry out some research and prepare a poster showing:

- diagrams of the procedure used to recover fingerprints using fingerprint powder
- three different types of fingerprint powder and examples of when each might be used
- at least one other method of recovering fingerprints.

Blood tests

It is important to make sure that a red stain is blood and whether it is human blood. It may then be possible to find out whose blood it is by using blood grouping techniques and comparing results with known information. Sometimes tests are needed to reveal the presence of blood where it is not clearly visible, for example when it is in very small amounts or somebody has tried to clean it up. These tests are called **presumptive tests**. One example is the use of phenolphthalein as shown in the diagram.

Phenolphthalein can turn pink in the presence of other substances such as potato. So if a pink colour is seen, other tests still need to be done.

If a presumptive test indicates blood then **conclusive tests** are carried out. These include tests to identify the presence of proteins specific to humans, blood grouping tests and DNA analysis.

Other forensic analysis techniques

Type of evidence	Possible analysis techniques
Hair, blood or other body fluids	Extraction of DNA, use of the polymerase chain reaction (PCR) process to multiply up small fragments of DNA, DNA profiling and comparison with national database
Teeth	Odontology – a comparison of structure and layout of teeth (including bite marks) with dental records
Bones or skull	DNA analysis (as above) **Anthropology** – analysing human remains to estimate age at death, height, sex; can include facial reconstruction based on measurements of the skull
Paint, alcohol, drugs or poisons	Chemical tests, chromatography, colorimetry

(a)

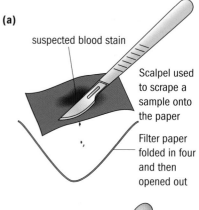

suspected blood stain

Scalpel used to scrape a sample onto the paper

Filter paper folded in four and then opened out

(b)

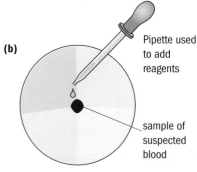

Pipette used to add reagents

sample of suspected blood

(c)

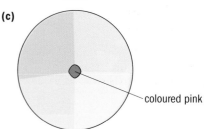

coloured pink

Testing for the presence of blood
(a) The test material is scraped onto a piece of filter paper. **(b)** A drop of phenolphthalein is added, followed by a drop of hydrogen peroxide. **(c)** A pink colour indicates the presence of blood.

Grading tip

The **P3** and **M2** tips from page 249 apply here too. If your plan also considers possible risks involved with your chosen techniques then this will help you gain criterion **D2**.

Successfully following your plan will help you achieve criterion **P4**. If you can see and describe patterns in the results of your analysis and explain these patterns it will help you towards **M3** and **D3**.

BTEC **Assessment activity 13.4**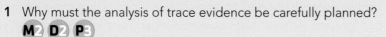

1 Why must the analysis of trace evidence be carefully planned? **M2 D2 P3**

2 If you are carrying out comparison techniques in the analysis of your evidence, why do you think it is important to keep samples separate and carefully labelled? **M3 D3**

13.5 Forensic science and the law

In this section: M4 D4 P5

Key term

Expert witness – somebody with special skills or knowledge who attends court to give factual information.

Courtroom interior.

Science focus

A written forensic statement contains these standard sections

- The name, qualifications and experience of the person making the report.
- A signed and dated declaration of the truth of the statement.
- An outline of the circumstances of the case (or investigation).
- A list of exhibits collected and analysed.
- A description of the experiments carried out and why they were carried out.
- The results obtained and what they indicate.
- A summary of the conclusions reached.

The role of forensic scientists

Forensic scientists have an important role in the recovery of evidence from a crime scene and in the analysis of that evidence. Another role of forensic scientists is to present information in court. All aspects of the work of the forensic scientist must be carried out to the highest possible standards. They must ensure that what they do is honest and correct. The results of their work can mean that somebody is imprisoned. This means that it is important that they agree to work in this ethical manner.

Most UK forensic science services are provided by the government-owned Forensic Science Service (FSS). There are also other, independent, organisations that provide forensic services. Some specialise in certain types of analysis, for example DNA profiling or document examination, while others offer a wide range of services.

Activity A

Working in pairs, use the Internet to research different organisations that provide forensic science services in the UK. Produce a table with the names of three providers of a wide range of services and three which provide specialist services. Indicate in your table the type of services they offer.

Reporting the results of an investigation in court

A forensic scientist must write a report of their findings following the analysis of evidence. This report is then submitted to the court. The forensic scientist may also be asked to appear in court as an **expert witness**. They must remain impartial and must report all their findings.

As a part of your assessment for this unit you will need to recover evidence from a mock scene-of-crime, analyse and draw conclusions about the evidence and write a report. You must be ready to present the findings from your investigations and express your opinion on the interpretation of these findings to a mock court. Your oral presentation should not contain too much technical detail. After all, you are the expert and your audience (for example, the members of the jury) are unlikely to have a scientific background.

You should prepare yourself to answer questions about the evidence you put forward. Answer each question clearly and make sure that you answer the question which has been asked.

BTEC Assessment activity 13.5 (M4) (D4) (P5)

In groups of four to six people, practise presenting evidence from one of your practical investigations. (It does not necessarily have to be for this unit.)

1 Take it in turns to act as the expert witness, prosecution lawyer and defence lawyer while the rest are members of a jury. **P5**

2 At the end of each person's turn to present their evidence, talk about how well they did.

- Was the information presented clearly and was it understandable?
- Were questions answered confidently?
- Were opinions justified by the evidence presented?
- What have you learned from this experience?
- What do you need to improve before your assessment?

PLTS

Effective participators

Writing and presenting a forensic report will help you develop a range of effective participator skills, particularly when you need to persuade others about your interpretation of results.

Functional skills

ICT and English

You will be able to develop your ICT skills and English skills in the production and presentation of a report to a mock court.

Grading tip

When you write a report of your investigations, include enough detail so that other people can understand exactly what you did, the result you got and what you think the results indicate. This will help you gain **P5**.

If your report explains your interpretation of what these results indicate, then this will help you to achieve **M4**. If you can give reasons for why you have interpreted the results as you have, then this will help towards **D4**.

Elaine Smyth
Forensic Scientist, Forensic Science Service

I am responsible for analysing the samples sent to me in the laboratory as quickly as possible.

A typical day for me would be to compare samples of DNA from a crime scene and from possible suspects. I have to write a detailed report on my findings and this would be used if the case goes to court. Occasionally I have to appear in court myself as an expert witness and make sure my report is easily understood by the jury and the judge. I must remain impartial when I present my evidence. I explain in simple terms what I have found and give an informed opinion on the interpretation of the test results. For example, I could indicate the probability of a match between a sample from a suspect and the sample at a scene of crime.

Think about it!

1 Why is it important for an expert witness to remain impartial?

2 Why does an expert witness give details of his or her qualifications and experience when in court?

13.6 The Criminal Justice System

Key terms

Closed question – likely to receive a single word or short phrase as an answer.

Open question – likely to receive a long answer.

Activity A

Think about different types of crime. Draw up a large poster-sized table with the headings 'Crime against property', 'Crime against the person' and 'Other crime'. Put the crimes that you have identified into the correct category.

Science focus

Quality of evidence

If the only evidence which placed a suspect at the site of a burglary was a smudged partial fingerprint found on the gate to the property, then it is unlikely that the suspect would be charged and brought to court.

PLTS

Independent enquirers

In researching a suitable criminal case you will have the opportunity to develop this skill.

The Criminal Justice System (CJS) makes sure that people guilty of a crime are punished and innocent people are protected.

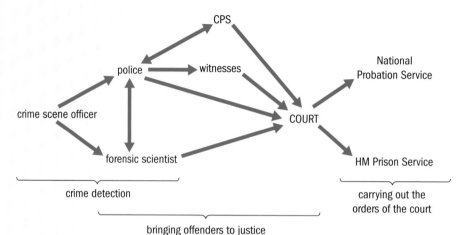

The Criminal Justice System at work. Different organisations work together to detect crime, bring offenders to justice and carry out the orders of the court.

Types of crime

Different types of crime are often grouped into general categories including: crimes against property, crimes against the person, knife crime, gun crime and vehicle crime.

Investigating a crime

Police investigate a reported crime to identify any suspects and gather evidence. The evidence gathered will include forensic evidence and interviews with witnesses, victims and suspects.

Interviewing techniques

Different types of questions can be used in interviews. **Closed questions** are quicker and designed to find out facts. They are more likely to be used when interviewing a suspect. **Open questions** get a more wordy answer and are more likely to be used when interviewing a victim.

Evidence

The police will consider the evidence but it is the role of the Crown Prosecution Service (CPS) to decide whether to charge a person with a criminal offence (see flowchart above). The quality and importance of the evidence will help the CPS make this decision.

The court system

If the CPS decides to prosecute, the suspect will be arrested and charged with the offence. The Crown Prosecutor will also consider alternatives to prosecution, which could include a caution or penalty notice (fine).

Table: Different criminal courts.

	People	Types of cases
Magistrates' court (including the youth court)	Three magistrates or Justices of the Peace (JPs) (volunteer members of the public without formal legal qualifications) advised by a legally qualified justices' clerk	Virtually all criminal cases start here 95% of all criminal cases are handled completely in this court
Crown Court	Judge and jury of 12 members of public. Jury listen to evidence presented and decide whether defendant is guilty or not. Judge responsible for sentencing if defendant found guilty	More serious offences
Court of Appeal	Three judges hear the appeal and reach a decision by majority. A judge in the Court of Appeal is called 'Lord Justice'	Suspected cases of miscarriage of justice: Prosecution may appeal if they think the sentence is too lenient Defence may appeal against the conviction or sentence

Trial process and rules of evidence

Not all evidence is allowed to be presented at a trial. The main categories of evidence allowed are:

- Real or physical evidence such as a murder weapon.

- Documentary evidence such as statements or video recordings.

- Witness evidence such as that presented by an expert witness.

Crime prevention

It is better if we can concentrate on prevention of crime rather than treating the end results of crime. Crime prevention methods in use throughout the country include neighbourhood watch schemes, surveillance cameras, regular patrols by the police, marking personal property, shredding personal documents to prevent identity fraud and increasing lighting in certain areas of a community.

Prosecution opens case
↓
Witnesses for prosecution are questioned to give their evidence
↓
Witnesses may be questioned by defence
↓
Defence opens their case
↓
Witnesses for defence are questioned to give their evidence
↓
Witnesses may be questioned by prosecution
↓
Closing speeches

Flowchart to show the trial process.

BTEC Assessment activity 13.6

1 How is the Forensic Science Service linked to the Criminal Justice System? **P6**

Choose a well-known criminal case. Then use books and the Internet to research the evidence presented during the trial.

2 Identify the evidence which was presented during the trial. Briefly show the role that each piece of evidence played in the conviction or acquittal of the defendant. **M5**

3 How were the Forensic Science Service involved with each type of evidence and what role did they play in the court? **D5**

Grading tip

To achieve **P6** you must clearly show how the role of the Forensic Science Service works with the Criminal Justice System. If you can identify and then explain the links that exist between the two by using an example of a criminal investigation then this will help you achieve **M5** and **D5**.

Just checking

1 What is a SOCO?
2 Identify three types of risks that a SOCO might come across.
3 Why is it important to search a crime scene in a particular pattern?
4 Explain the importance of using the correct packaging when recovering forensic evidence.
5 Name two techniques that could be used to analyse a sample of blood.
6 How can insects help the forensic scientist?
7 What technique could be used to analyse a sample of paint from a car?
8 What is the National Automated Fingerprint Identification System?
9 Name two different types of court and identify the major difference between them.
10 Watch an episode of CSI on television and make a list of all the things that you think are not done as they would be in real life.

edexcel ▦

Assignment tips

- Remember that for **P1** you will need to work closely with other members of your team when you are collecting, recording and packaging the evidence from a mock crime scene. Each team member must take responsibility for agreed tasks.

- Don't rush your initial planning of how you will investigate the crime scene. It is important that everybody in your team knows what they need to do and how to do it. Talk to each other as much as possible as this is what real investigators of a crime scene would do and it will help you meet **P1** and **P2**.

- The more detailed notes you make, and as soon as possible after the investigation, the more clearly you will be able to describe and evaluate the processing of the crime scene to meet **M1** and **D1**.

- Remember that for **D1** evaluating will mean identifying good things as well as things that could have been done better. You will also need to clearly identify the purpose of collecting the pieces of evidence.

- The better you plan how you will analyse the evidence collected the easier it will be to analyse and the easier it will be to decide what the results mean and to achieve **P3** and **M2**.

- When researching information on the Internet, be aware that many websites are American or Australian. You need to be careful that the information you find and use is relevant to the United Kingdom.

- Remember that not all crime scenes involve murder.

14 Science in medicine

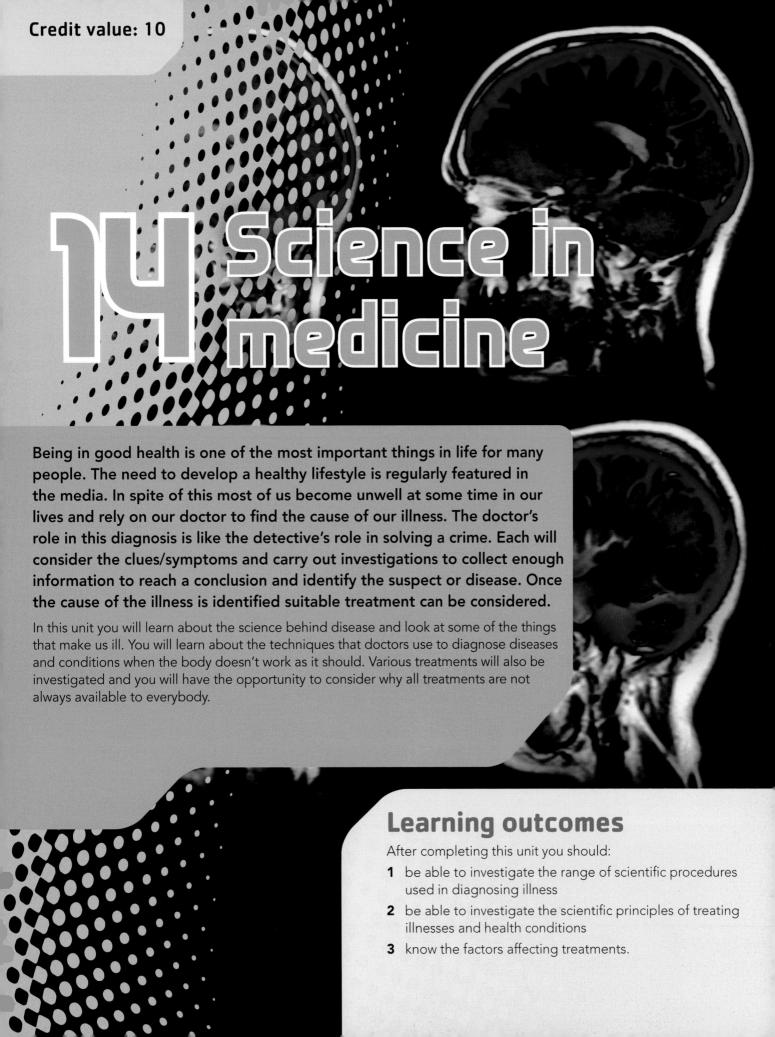

Being in good health is one of the most important things in life for many people. The need to develop a healthy lifestyle is regularly featured in the media. In spite of this most of us become unwell at some time in our lives and rely on our doctor to find the cause of our illness. The doctor's role in this diagnosis is like the detective's role in solving a crime. Each will consider the clues/symptoms and carry out investigations to collect enough information to reach a conclusion and identify the suspect or disease. Once the cause of the illness is identified suitable treatment can be considered.

In this unit you will learn about the science behind disease and look at some of the things that make us ill. You will learn about the techniques that doctors use to diagnose diseases and conditions when the body doesn't work as it should. Various treatments will also be investigated and you will have the opportunity to consider why all treatments are not always available to everybody.

Learning outcomes

After completing this unit you should:

1 be able to investigate the range of scientific procedures used in diagnosing illness

2 be able to investigate the scientific principles of treating illnesses and health conditions

3 know the factors affecting treatments.

Assessment and grading criteria

This table shows you what you must do in order to achieve a **pass**, **merit** or **distinction** grade, and where you can find activities in this book to help you.

To achieve a **pass** grade, the evidence must show that the learner is able to:	To achieve a **merit** grade, the evidence must show that, in addition to the pass criteria, the learner is able to:	To achieve a **distinction** grade, the evidence must show that, in addition to the pass and merit criteria, the learner is able to:
P1 Carry out investigations into biological and physical procedures used to diagnose illness **See Assessment activities 14.1 and 14.2**	**M1** Explain the scientific principles underlying the biological and physical procedures used to diagnose illness **See Assessment activities 14.1 and 14.2**	**D1** Evaluate the advantages and disadvantages of using biological and physical procedures to diagnose illness **See Assessment activities 14.1 and 14.2**
P2 Carry out investigations into the scientific principles of therapeutic drugs used to treat given illnesses **See Assessment activity 14.3**	**M2** Describe how the therapeutic drugs would be used to treat given illnesses **See Assessment activity 14.3**	**D2** Explain why the actions of therapeutic drugs are used to treat given illnesses **See Assessment activity 14.3**
P3 Carry out investigations into the scientific principles of physical therapies used to treat given conditions **See Assessment activity 14.4**	**M3** Describe how physical therapies would be used to treat given conditions **See Assessment activity 14.4**	**D3** Assess, using scientific and other evidence, which physical therapies are effective in the treatment of conditions **See Assessment activity 14.4**
P4 Identify the general risks of specified treatments **See Assessment activity 14.5**	**M4** Describe, using scientific evidence, the particular risks involved in specified types of treatments **See Assessment activity 14.5**	**D4** Explain why some individuals may choose not to take advantage of all types of available treatments **See Assessment activity 14.5**
P5 Identify other factors affecting the choice and availability of treatments to patients. **See Assessment activity 14.6**	**M5** Describe controversial decisions in prescribing treatments. **See Assessment activity 14.6**	**D5** Explain why decisions to give prescription drugs to some and not to others are always controversial. **See Assessment activity 14.6**

How you will be assessed

Your assessment could be in the form of:

- a practical demonstration or essay e.g. on procedures used to diagnose illness

- a table e.g. showing the uses/administration of therapeutic drugs

- a dialogue or report e.g. on alternatives to drug treatments

- notes e.g. on the risks of different treatments

- a presentation e.g. on why a certain treatment isn't available in a particular area.

Hayley, 16 years old

I always wanted to work in a hospital and being able to do this unit allowed me to see a little bit about what it would be like. I particularly liked the visits. We went to a small hospital close to our school, a very big city hospital which had operating theatres and a company that made drugs. Learning about the different ways in which illness is diagnosed was interesting and it was good to find out how different people do different jobs at the doctors' and in hospitals. I thought that it was mainly doctors and nurses working in the hospital; I didn't realise there are so many technicians carrying out different tests. Seeing all the equipment in the hospital helped me to understand what we had done in class. This really helped me when I had to do my assignments and use the Internet to find out all of the information that I needed to write my report. Our class also had to do a presentation about different types of drugs and treatments. Preparing the slides was really good but I found it difficult to explain why some drugs aren't available for everybody. After we finished the unit I decided that I want to work in a big hospital because there are more things using science going on there.

Catalyst

Science in medicine

1 Make a list of 10 different reasons for being admitted to hospital.

2 Give a very brief description of six different job roles within a hospital.

3 Make a list of signs that might be used to identify that somebody is ill.

4 Identify three different ways in which illness may be treated.

5 Give two possible side effects of drugs.

14.1 Diagnosing illness

In this section:

Key terms

Symptom – a sign of disease felt by a patient.

Diagnosis – the process by which a doctor identifies a disease.

Infection – the invasion of living tissue by harmful microorganisms.

Disease – an abnormal condition of the body that has a specific cause.

Disorder – a type of disease caused by a part of the body not working as it should.

Medical consultation between doctor and patient.

Some diseases are caused by harmful microorganisms such as bacteria, viruses or parasites. Some are **disorders** or parts of the body not working properly. For example, in diabetes the part of the body that produces insulin, a hormone which controls blood sugar levels, doesn't produce enough of this chemical.

Unit 14: See pages 262–263 for more information on diseases caused by microorganisms.

Symptoms of illness can be noticed by the patient but very few signal a specific disease. We will look at some of the techniques that are used by healthcare professionals to **diagnose** illness.

Physical diagnosis

Temperature

Measuring body temperature can help to tell us if someone is ill or not. It does not tell us why somebody is ill. The human body has a normal body temperature of about 37°C. A temperature outside the range of 36.2 to 37.5°C indicates illness or disease.

Blood pressure

A blood pressure measurement indicates if somebody is ill or likely to suffer illness. Normal blood pressure is in the range between 120 over 80 and 140 over 90 mmHg. The first number is the highest pressure reached when the heart beats and the second number is the lowest pressure reached when the heart relaxes between beats. The higher a person's blood pressure the more likely they are to suffer a heart attack or stroke. It is better to have blood pressure readings around 120 over 80 mmHg to reduce the risk of health problems. It is also possible to have health problems as a result of blood pressure readings which are too low, for example below 90 over 60 mmHg, but these are usually minor problems.

Activity A

Working in pairs, use the Internet and books to help you research different types of diseases. Produce a table that identifies at least three examples of diseases caused by bacteria, at least three that are caused by viruses, at least two that are caused by larger parasites and at least three that are not caused by microorganisms.

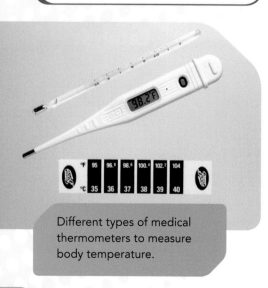

Different types of medical thermometers to measure body temperature.

External symptoms

The external appearance of parts of the body can indicate that something is wrong. Examples are:

- a part of the body can become swollen due to injury or **infection**
- a rash can indicate an infection or an allergy.

Outward symptoms do not always indicate a specific cause of **disease** so the doctor often needs to investigate inside the body.

Internal images

Endoscopy is a technique that allows a doctor to see inside a patient's body. A tiny fibre-optic light and camera are inserted into the body and the camera picture is seen on a screen.

Another way to get an image of the inside of a patient is to take a body scan. The most common type is an X-ray. X-rays are often used to diagnose broken bones but can also be used to photograph areas such as the stomach. In this case the patient would first need to eat a barium porridge. X-rays will not pass through barium so it will show up on the X-ray film as it flows into parts of the gut.

A computerised tomography (CT) scanner allows a doctor to see a slice through the body without surgery. The machine takes a series of X-rays from which a computer produces an image. A magnetic resonance imaging (MRI) machine uses radio waves from a strong magnetic field to show images from inside the body.

Did you know?

Some illnesses produce a specific kind of rash which can help in the diagnosis of the illness. Meningitis is an example of this.

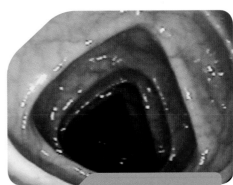

Endoscope view of the large intestine.

BTEC Assessment activity 14.1

As a hospital technician you have been invited to speak to a visiting school group about some of the physical methods used to diagnose illness.

1 Carry out any research necessary to help you prepare a presentation on methods used to see inside the body and help diagnose illness. **P1**

2 In your presentation explain what information about the body these methods give and how this helps in the diagnosis. **M1**

3 Identify the advantages and disadvantages of each of the methods and give an example of an illness for which each method might be most useful. **D1**

Grading tip

For **P1** your presentation should describe body scan and endoscopy methods.

When working towards **M1** think about how each method works.

To achieve **D1** you need to be careful not to concentrate just on the advantages but clearly highlight any disadvantages as well. Finish by making a judgement about whether each method is a good idea.

Safety and hazards

Exposure to too many X-rays can cause serious illness, so the health professionals who carry out X-rays usually go into a side room while the X-ray is being taken.

Activity B

What advantage do you think an MRI machine using radio waves might have over X-rays?

14.2 Biological diagnosis

In this section: P1 M1 D1

Key terms

Pathogen – an organism which is harmful to the human body.

Parasite – an organism that lives on or in another living thing (the host) and obtains nourishment from the host. The parasite harms the host.

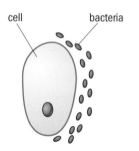

cell   bacteria

(a) Bacteria surround the cell

(b) Bacteria produce substances which damage the cell

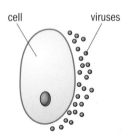

cell   viruses

(a) Viruses surround the cell

(b) Some of the viruses enter the cell

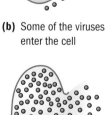
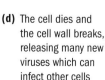

(c) The invading viruses change the cell and make the cell produce more viruses

(d) The cell dies and the cell wall breaks, releasing many new viruses which can infect other cells

How bacteria and viruses invade cells in the body.

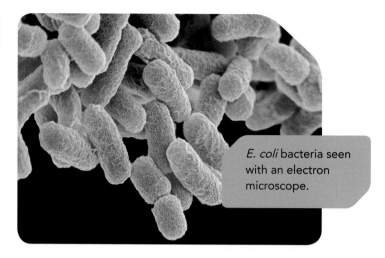

E. coli bacteria seen with an electron microscope.

Doctors can investigate parts of the body to help them diagnose an illness. This can include blood and urine as well as looking at cells from the body under a microscope or analysing DNA.

Non-infectious diseases are caused by lifestyle, ageing or genetics. Where an illness is caused by a **pathogen** it is called an infectious disease.

Unit 6 looks in more detail at these two types of diseases.

Identifying the microorganism that is causing an illness can help decide the treatment needed.

Infectious diseases

Bacteria, fungi, viruses and larger **parasites** cause disease in different ways. In all cases the first stage is for the pathogen to enter the body.

Activity A

Identify one parasite that lives within the intestines and one that lives in the blood. Describe how each parasite may enter the body and how it causes illness.

Non-infectious diseases

Your lifestyle can affect your chances of suffering a number of non-infectious diseases. Poor diet, smoking, alcohol or drug abuse and lack of exercise can all increase your chances of suffering from heart disease, stroke, cancer or liver damage.

Your genes can also affect your chance of suffering from certain conditions. Cystic fibrosis is an example of a condition that is genetic and can be inherited. DNA analysis can show if an individual has inherited the gene that causes a specific condition. A person may be a carrier of a gene but not show the signs of the condition themselves. Genetic counselling is often offered to help parents consider the risk of any children they have inheriting the faulty version of the gene and suffering from the condition.

Diagnosis

There are many biological techniques that can be used in diagnosis.

- Cytology is the study of cells. Medical professionals look directly at a sample of cells through a microscope to see if there are any changes happening. For example, a cervical smear test helps in the early detection of diseases like cervical cancer.

- Haematology is the study of blood. An understanding of the normal structure of blood helps identify abnormalities. For example, an increase in the number of white blood cells in the blood could indicate leukaemia.

- Chemical tests on the blood or urine can show abnormal levels of substances. For example, people with diabetes often have to carry out their own regular checks on sugar levels to monitor the condition. This can be done by test strips which use a biological enzyme reaction or an electronic meter.

Did you know?

Not all microorganisms are pathogens; in fact we rely on beneficial bacteria to help the normal functioning of our bodies.

Activity B

Make a list of five inherited diseases and conditions and provide a short description of the symptoms.

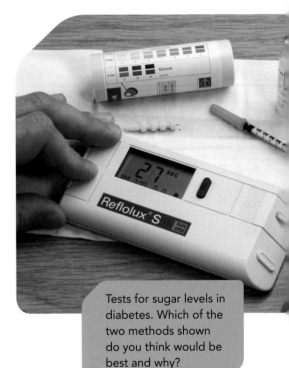

Tests for sugar levels in diabetes. Which of the two methods shown do you think would be best and why?

BTEC Assessment activity 14.2 **P1** **M1** **D1**

The young people that attended your presentation for page 261 were so impressed that they have now asked you to produce a poster to show some biological methods used to diagnose disease.

1 Research the normal structure of blood and how changes in this can indicate illness. Show named illnesses and the changes in blood that they cause. **P1** **M1**

2 How effective are the tests used to identify the changes in blood? **D1**

Grading tip

To meet the higher criteria you need to show an understanding of the procedures used to identify changes or abnormalities in the blood. For **M1** show that you know how tests identify the changes and what these changes can mean.

For **D1** you need to show advantages and disadvantages of these tests. Be careful not to concentrate just on the advantages. Clearly highlight the disadvantages as well. Finish by making a judgement about whether the tests are a good idea.

14.3 Using therapeutic drugs

In this section: P2 M2 D2

Key terms

Therapeutic drugs – drugs used for treating disease.

Analgesic drugs – a drug taken to stop pain. It does not cure the problem but reduces the feeling of pain.

Nurse preparing medication from a hospital drugs trolley.

A range of drug types is available by prescription from a doctor or can be bought 'over the counter' (OTC) at a pharmacy. Different types of drugs are used to treat different types of illnesses. We will look at the use of some of the more common drugs.

Types of drugs

Therapeutic drugs usually have specific uses. For example, **analgesic** drugs are used to reduce pain. You may know of the painkillers paracetamol and aspirin that can be bought over the counter, although there may be limits on how many you can buy. Doctors can prescribe stronger codeine- or morphine-based painkillers. These painkillers block the chemical and electrical signals being sent from nerves that tell the brain there is pain.

Other types of drugs include:

- Anti-inflammatory drugs: act to reduce inflammation due to an injury or an infection. They block the effect of the enzyme that produces the substance which causes the pain and swelling. An example is ibuprofen.

- Antibiotics: substances capable of stopping the growth of bacteria, e.g. penicillin.

- Antihistamines: drugs which counteract the effect of histamine. Histamine is produced by the immune system to fight a substance the body sees as harmful, for example pollen in hay fever sufferers. An example of an antihistamine is acrivastine.

- Chemical replacements: used when the body is not producing enough of a substance for itself. Examples are insulin, growth hormone and thyroxine.

Others drugs include those used for chemotherapy, antidepressants, heart drugs and blood pressure drugs.

Activity A

Carry out research at a pharmacy or on the Internet to help you complete the table below.

Drug name	Use	Type of drug
Paracetamol	Painkiller. Also found in cold remedies	
	Reducing inflammation	
		Antihistamine
Penicillin		

Drugs can be given or used in a number of different ways and so come in different forms. For example, some drugs are given by injection while others have to be inhaled. Some drugs need to injected directly into the blood (intravenous) while others need to be injected just under the skin (subcutaneous).

Activity B

For each method of giving or applying a drug listed below select the appropriate form or forms of drug from this list.

cream   patch   ointment   tablet   capsule   liquid   spray

(i) injection – inserted into the blood using a syringe

(ii) oral – taken by mouth

(iii) inhalation – breathed in

(iv) topical – applied directly to the skin.

Pharmaceutical companies often produce drugs under a 'brand' name. Some common OTC drugs are available as different products from different companies. Even with prescription drugs, your doctor can often prescribe a number of different products containing the same drug.

BTEC Assessment activity 14.3 P2 M2 D2

You work for a health centre and have been asked to produce some leaflets that will help members of the public understand the use of therapeutic drugs.

1 Choose six different illnesses and prepare a leaflet on each to show:

(a) the main symptoms and drugs that might be used to treat the illness **P2**

(b) how each of the drugs would be used, including how it would be given **M2**

(c) how the drugs help to treat the illness. **D2**

Grading tip

To achieve **P2**, **M2** and **D2** make sure that the illnesses that you choose cover the full range of different types of drugs.

For **P2** your leaflets must correctly match the types of drugs used for treatment of the identified illness.

You need to describe the form of each of the drugs you have identified and how they would be given or applied to meet **M2**.

For **D2** your leaflets should include a clear explanation of the effect of the drugs on the illness. You will probably find that including diagrams will help your explanations.

Did you know?

Sometimes the same drug comes in different forms. For example, ibuprofen is available as tablets, capsules and as a gel.

Child receiving an injection of antibiotic. What other reasons could there be to give a child an injection?

An inhaler used to treat asthma.

14.4 Using physical therapies

In this section: **P**3 **M**3 **D**3

The use of drugs is not the only way to treat illness. There are a number of physical therapies that can also be used. We will look at some examples of these.

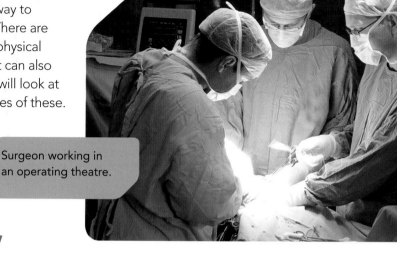

Surgeon working in an operating theatre.

Key terms

Organ transplant – a healthy organ donated from one person to replace a faulty organ in a patient.

Preventative therapy – treatment used to reduce the chance of a specific illness happening.

Did you know?

Although there are two kidneys in the human body it is possible for somebody with one properly functioning kidney to lead a full and healthy life. However, this can put a strain on the remaining kidney.

Surgery

Surgery can be used to correct defects in the patient's body. For example, surgery on the eye can correct a squint caused by some muscles in the eye being stronger than others.

Parts of the body can become so damaged that the only effective treatment is to remove the damaged tissue. The part of the body to be removed might not be essential to the normal working of the body. An example is the removal of the appendix (appendectomy).

If a whole organ is damaged or not working correctly then an **organ transplant** may be needed. For example, where a kidney, liver or heart fails to function correctly a transplant would help the patient to live a longer and more fulfilling life.

Transplanting organs depends on organs being donated from the body of another person. The organs must be transported very quickly after removal to give the person receiving the transplant the best chance of survival.

More people are waiting for a transplant than there are organs being donated, so scientists are working to develop other ways of obtaining organs and tissue. One way that this is being done is through stem cell research.

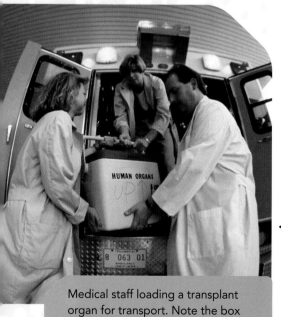

Medical staff loading a transplant organ for transport. Note the box designed to keep the organ cool. Why do you think the organ needs to be kept cool and why does it need to be transported quickly?

Activity A

What are stem cells and how may they be used to help treat people with organ damage?

Replacement therapy

It is not just organs that are transplanted. Joints such as the hips or knees can be replaced with artificial joints during surgery. Tissues such as skin or heart valves can also be replaced.

Blood transfusion is another example of a replacement therapy. Like organ transplant, it is very dependent on people donating their blood. A person receiving donated blood has to have a blood group compatible with that of the donated blood. Some people have a blood group that is very useful because it can be given to a large number of people.

Vaccinations

Vaccinations can be classed as a form of **preventative therapy**. Vaccination involves injecting a person with the pathogen that causes a disease in order to protect them from that disease in the future. This therapy works by stimulating the body to produce antibodies against the particular pathogen. If the patient is then exposed to the pathogen the body is prepared to fight any infection.

In some cases the vaccine that is used is a small amount of the live, but weakened, pathogen. Sometimes dead pathogens are injected and in other cases just the antigens from the pathogen are injected so that it causes the body to produce the antibodies but it cannot cause the disease.

Other therapies

The table below shows some other physical therapies that you might come across.

Therapy	Description	Example of use
Radiotherapy	Use of radiation to kill growing cells	Cancer treatment
Physiotherapy	Exercise, massage and manipulation	Arthritis treatment
Laser therapy	Use of a concentrated beam of light	Correcting short-sightedness
'Complementary/ alternative'	Acupuncture – use of needles	Rheumatoid arthritis
	Herbalism – use of herbs and plants	Allergies
	Osteopathy – non-surgical manipulation of bones	Back injuries

 **Assessment activity 14.4** P3 M3 D3

You work for Patient UK, a patient support organisation, and have been asked to produce a booklet to be issued to patients to help them understand the use of physical treatments.

1 Carry out research to allow you to prepare the booklet on different types of physical therapy. P3

2 Describe an example of when each therapy can be used. M3

3 How effective is each of the therapies that you have identified? D3

Unit 6: See page 156 for more information on blood groups.

 Activity B

Why are people who donate blood able to do this on a regular basis?

What other useful substances can be taken from donated blood?

Unit 6: See pages 150–151 for more information on vaccinations

 PLTS

Independent enquirers and Self-managers

Completing this activity will help you to develop research skills and self-management skills by finding and selecting the best information and presenting it in an appropriate way.

Grading tip

For P3, make sure that your booklet covers the range of therapies. Including some diagrams or pictures will help people understand the therapies and is a good way of showing your knowledge.

You must use suitable examples of illnesses or conditions to describe how each therapy is used in treatment for M3.

To gain D3 identify any risks or side effects from each treatment and any other disadvantages and then balance these against the advantages of the treatment. This will help you to assess which therapies are effective for which illnesses or conditions.

14.5 Factors affecting treatments

Have you ever wondered why different people with similar illnesses receive different treatments? Why some people choose to have surgical procedures while others don't or why people have different types of drugs?

Risks

Many types of treatment involve some risk. Some people are more likely to be affected by these risks, or the risk may be greater in some circumstance than in others. For example, some drugs may be more suitable for an adult than for a young child. Also the risks of surgery may be higher if a person already has another illness such as heart disease.

Key terms

Side effect – effect that occurs in addition to the intended therapeutic effect of treatment.

Allergic reactions – various effects caused by the immune system acting on normal substances as if they were harmful.

 Activity A

Why do you think some drugs may not be suitable for young children?

Physical treatments

Some physical treatments use potentially harmful radiation, but the way these treatments are carried out minimises the risks. For some people, certain treatments may not be suitable for other reasons. For example, an MRI scan does not have any particular **side effects** or use potentially harmful X-rays. Even so, somebody who has a heart pacemaker or metal implants cannot have an MRI scan because of the risk of the strong MRI magnet moving the metal. This risk is based on the scientific evidence of the attraction of metal to a magnet.

Some patients decide against having a surgical procedure because they feel that the risks are greater than the potential benefits.

A pharmacist explaining the possible side effects to a patient.

Drug treatments

Most drugs have some potential side effects. Aspirin, for example, is a useful painkiller, anti-inflammatory and antipyretic drug but it has a side effect of sometimes causing bleeding in the stomach.

Some people are **allergic** to certain drugs. The most common drug allergy is to the antibiotic penicillin. This can produce a rash and inflammation (see photo) but it can also cause more serious reactions such as difficulty in breathing.

There are other risks of drug treatments: the patient could become addicted to the drug, or the drug could interfere with another drug the patient is taking.

This is why it is very important that all drugs are carefully prescribed by a qualified healthcare professional who takes account of all the possible risks. These risks are based on scientific evidence from the study of the chemical nature of the drugs.

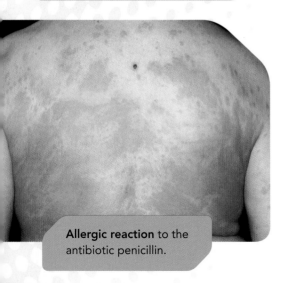

Allergic reaction to the antibiotic penicillin.

Religious views

In a multicultural society there are people of many different religions and beliefs. For some people certain treatments are not acceptable within their religion. For example, a Jehovah's Witness would find it unacceptable to receive a blood transfusion. Whether you agree with this thinking does not matter, as every individual has the right to their own beliefs. However, the situation can be more complicated. If doctors decided that a child needed a blood transfusion, but the parents refused because they were Jehovah's Witnesses, what would be the right thing to do? Healthcare professionals such as doctors and nurses must be aware of different religious views and may have to look at alternative treatments in these cases.

Ethical and social factors

Patients can have ethical concerns about types of treatment based on whether they see the treatment as right or wrong. For example, some people believe that abortion is wrong. Different cultures have very different opinions on aspects of healthcare, for example how to look after old and infirm people.

Patients in a coma can be kept alive on a life-support machine for a long time. If it is turned off they will die. But if there is no hope of the patient recovering, family members and doctors have to make a decision about whether to turn off this support. This is another example of an ethical decision about treatment.

Did you know?

Some allergic reactions to foods like nuts or to insect bites or stings can be so severe that they cause the airways to narrow, so the person finds it difficult to breathe. Their blood pressure can suddenly drop, which can make them faint. This is known as anaphylaxis or anaphylactic shock and is life threatening.

Activity B

Why might a Jehovah's Witness not want to have a blood transfusion? Research their belief about blood.

Activity C

With a partner, discuss how long you think somebody in a coma should be on a life-support machine before it is turned off. Should it be the same length of time for everyone?

BTEC Assessment activity 14.5 P4 M4 D4

You are a trainee healthcare worker and have been asked to put together a recorded audio tape for patients explaining the risks of various treatments.

1 Work in a small group to research the risks of the range of treatments covered on pages 264–267. Select two drug treatments and three physical therapies. Identify and describe the risks for these treatments on your audio tape. **P4 M4**

2 Produce an additional section to your tape to explain the reasons why some people may choose not to have certain types of treatment. **D4**

Grading tip

To achieve **P4** you need to identify the risks of named treatments. To reach **M4** make sure you include scientific evidence to describe the risks.

Your tape will cover some of the ethical and social issues in addition to balancing risks and benefits of named treatments to meet **D4**.

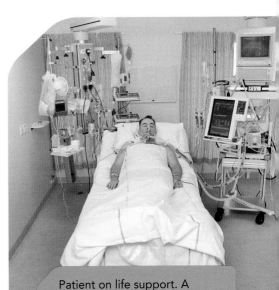

Patient on life support. A ventilator helps him breathe and a tube through the nose into the stomach feeds the patient. He is also linked to a vital signs monitor.

14.6 Finance and the National Health Service

In this section: P5 M5 D5

Key terms

Primary care – care provided by people you see when you first have a health problem, e.g. your GP, dentist, optician.

Secondary care – specialist medical care such as surgery or emergency care, usually given in a hospital.

Department of Health
Government department responsible for the NHS

10 Strategic Health Authorities
manage the local NHS

152 Primary Care Trusts	150 Hospital Trusts
provide front line services	(more than 100 of which are Foundation Trusts)

60 Mental Health Trusts	12 Ambulance Services
provide services for mental health	provide emergency and patient transport

Simplified structure of the NHS.

Did you know?

NHS North Yorkshire and York has a budget of over £1 billion for the year 2010/11. From this they must buy all NHS healthcare for the 800 000 people in the area. Can you find out the name of your local primary care trust and how much their budget is for this year?

Structure of the NHS

The British government Department of Health is in charge of the NHS in England. Scotland, Wales and Northern Ireland run their local NHS services separately. The NHS provides healthcare to residents of the United Kingdom.

Hospital nurses.

Funding and cost of treatment to the NHS

Most NHS services are free for the patient. Some services have a charge, for example eye tests, dental care and prescriptions, but the charge to the patient does not cover the full cost of the services. The NHS pays the remaining cost.

Money to fund the NHS comes from taxpayers.

Primary care trusts control about 80% of NHS funding. They are allocated money each year and are responsible for deciding how to spend these funds to best serve the people in their area. They must provide enough services such as GPs, dentists, opticians and pharmacists in their area.

Primary care trusts also commission or 'buy' services for **secondary care** from hospital trusts.

A new type of hospital trust is called a foundation trust. This type of hospital is still within the NHS, but is run in a different way. There are currently over 100 foundation trusts in the UK.

Regional variation

Because different parts of the country are run by different health authorities, there can be regional differences in how hospital trusts and primary care trusts spend their money.

Waiting times vary between hospitals and in different regions but the NHS is trying to make sure that nobody waits longer than 18 weeks before starting treatment.

You may have heard of the 'postcode lottery' of healthcare. This means that patients in some parts of the country do not get the same access to drugs (particularly new and more expensive ones) and treatments as others. Primary care trusts can only provide services and treatments within the budget that they are given, i.e. they have a finite amount of money and have to pay for all the healthcare needs of the population with it.

Private medicine

Some people choose to have private medical care instead of NHS care. This means that they have to pay for the treatment themselves. One advantage of this is that they do not have to wait as long or can arrange treatment at a time they wish. They may also be able to pay for particular drugs that may not be available from the NHS in their area. Because the treatment has to be paid for by the individual, some people take out health insurance to help pay for private treatment.

Activity A

Discuss the differences in health-care in different areas with two or three people in your class. Do you think that all treatments should be funded everywhere? Should waiting times be shorter and the same in all areas? Would you or your parents be prepared to pay more tax to fund improvements in the NHS?

BTEC **Assessment activity 14.6** **P5** **M5** **D5**

You are working as a trainee science reporter for a newspaper and have been asked to prepare an article about healthcare treatments available in different regions.

1 Use the Internet or newspapers to find examples of when people have been denied a particular drug or treatment while others in a different area have been given the drug or treatment. Use this information and any information from pages 268–269 to make a list of the things which affect choice and availability of treatments. **P5**

2 Discuss with a member of your class (your editor) the reasons why treatments are not always available to everybody. **M5** **D5**

Grading tip

To reach **M5** you must be able to describe examples of decisions about availability of treatments. You will need to show your understanding of these issues by being able to explain why these decisions are open to argument to meet **D5**.

WorkSpace

Sujata Patel
Chief Technician, Aseptic Unit, Hospital Pharmacy Department

STEM AMBASSADORS ILLUMINATING FUTURES

Nationally coordinated by STEMNET

I am responsible for the unit that makes intravenous drugs for adults with cancer and nutrition for premature babies who cannot feed.

I receive prescriptions from the doctors and then prepare worksheets so that the chemotherapy drugs are made to the correct formula. The ingredients are assembled and checked before the mixture is made in a sterile room. I monitor the air in the sterile rooms for microscopic particles, including bacteria.

Patients come to the chemotherapy day unit to receive their chemotherapy. They have a blood test before the drugs can be given. If the results are low, then the patient's treatment may be delayed or dosage reduced. We tailor each treatment while the patient waits. I have to gown up in specialised clothing including hood, mask and boots before preparing chemotherapy. Checks are performed at each stage of the preparation process. The treatment might have to be given to the patient over several hours so it is important to work fast.

Think about it!

1 Why are chemotherapy drugs and nutrition for infants prepared in sterile rooms?

2 What other checks might be performed?

3 What would you say to a patient who is unhappy about how long their treatment takes to prepare?

Just checking

1 What is illness? Give three examples of causes of illness.
2 Describe four biological diagnosis methods and four physical diagnosis methods.
3 List five different types of drugs and give one example of each.
4 Describe four different methods of giving drugs and four different forms of drugs.
5 Why do people taking X-rays need to take extra precautions like moving behind a screen or leaving the room?
6 Give two examples of when radiation may be useful in diagnosis and one example of when it may be useful in treatment.
7 List four different 'complementary' or 'alternative therapies' and give one example of when each might be used.

Assignment tips

To get the grade you deserve in your assignments remember to do the following.

- Explain and evaluate the procedures of diagnosis and the actions of treatments.

- In your evaluations clearly display in your work the advantages and disadvantages and draw a conclusion from these to make a statement on the overall effect.

- When you draw conclusions or make statements back these up with scientific principles rather than just stating your opinion.

- Be sensitive when you consider the reasons why people may not choose treatments. Remember that not everybody believes the same thing but that does not mean that you are right and other people are wrong.

The key information you'll need to remember includes the following.

- Both biological and physical methods are often needed to allow medical professionals to diagnose illness.

- Not all illness is caused by bacteria or viruses.

- Therapeutic drugs can be grouped by their uses, for example painkillers or antibiotics, but remember that the same drug is often available under different manufacturers' names.

- Most drugs have some side effects and may affect people differently – remember that doctors sometimes have to try different types of the same group of drugs.

- Where you live may mean that certain treatments are not available to everybody.

- Some 'complementary therapies' are now available through the NHS, but this may also depend on the area where you live.

You may find the following websites useful as you work through the unit.

For information on...	Visit...
all aspects of the UK National Health Service	UK NHS
health advice and information	NHS direct

Credit value: 5

15 Using mathematical tools in science

Mathematics is an important tool in science. Whether you are an assistant microbiologist working out the percentage of samples that have been contaminated with *E. coli* bacteria, an assistant pharmacist calculating the correct concentration of a medicine or a trainee engineer designing new products such as efficient engines, you will need to work with numbers. Scientific results are usually numerical and need to be displayed, sometimes as charts, sometimes as linear or non-linear graphs. Displaying data like this makes the results understandable to other scientists and engineers.

This unit introduces you to the basic mathematical tools needed for science. We will look at how accurate measurements are, and how to give results to the right number of significant figures or decimal places. You will learn how standard form simplifies calculations, and how algebra is used to solve science problems. The last part of this unit introduces methods for collecting and displaying scientific data so that other scientists and engineers can better understand your experimental results. It also looks at the importance of identifying the types of errors that occur in scientific experiments.

Learning outcomes

After completing this unit you should:

1. be able to use mathematical tools in science
2. be able to collect and record scientific data
3. be able to display and interpret scientific data.

Assessment and grading criteria

This table shows you what you must do in order to achieve a **pass**, **merit** or **distinction** grade, and where you can find activities in this book to help you.

To achieve a **pass** grade, the evidence must show that the learner is able to:	To achieve a **merit** grade, the evidence must show that, in addition to the pass criteria, the learner is able to:	To achieve a **distinction** grade, the evidence must show that, in addition to the pass and merit criteria, the learner is able to:
P1 Carry out mathematical calculations using suitable mathematical tools **See Assessment activities 15.1, 15.2, 15.3 and 15.5**	**M1** Use standard form to solve science problems **See Assessment activities 15.1, 15.2 and 15.5**	**D1** Use ratios to solve scientific problems **See Assessment activity 15.3**
P2 Carry out mathematical calculations using algebra **See Assessment activities 15.4 and 15.5**	**M2** Use mensuration to solve scientific problems **See Assessment activity 15.5**	**D2** Use algebra to solve scientific problems **See Assessment activities 15.4 and 15.5**
P3 Collect and record scientific data **See Assessment activity 15.6**	**M3** Describe the process involved in accurately collecting and recording scientific data **See Assessment activity 15.6**	**D3** Compare methods of data collection **See Assessment activity 15.6**
P4 Identify errors associated with collecting data in an experiment **See Assessment activity 15.6**	**M4** Calculate any errors associated with scientific data collected in an experiment **See Assessment activity 15.6**	**D4** Explain how errors can be minimised in data collected in the experiment **See Assessment activity 15.6**
P5 Select the appropriate formats to display the scientific data that has been collected **See Assessment activity 15.7**	**M5** Interpret the trend in the scientific data collected in an experiment. **See Assessment activity 15.7**	**D5** Calculate scientific quantities from linear and non-linear graphs. **See Assessment activity 15.7**
P6 Interpret scientific data. **See Assessment activity 15.7**		

How you will be assessed

Your assessment could be in the form of:

- worked evidence showing you have solved scientific problems e.g. using standard form, ratios, algebra and mensuration

- an experiment, including plan and results showing you have collected scientific data

- evidence to show that you can display and analyse data from an experiment e.g. description, charts, graphs, calculation of errors, comparison.

Georgia, 16 years old

This unit is useful regardless of the area of study you go into. I learned maths skills that are related to the real world; it wasn't just maths without a purpose. I found the way the maths tools were linked to experiments interesting, as I could see how the maths was being used and it made sense to me.

Catalyst

Maths in our lives

In groups of three discuss how you have used maths in your life. For example, it could be when buying cinema tickets with friends, going on holiday abroad or buying DVDs on special offer. In your groups, list what kind of maths is involved (for example fractions, money conversions etc). Now compare these calculations with the maths you have used in the science you have studied so far. Present your results in a poster.

15.1 Standard form and SI units

In this section: P1 M1

Key terms

Standard form – a way of writing down small and big numbers easily as powers of 10.

Unit – a symbol that follows a number; it tells us about the quantity.

SI units – units that have been agreed internationally.

Imperial units – an old system of units that is mainly used in the UK and USA.

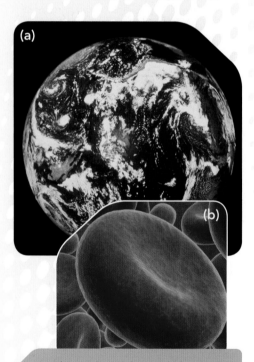

(a) The Earth and (b) a red blood cell.

Did you know?

When solving problems in standard form, remember that a negative power means a small number and a positive power means a large number.

Scientists have to handle very small and very large numbers, for example measuring the size of cells or calculating the memory size of a memory stick. They use **standard form** to make these numbers easy to handle.

Standard form

The mass of the Earth is about 6 000 000 000 000 000 000 000 000 kilograms, or 6 million billion billion kilograms. The distance between the Sun and Earth is about 149 000 000 000 metres; that's 149 billion metres. These numbers are difficult to use written like this, so scientists change them into something called standard form.

Numbers in standard form are written as numbers between 1 and 10 multiplied by a power of 10. So the mass of the Earth would be written as 6.0×10^{24} kg (count the zeros) and the distance between the Sun and the Earth as 1.49×10^{11} m.

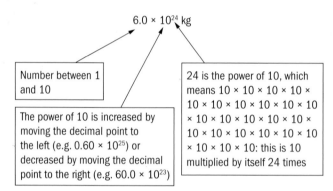

6 000 000 000 000 000 000 000 000 kilograms

6.0×10^{24} kg

Number between 1 and 10

The power of 10 is increased by moving the decimal point to the left (e.g. 0.60×10^{25}) or decreased by moving the decimal point to the right (e.g. 60.0×10^{23})

24 is the power of 10, which means 10×10: this is 10 multiplied by itself 24 times

Writing large numbers in standard form.

Very small numbers, like the mass of an electron (0.000 000 000 000 000 000 000 000 000 000 981 kilograms) or the diameter of a red blood cell (about 0.000 007 metres) are just as difficult to use as very large numbers. We write them in standard form too, so the mass of the electron is 9.81×10^{-31} kilograms, and the diameter of a red blood cell will be 7.0×10^{-6} metres.

Worked example

The average width of a human hair is 0.000 07 metres. Change this to standard form.

Move the decimal point to the right of the first number that is not a zero. The decimal point has moved five places to the right, so the power of 10 will be −5. The width of the hair is 7.0×10^{-5} m.

0.000 000 000 000 000 000 000 000 000 981 kilograms

9.81×10^{-31} kg

Number between 1 and 10

A negative power of 10 means 'one divided by', so 10^{-31} means $1/10^{31}$

Here, the decimal point has been moved 31 places to the right, so there is a power of –31

Writing small numbers in standard form.

SI (metric) units

In science the quantities that describe what is going on in the world are likely to have a number with a **unit** after it; for example, for a bag of potatoes of mass 5 kg, the unit is the kilogram (kg). **SI units** like these are used throughout the world. **Imperial units** are used mainly in the UK and the USA, but are no longer common. Some SI units you may have met are kilograms (kg), metres (m), seconds (s) and kelvin (K); imperial units still in use include ounces (oz), pounds (lb) and stone (st), inches (in), feet (ft) and yards (yd), pints and gallons, and degrees Celsius and Fahrenheit.

Worked example

Ikram has a mass of 80 kg; what is his mass in stone?

1 st = 6.35 kg, so 1 kg = 1/6.35 st = 0.16 st

and so 80 kg = 80 × 0.16 = 12.8 st.

BTEC Assessment activity 15.1   (P1) (M1)

1 A mirror designed in the USA for use in solar cell applications has dimensions of 2.0 ft × 1.5 ft. Will it fit in a frame of 0.60 × 0.40 m? **P1**

2 An optical microscope revealed that a plant cell had a diameter of 0.000 0032 m. Give this in standard form. **M1**

3 You must prepare one litre of a solution containing 4×10^{-3} lb of salt to 0.5 litre of water. How many kilograms of salt do you need? Give your answer in standard form. **M1**

Activity A

In groups of three, write your ages in minutes. Then re-write your answer in standard form.

Table: Converting between SI and imperial units

Quantity	Conversion factor	
Mass	1 kg	~2.2 lbs
	1 ton	160 st
	1 st	~6.35 kg
Length	1 ft	~30 cm
	1 yd	3 ft
	1 ft	12 in
Volume	1 litre	~ 1.75 pints
	1 gallon	8 pints

PLTS

Independent enquirers

This type of question will allow you to develop your skills as you identify a mathematical problem and display the data in an appropriate format.

Grading tip

For **P1** remember that in science most quantities have a unit; without the unit you will not know what the quantity is.

To achieve **M1**, use standard form; remember the power is related to the number of places of the decimal point.

15.2 More mathematical tools for science

In this section: **P1** **M1**

Key term

Prefix – letters added to the start of the name of a unit to represent powers of 10 (both positive and negative).

Prefix	Symbol	Factor
giga	G	10^9 (1 000 000 000)
mega	M	10^6 (1 000 000)
kilo	k	10^3 (1000)
deci	d	10^{-1} (0.1)
centi	c	10^{-2} (0.01)
milli	m	10^{-3} (0.001)
micro	μ	10^{-6} (0.000 001)
nano	n	10^{-9} (0.000 000 001)

Prefixes

As well as standard form, there is another way to make large and small numbers neater. Scientists often use **prefixes** in front of the base unit to represent powers of 10. The table shows the prefixes that you need to know. You will recognise some of them from words like 'microwave' and 'gigabyte'.

Activity A

Viruses have a diameter of about 1.5×10^{-7} m. Rewrite the number with a different power of 10 and an appropriate prefix.

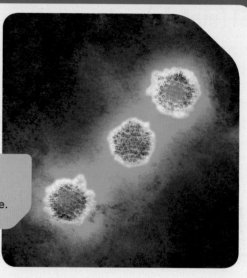

A virus magnified 83 125 times by an electron microscope.

Science focus

You can work easily with powers of 10 by remembering that when you multiply two numbers in standard form, you simply have to add the powers together: $100 \times 10^{-6} = 10^2 \times 10^{-6} = 10^{-4}$.

Did you know?

When you use a calculator you must make sure that you do each part of the calculation in the correct order, otherwise you'll get the wrong answer. Remember BODMAS (brackets, of, divide, multiply, add, subtract) to get the order right.

Using a scientific calculator

[EXP] is the button to press to key powers of 10 (on some newer calculators the [EXP] button is now labelled [10ˣ]). For example, the speed of light is 3×10^8 metres per second. To enter this number into a calculator, follow these steps:

- key in the number between 1 and 10, in this case [3]
- • press [EXP]
- • press the power of 10, in this case [8]
- • the calculator will show *300000000* or *3.0 EXP 8*.

Don't key [3] × [1] [0] [EXP] [8]. That would give you an answer that is 10 times too big, because [EXP] already means '× 10'.

Rounding off

Imagine you work in a plant science laboratory. You have to find the average mass of grains from a new breed of wheat. To increase accuracy you collect 500 grains. If you find their combined mass is 34.23 g, you can calculate that one grain has a mass of 0.06846 g. But if your scales can only measure the combined mass to two decimal places, your answer for one grain can't be accurate to five decimal places. You must round off the answers to a specific number of decimal places or significant figures. In this case the answer will be 0.07 g.

Decimal places (d.p.)

Worked example

The mass of a neutron is 1.6752 × 10⁻²⁷ kg. What is it to two decimal places?

1.6752×10^{-27} kg $\rightarrow$ = 1.68×10^{-27} kg (2 d.p.)

| Last digit | Digit after last digit you want is 5, so round up |

Significant figures (s.f.)

Worked example

Write 6.025301 to three significant figures.

$6.025301 \rightarrow$ = 6.03 (3 s.f.)

| 1st | 2nd | 3rd |

Write 0.000120543 to four significant figures.

$0.000120543 \rightarrow$ = 0.0001205 (4 s.f.)

| 1st | 2nd | 3rd | 4th |

In this case the first significant figure is the first digit that isn't a zero. The second, third, etc. significant figures follow from the first, whether they are zero or not.

Activity B

Use your calculator for the following calculation. Give your answer in standard form.

$(4.3 \times 10^{-4}) / (3.2 \times 10^{3})$

Science focus

When rounding up or down, find the last digit you want and look at the next digit to the right. If it is 5 or more, then round up the last digit; if it is less than 5, then leave the last digit as it is.

For example 2.78 would be rounded up to 2.8, while 2.72 would be rounded down to 2.7.

BTEC **Assessment activity 15.2**

You are helping to design a microbiology laboratory and would like the dimensions to be similar to those of the science laboratory in your school or college.

1 Measure and record the dimensions of the doors, windows and walls of your laboratory and calculate their areas. Leave your answers to the appropriate significant figures. **P1**

2 What is the volume of the laboratory? Give your answer in standard form. **M1**

Grading tip

When attempting **P1** the number of significant figures in the areas should not be any more than those you give in your measurements. When working towards **M1** don't get confused with the notation given by the calculator; that may not be in standard form.

15.3 Using fractions, percentages and ratios in science

In this section: **P D**

Key terms

denominator – the bottom number of a fraction.

numerator – the top number of a fraction.

ratio – a comparison of the sizes of different numbers.

Scientists often need to compare numbers. For example, an ecologist may look at the percentage of an area covered by a particular plant species.

Activity A

Work out the following problems without a calculator. Check your answers with a calculator.

$\frac{3}{5} + \frac{1}{5}$

$\frac{3}{8} \times \frac{4}{6}$

$\frac{5}{9} \div \frac{3}{8}$

$\frac{3}{4} - \frac{1}{3}$

Fractions

Below is a reminder of how to use and simplify fractions.

- Multiplying and dividing:

  Multiply top and bottom separately

  e.g. $\frac{2}{7} \times \frac{3}{4} = \frac{2 \times 3}{7 \times 4} = \frac{6}{28} = \frac{3}{14}$ (simplifying)

  Divide by turning the second fraction upside down and then multiplying

  e.g. $\frac{5}{8} \div \frac{1}{2} = \frac{5}{8} \times \frac{2}{1} = \frac{10}{8} = \frac{5}{4}$ (simplifying)

- Adding and subtracting

  When the **denominators** (bottom numbers) are the same, add or subtract the **numerators** (top numbers):

  e.g. $\frac{1}{8} + \frac{6}{8} = \frac{1+6}{8} = \frac{7}{8}$

  When the denominators aren't the same then you need to cross-multiply:

  $\frac{2}{5} + \frac{1}{4} = \frac{2 \times 4 + 1 \times 5}{5 \times 4} = \frac{8+5}{20} = \frac{13}{20}$ (This can't be simplified as there are no whole numbers that both 13 and 20 can be divided by.)

Percentages (%)

Percentages are fractions, but the denominator is always 100. Below are some examples of percentages and fractions used in science.

Worked example

1 In a biology experiment, 10 out of 29 students had a pulse rate between 69 and 79 beats per minute. Write down this finding as a fraction and a percentage.

$\frac{10}{29}$ or $(10 \div 29) \times 100 = 34\%$ (2 s.f.)

2 A chemistry technician found that an unknown compound contains 26.4% nickel by mass. How much nickel is present in 0.124 kg of the compound, to three significant figures?

26.4% means $26.4 \div 100$

so 26.4% of 0.124 will be $26.4 \div 100 \times 0.124 = 0.032736$ kg

so there were 0.0327 kg of nickel in the compound (3 s.f.)

The Earth's mass is six times the Moon's mass.

Ratios

Ratios are another form of fractions used in solving science problems. For example, ratios are very important in genetics.

Worked example

In an experiment on air quality, it was found that the ratio of the volume of nitrogen to the volume of other gases in the air was 72:15, read as '72 to 15'. An air sample contained 5.2 dm³ of nitrogen. What was the volume of other gases?

Step 1: Let us call the volume of nitrogen $V_{Nitrogen}$ and the volume of other gases V_{Other}.

$$\frac{\text{volume of nitrogen}}{\text{volume of other gases}} = \frac{V_{Nitrogen}}{V_{Other}} = \frac{72}{15} = 4.8$$

Step 2: Now make V_{Other} the subject:

$V_{Nitrogen} = 4.8 \times V_{Other}$

$V_{Other} = V_{Nitrogen}/4.8$

Step 3: Substitute the value for $V_{Nitrogen}$ (5.2 dm³)

$V_{Other} = 5.2/4.8 = 1.1$ dm³. Note that the answer was left to the same number of significant figures as $V_{Nitrogen}$.

Assessment activity 15.3 P1 D1

1 As an environmental officer for the council you have been carrying out field observations. Your data show that the number of birds in one area of town has increased by 8% each year. If the number of birds at the start of the study was 50, how many birds would be counted after the first year? **P1**

2 You are an assistant engineer working on a Moon mission. Your task is to calculate the baggage weight allowance. The ratio of the weight of an object on the Earth to that on the Moon is 6:1, and the mass of baggage can't be more than 2.5 kg.

Weight, in newtons (N), is given by the equation $W = mg$, where m is the mass in kg and $g = 9.81$ N/kg on Earth.

From the equation, work out the maximum weight of baggage on Earth. **P1** Using the ratio, work out its weight on the Moon. **D1**

Grading tip

For **P1**, remember when you are using a percentage in a calculation you need to change the percentage to a decimal or a fraction. To change a percentage to a fraction, just divide it by 100.

Learners often make the mistake of trying to solve ratio questions in one step. To achieve **D1**, learn to follow the three steps in the worked example and remember there is likely to be a unit to include in your final answer.

15.4 Scientific problems involving algebra

In this section: P2 D2

Science focus

The relationship between density, mass and volume can be expressed as a word equation: density = $\frac{\text{mass}}{\text{volume}}$

or in symbols as: $\rho = \frac{m}{V}$

where ρ is the density, m is the mass and V is the volume.

Did you know?

ρ, the symbol for density, is a Greek letter pronounced 'row'.

What is algebra?

Scientists try to understand the world by finding the mathematical rules that describe it. We can write these rules down without using actual numbers. Instead we use symbols, usually letters, to represent quantities and how they link with each other.

We can write an equation in words or in symbols. You need to remember the following points.

- There is an equals sign between the left-hand side (LHS) and the right-hand side (RHS) of the equation. What is on the LHS should always be equal to what is on the RHS.

- We can **transpose**, or rearrange, equations, or we can substitute numbers into them.

- In science, most answers will have a unit as well as a number.

Substitution

This means replacing the symbols with numbers. The worked example will help you understand this.

Worked example

You have a sample of an unknown metal that has a mass of 100.2 kg and a volume of 0.023 m³. Find the density. Give your answer to two significant figures.

Step 1: Write down the equation for density: $\rho = \frac{m}{V}$.

Step 2: Substitute (put in) the values for mass and volume into the equation: $\rho = 100.2/0.023$.

Step 3: Use a calculator to find an answer: $\rho = 4357$.

Step 4: Give the answer to two significant figures: $\rho = 4400$.

Step 5: Include the correct unit (we always need to do this in science): $\rho = 4400$ kg/m³.

Activity A

Weight $W = m \times g$, where m is the mass and g is the acceleration due to gravity. The unit of W, weight, is the newton (N).

Calculate the weight of a girl if her mass is 54.7 kg and $g = 9.81$ m/s². Give your answer to three significant figures and in standard form.

Transposition of equations

Transposing means rearranging the equation. For example, if we wanted to calculate the volume of an object for which we know the density and mass, we would have to rearrange $\rho = m/V$. Remember that if you do anything to the LHS of an equation, you must do the same to the RHS.

- If the symbol that is required is multiplied by another symbol, then divide both sides by that symbol.

For example: voltage = current × resistance, $V = IR$. If we want I as the subject then we divide both sides by R:

$\dfrac{V}{R} = \dfrac{IR}{R}$, which simplifies to $\dfrac{V}{R} = I$

- If the symbol that is required is divided by another symbol, then multiply both sides by that symbol.

For example: number of moles = mass/molecular mass, $n = \dfrac{m}{M_r}$. If we want m as the subject then we multiply both sides by M_r:

$nM_r = \dfrac{m}{M_r} \times M_r$, which simplifies to $nM_r = m$.

- When the equation is of the form $v = u + 2$, and we want to make u the subject, we *subtract* the value we don't want (in this case 2) from both sides:

$v - 2 = u + 2 - 2$
$v - 2 = u$

- The same rule applies if the equation is of the form $v = u - 2$, and we want to make u the subject; we *add* the value we don't want to both sides.

Algebra is involved in investigating how force, mass and acceleration affect the design of a car.

BTEC Assessment activity 15.4 P2 D2

As a technical engineer working for a car manufacturing company, you have to investigate how force, mass and acceleration affect a design. The equation for force (F), mass (m) and acceleration (a) is $F = ma$.

1 If $m = 400$ kg, $a = 5$ m/s², find F. **P2**

2 If the mass is changed to 800 kg, what will the new acceleration be? **D2**

3 The speed of the car is given by the equation $v = \dfrac{s}{t}$, where v = speed, s is the distance and t is the time. Hint: make sure your units are consistent. Calculate the distance travelled in 2 minutes at a speed of 110 km/h, giving your answer to two significant figures and inserting the appropriate prefix. **D2**

Grading tip

For **P1**, remember if you divide by a number on one side of the equation to do the same thing to the other side. For **D2**, make sure you use the SI units of distance (change from kilometres to metres), time (change from minutes to seconds) and speed (change from km/h to m/s) before doing any calculations.

15.5 Mensuration

In science you will have to calculate areas and volumes of different shapes; this is called **mensuration**. For example, the Earth can be thought of as a sphere so we can calculate its volume. From this we can work out its mass and density. Similarly, the cross-sectional area of an artery is nearly circular, so the area through which blood flows can be calculated.

Key term

Mensuration – calculation of areas and volumes.

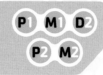

The Earth is nearly spherical.

Blood vessels have a nearly circular cross-section.

Did you know?

Geologists need to be able to work out the volume of the layers inside the Earth to investigate the planet.

Functional skills

Mathematics

This type of question allows you to develop skills in selecting mathematical methods to solve an unfamiliar problem.

Surface areas of some objects

The surface area and volume of a sphere are given by the formulas:

$$\text{Surface area} = 4\pi r^2 \qquad \text{Volume} = \frac{4}{3}\pi r^3$$

where r is the radius of the sphere.

Worked example

The radius of the Earth is 6.4 x 10⁶ m. Calculate its surface area and volume. Give your answer in standard form to two s.f.

Surface area = $4\pi r^2$

Surface area of Earth = $4 \times \pi \times (6.4 \times 10^6)^2$

= 5.2×10^{14} m²

Volume = $4/3 \times \pi \times (6.4 \times 10^6)^3$

= 1.1×10^{21} m³   Note that the units must be included.

The total surface area and volume of a cylinder are given by the formulas:

$$\text{Total surface area} = 2\pi rh + 2\pi r^2 \qquad \text{Volume} = \pi r^2 h$$

where r is the radius of the cylinder and h is the height (or length) of the cylinder.

Worked example

As part of an experiment to investigate blood flow in arteries, a 4 cm long section of an artery of cylindrical shape is found to have a diameter of about 0.44 mm. Calculate the volume of the artery.

Step 1: State the formula for volume of a cylinder:

Volume $V = \pi r^2 h$

Step 2: Change the units to SI units (m)

Length of artery = 4 cm = 0.04 m

Diameter of artery = 0.44 mm = 0.00044 m

Change diameter to radius: radius = diameter/2 = 0.00022 m

Step 3: Substitute values of radius and length in the formula:

$V = \pi \times (0.00022)^2 \times 0.04$

Volume = 6.1×10^{-9} m³

BTEC Assessment activity 15.5 (P1)(M1)(P2)(M2)(D2)

You are working at the National Blood Centre, investigating blood flow in arteries.

If the average mass of blood in the artery section above is estimated to be 6.9×10^{-3} g, use the volume calculated in the Worked example to determine the density of the blood. (P1)(P2)(M2)(D2)

Leave your answer in standard form and to the correct significant figures (M1). Density (ρ) is given by the formula

$$\rho = \frac{m}{V}$$

Grading tip

For this sort of problem, first identify the shape (M2) and write the equation out, then use the basic rules of algebra you learned on pages 282–283 to rearrange the equation to get the subject if necessary (P2). Substituting the values correctly and leaving your answer to the appropriate significant figures and with a unit will go towards meeting (P1), (M1) and (D2).

Activity A

If the average density of the Earth is 5500 kg/m³, use the volume calculated in the Worked example to calculate its mass. Density (ρ) is given by the formula:

$$\rho = \frac{m}{V}$$

PLTS

Independent enquirers

In this type of problem you have to identify a mathematical problem that needs solving, and plan and carry out the calculations.

WorkSpace

Anji Patel
Health Visitor, local primary care trust

I visit new parents and their babies and monitor any problems which crop up, particularly in the early weeks.

We talk to parents about how to feed and care for their babies, and check that the babies are developing normally. We weigh the babies regularly and help the parents to plot the weights on a growth curve. If a baby isn't putting on weight normally this will show up on the curve. I carry accurate scales that can weigh a baby to the nearest 10 g.

Most parents talk about the weight of their babies in pounds and ounces. My scales read weights in kilograms because it is much easier to work in these units – try working out a 5% weight loss in pounds and ounces for example! So I often need to convert between the different units. I'm pretty good at plotting the weights accurately and I teach the parents how to do this as well. Sometimes, worried parents want me to come and weigh a newborn baby almost every day – I have to explain that even my equipment has small measurement errors, so if you are hoping to measure the small change that you get in a baby's weight in a day it won't tell you much.

Think about it!

1 Why is it much harder to work out a 5% weight loss in pounds and ounces?

2 Does it matter that the scales only read to the nearest 10 g?

15.6 Collecting and recording scientific data

In this section: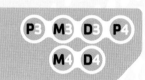

Key terms

Data logger – automated system used to make measurements.

Precision – the smallest change in a quantity that an instrument can measure.

Systematic error – errors in the experimental technique or instrument.

Random errors – chance occurrence in any experiment.

Data collection in space.

Did you know?

The Large Hadron Collider experiment at CERN (the European Organization for Nuclear Research) will collect enough data to fill about 100 000 dual-layer DVDs every year.

Collecting data

Data can be collected manually, for example when an assistant pharmacist uses a pressure chamber and piston to measure the volume of gas given off during a chemical reaction. She records the data as a table in a log book. Alternatively, she could connect a **data logger** to the apparatus. The data logger has a digital sensor that automatically measures the time and the volume of gas and stores the data in a computer.

- When someone collects data themselves they call the data primary data.

- When data are taken from published sources they are called secondary data.

Activity A

Give one example of **(a)** primary data and **(b)** secondary data that you have used.

Errors

The accuracy of any measurement depends on both the skill of the experimenter and the equipment used. The **precision** of an instrument is the smallest change in quantity that the instrument can measure. For example, a string used in a pendulum could be 12.3 cm long according to a 30-cm rule with a precision of 0.1 cm (1 mm). Even if the string is in fact 12.28 cm or 12.34 cm long, you will still read about 12.3 cm with the 30-cm rule. We say that there is a maximum absolute uncertainty or error of half of the precision, in this case 0.05 cm (or 0.5 mm). We can write this error as ±0.05 cm. The string has a maximum length of 12.35 cm and a minimum length of 12.25 cm, so the measurement is written as (12.30 ± 0.05) cm.

Sometimes it is useful to compare measurement errors as a percentage error.

$$\text{Percentage error} = \frac{\text{maximum error}}{\text{measurement}}$$

Experimental results may have errors related to the apparatus (for example, a mass balance that is not zeroed); these are called **systematic errors** and can be eliminated by correctly calibrating or redesigning the instrument. There are other errors known as **random errors** introduced by the experimenter (for example, human reaction time when starting a stopwatch, or failure to let a balance settle before noting the value). Random errors can be minimised by taking averages.

Instrument	Precision	Maximum absolute error
30-cm rule	1 mm	0.5 mm
Measuring cylinder	10 ml	5 ml
Balance	0.1 g	0.05 g
Micrometer	0.01 mm	0.005 mm

Science focus

In science, 'error' does not mean 'mistake'. Error tells us how far a result might be from the true value – the difference is caused by the limitations of the equipment or the technique used.

Activity B

The mass of a sample of calcium carbonate is found to be 2.34 g.

(i) What is the precision of the instrument?

(ii) Write down the maximum and minimum values of the mass.

(iii) What is the percentage error?

BTEC Assessment activity 15.6 P3 M3 D3 P4 M4 D4

As a trainee engineer working for a car manufacturer, you have to investigate the motion of a ball bearing in oil by measuring the distance it falls in a measuring cylinder and the time. You have a rule and a stopwatch to carry out the investigation. You have been told that the mass of the ball bearing is 23.4 g and its average diameter is 1.8 cm.

1 (a) Carry out a similar experiment and display your data in a table. **P3**

   (b) Describe the stages you followed in collecting the data. **M3**

   (c) How would the method of collecting the data be different if you used a data logger? **D3**

2 (a) Write down some of the errors that could occur in the experiment. **P4**

   (b) Calculate the percentage error in the mass and the diameter of the ball bearing using the numbers above. **M4**

   (c) How can the errors you listed be minimised? **D4**

Grading tip

When attempting **P3**, don't forget to include both primary and secondary data. For **M3** and **D3** you need to consider methods of collecting data and refer to how you collected both primary and secondary data. To meet **P4** you must include both random and systematic errors. For **M4**, show your working. A common mistake when attempting **D4** is to generalise and not refer to the experiment.

15.7 Displaying and interpreting scientific data

In this section: P5 M5 D5 P6

Key terms

Continuous data – data that can take any value.

Discrete data – data which can take certain values only.

Tangent – a line that passes through a point on a non-linear graph and has the same gradient as the graph at that point.

Types of data

Data may be **continuous**, such as the thickness of a material, or **discrete**, such as the number of earthworms in a soil sample.

Other data can be a mixture of numbers and names, for example the types of fingerprints found on a glass of water at the scene of a crime (loops, whorls and arches).

Activity A

What kind of data was that measured by the assistant pharmacist on page 286?

Displaying data

A forensic scientist analysing fingerprints needs to show her results to others. She can do this by plotting the data in a bar chart or a pie chart.

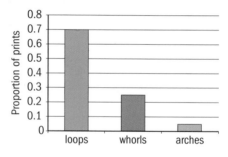
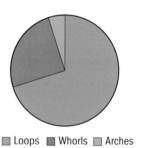

Two ways of showing results of fingerprinting tests. In the bar chart, height represents the numbers of type of fingerprints found at the crime scene. In the pie chart, the fraction of the 'pie' represents the fraction of type of fingerprints.

For some data, such as the experiment carried out by the assistant pharmacist on page 286, a chart would not be appropriate, as we want to see how the gas volume varies with time. A graph is normally plotted for measurements showing how one quantity varies with another. A plot of that experimental data is shown here.

Did you know?

The type of graph shown in the picture is called non-linear, as it is not a straight line.

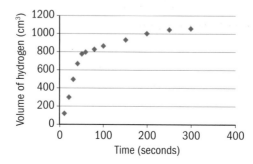

A plot of hydrogen gas versus time for the chemical reaction. The axis must be labelled with the correct units. The data should cover at least 50% of the graph paper. A tangent is drawn at $t = 0$ seconds and a gradient calculated by measuring the change in the y-axis and then dividing by the change in the x-axis. Gradient $= (670 – 0)/(40 – 0) = 17$ cm^3/s.

In this graph, the amount of hydrogen gas increases rapidly at the start, but after 200 seconds the gas is increasing only gradually. We find out how fast the chemical reaction gives off the gas at first by drawing a **tangent** at 0 seconds and measuring the gradient.

The first linear graph shows data for a car travelling at constant speed. In 1 second, the distance travelled was 30 m, and after 2 seconds the distance was 60 m. We say that the distance is proportional to the time. From the gradient we can find out the actual speed travelled over this time. In the second linear graph you can see how area can also give us useful information.

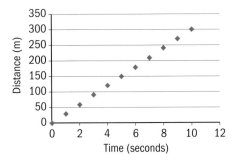

A linear graph (a straight line) for a car travelling at constant speed. The gradient (y/x) shows the speed: 90 m/3 s = 30 m/s.

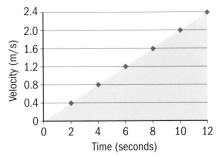

A linear graph showing how velocity varies with time for an accelerating car. The area under the graph represents the distance covered. The shape is a triangle, so area = ½ × base × height = ½ × time × velocity. In this case the area will be ½ × 12 × 2.4 = 14.4 m.

Now consider the non-linear plot showing a radioactive substance decaying with time (on the right).

Science focus

Scientists handling experimental data need to know the definitions of mean, median and mode.

e.g. For the nine values 1.28, 1.24, 1.23, 1.27, 1.23, 1.24, 1.29, 1.23, 1.21

Mean: average value = (1.28 + 1.24 + 1.23 + 1.27 + 1.23 + 1.24 + 1.29 + 1.23 + 1.21)/9 = 1.25 to 2 d.p.

Median: middle number = 1.24 (the fifth number in order)

Mode: number that occurs most often = 1.23

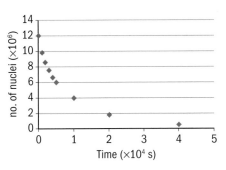

Radioactive decay: At the start of the experiment there were 12 million nuclei. In 10 000 seconds (nearly 3 hours) the figure has gone down to 4 million.

BTEC Assessment activity 15.7 (P5)(P6)(M5)(D5)

You are working as a radiographer in a local hospital. Part of your work is to monitor the age and gender of the patients that visit your department.

1 Prepare a presentation showing how you would display your data, and explain why you chose that format. **(P5)**

One of the radioactive substances you use has an activity curve as shown on the right.

2 (a) Interpret the data displayed in the radioactivity graph. **(P6)**

(b) Interpret the trend in the radioactivity graph, giving reasons. **(M5)**

(c) The area of the plot gives the total number of nuclei that have decayed. Estimate this value. **(D5)**

Grading tip

When displaying data make sure that you label your graphs or charts, otherwise you will not meet **(P5)**. **(P6)** requires you to cover the full timescale of your experiment, not just the start or the end. To meet **(M5)** you must explain the reasons for the trend in your graph. For **(D5)**, make sure that you include the units and give your answer to the correct significant figures.

Just checking

1 What are the SI units for:
   (a) mass   (b) length?
2 Convert
   (a) 10 stones into kilograms   (b) 4 kilometres into miles.
3 Use a calculator to solve the following
   (a) $1.2 \times 10^5 \times 3.2 \times 10^6$ (b) 0.00034/0.023.
   Give your answers to (a) and (b) in standard form.
4 What physical quantity can you find from the slope of a distance–time graph?
5 Give an example from science of a graph that is
   (a) linear   (b) non-linear.
6 Give an example of a random error and a systematic error.
7 The volume of a solution was measured as 32 ± 2 ml.
   (a) What is the precision of the instrument?
   (b) What is the percentage error of the volume?

edexcel

Assignment tips

- Remember most quantities in science have a unit. Without the unit or with the wrong unit the quantity will not make sense, and none of the grading criteria will be met.

- Make sure that when you use a calculator you make use of the brackets, as the order of the calculation will matter. Don't use calculator notation in your answer as that will not be standard form and **M1** will not be met.

- When you have used algebra, mensuration or ratios to solve a scientific problem, you have to show a step-by-step method in order to satisfy the grading criteria **P1**, **D1**, **P2**, **M2** and **D2**; don't just write the answer down without the working.

- When carrying out experiments, make a note of the precision of the equipment and any possible errors that may occur; these factors are important in meeting criteria **P4**, **M4** and **D4**.

- When you plot graphs, make sure that your data points cover at least 50% of the graph paper and that all axes are labelled correctly, with the correct units.

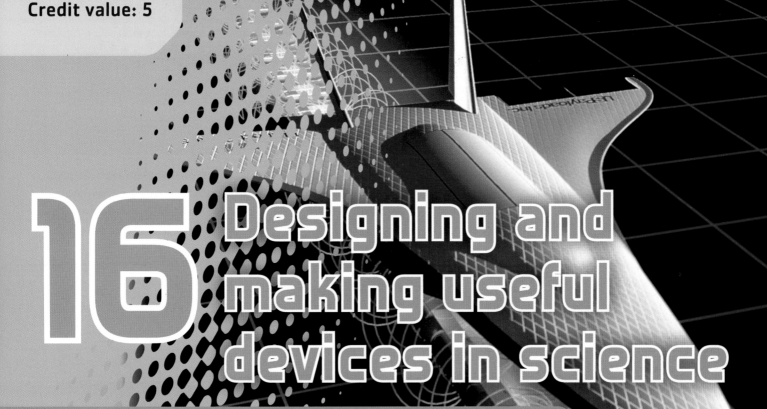

16 Designing and making useful devices in science

In this unit you will use your scientific knowledge to design and make four different devices. You will use a similar design process to that used by scientists, engineers and technicians who work across the whole range of manufacturing industry. The skills that you learn in this unit will provide a good foundation that could lead you to work on innovative new products in the future.

You will design and make:

- a pinhole camera
- a periscope
- a microbalance weighing scale
- batteries to generate electricity.

Just like real designers, you will need to consider how long it will take you to design and build your device so that you finish in time. You will also have to make sure that you work safely. As in real organisations, you will be limited by the materials that are available to you and how much they cost. You will also need to prove that your devices are effective.

Learning outcomes

After completing this unit you should:

1 be able to construct a device to record an image, with available resources

2 be able to construct a microbalancing device, with available resources

3 be able to construct a device for seeing around corners, with available resources

4 be able to construct devices that generate electricity, with various resources.

Assessment and grading criteria

This table shows you what you must do in order to achieve a **pass**, **merit** or **distinction** grade, and where you can find activities in this book to help you.

To achieve a **pass** grade, the evidence must show that the learner is able to:	To achieve a **merit** grade, the evidence must show that, in addition to the pass criteria, the learner is able to:	To achieve a **distinction** grade, the evidence must show that, in addition to the pass and merit criteria, the learner is able to:
P1 Design a pinhole camera **See Assessment activity 16.1**	**M1** Identify future improvements to the camera design	**D1** Explain the science behind the camera design
P2 Produce a pinhole camera providing evidence for its functioning ability		
P3 Design a microbalance **See Assessment activity 16.1**	**M2** Use the microbalance to determine the weight of objects, commenting on the accuracy **See Assessment activity 16.2**	**D2** Evaluate the design of the microbalance, suggesting improvements for the future **See Assessment activity 16.2**
P4 Produce and calibrate a microbalance **See Assessment activities 16.1 and 16.2**		
P5 Design a periscope **See Assessment activity 16.1**	**M3** Identify future improvements to the effectiveness of the periscope design	**D3** Explain the science behind the periscope design
P6 Produce a periscope providing evidence for its effectiveness		
P7 Design batteries which use various resources **See Assessment activity 16.1**	**M4** Explore ways to improve the effectiveness of batteries.	**D4** Assess the impact of batteries on the environment.
P8 Produce batteries using various resources.		

How you will be assessed

Your assessment could be in the form of:

- a portfolio of evidence for each device, to include initial design, completed product, evidence for end product and an evaluation of the effectiveness of the end product
- an article e.g. on the science behind the pin-hole camera and the periscope
- evidence of calibration improvements to your microbalance
- a leaflet e.g. describing the impact of using batteries on the environment and environmentally friendly alternatives.

Alistair, 15 years old

This unit looks fun and interesting. It involves lots of practical work which I will enjoy. It tells you what steps scientists, engineers and technicians work through when designing devices and products.

The designing part of the unit is something I have not done much of before. Making the devices is the part I am looking forward to the most and I think I will really enjoy it.

I think that this unit will help me to prepare for a job which involves either designing or making things. It will also help me decide whether to pursue a career in this area of science and technology.

Catalyst

Imagine you are a design engineer working for a large manufacturing company. You have been asked to design an egg box that will prevent the eggs from breaking, even when it is dropped from table height.

1 What should you consider about the materials you will use when designing or making the device in the future?

2 What health and safety considerations should you take into account?

3 How could you test the device to make sure it works and is effective?

16.1 Designing and planning

In this section: P1 P3 P4 P5 P7

Key terms

Porosity – the measure of how much fluid or gas can pass through a material due to tiny holes in its structure.

Malleability – the property of something that can be worked, hammered or shaped without breaking.

Ductility – a metal's ability to be drawn, stretched or formed without breaking.

Brittleness – having little elasticity; the material is easily cracked, fractured or snapped.

Calibrate – compare the output of a device with a known standard.

Safety and hazards

When you are writing your plan you need to think about any dangerous equipment or risky procedures that might be involved. You then need to decide how you can use preventative measures to keep yourself and others around you safe.

For example, when using a sharp knife to cut card there is a risk of the knife slipping. To prevent this, use a wooden strip as an edge to cut against and always cover the blade immediately after use.

Also, do not wear scarves or ties when using power tools and wear goggles to protect your eyes.

A builder would never start work on building a house without an architect's plan. How would they know how big to build the house or where to divide the rooms?

You are going to design four devices: a camera, a microbalance, a periscope and a battery. You need to consider various factors that will allow you to produce an end product which is functional, effective and safe.

1 Research each device to find what the design parameters are. Think through how the device will be used. For the microbalance this will include the range of masses to be weighed and the type of display.

2 Read about the science behind the principle of how a camera and periscope work. This will help you to specify the materials you need and to produce sketches of how the device will be made.

Unit 2: See pages 56–57 for the science behind batteries.

3 Decide what materials and equipment you will need. For example, for the periscope you will need some reflective mirrors and a tube. Where will you get these from? Also consider the impact of your device on the environment when it is disposed of.

4 Produce a sketch showing where the materials will be used and make some notes on how to assemble them.

5 Consider what health and safety rules need to be followed.

Knowing the science behind how things work is an essential part an engineer's job. This engineer is designing and testing a jet engine to make sure that it works efficiently and safely, e.g. being able to survive a bird-strike on take-off.

Materials

There is a wide range of different materials which you could choose when designing each device. Each material has different properties which makes them useful for some applications and not for others.

Activity A

Why wouldn't cardboard be good for transporting a large amount of liquid? What material would be better to use instead? Why?

Activity B

Why wouldn't you use glass to make a football? What would be a better material to use? Why?

Case study: Designing a miniature telescope

Jake is an optical engineer. He works for a large manufacturing company that makes products including microscopes and telescopes. He is designing a miniature telescope that is easily transported and can be used by small children.

List three things that Jake should consider.

 ## Assessment activity 16.1   **P3**

1   Describe the science that governs how a playground seesaw works. Working in pairs, discuss how you could use this science to design a microbalance. **P3**

2   How could you **calibrate** the microbalance? (Hint: think about weighing a large piece of graph paper with a top-pan balance and then working out the weight of one small square of paper.) **P3** **P4**

Did you know?

Batteries contain toxic materials such as lead and cadmium which will pollute the environment if improperly disposed of. In the UK, we throw away about 20 000 tonnes of general purpose batteries every year. By 2016, 45% of batteries will have to be recycled.

Some properties of materials you may need to think about	
Strength	Stiffness
Weight	Flexibility
Cost	
Malleability	Porosity
Brittleness	Ductility

PLTS

Creative thinkers

You will be able to show how much of a creative thinker you are when you generate your ideas and explore possibilities when planning and designing the devices. You will have to adapt your ideas as you identify what materials are available.

Grading tip

To meet the **P1**, **P3**, **P5** and **P7** grading criteria you need to:

- consider and list which materials and equipment you will use

- produce a labelled drawing of your proposed design, showing the materials and measurements of how big it will be when it is made

- produce clear step-by-step instructions explaining how you will make your device, including any health and safety considerations.

16.2 Producing and evaluating

In this section:

P2 M1 D1 P4 M2 D2

P6 M3 D3 P8 M4 D4

Key term

Evaluating – drawing conclusions by examining data or information gained from an investigation.

A manufacturing apprentice producing a prototype of a new design.

Did you know?

Airbus produced five test aircraft of the A380 'super jumbo'. One of these aircraft was used to make sure that it was safe in extremely cold countries. It was tested at minus 40°C in Canada.

Producing your devices

When producing your devices make sure that:

- your working space is clear and clean before you start

- all of your materials and equipment are neatly arranged to one side of your working space

- you follow your design instructions – remember, if you have any problems they can be adapted as you go along

- you calibrate your balance using known masses.

Evaluating your devices

Once you have made your devices you must test them to make sure that they do the jobs they are supposed to do. This will help you identify any improvements you could make. **Evaluating** a product is an important step for a manufacturing company. It will not stay in business very long if it sells products that do not work correctly or aren't safe. Imagine buying a camera which produces unfocused (blurred) photographs or a weighing scale that is inaccurate.

Device	How to evaluate it
Camera	Produce a photograph and assess the image for focus and brightness
Batteries	Show that electricity can be produced using your batteries and see how long they last before being discharged
Periscope	Test that your periscope can see items around corners or over walls. Is the image bright enough to be useable?
Microbalance	Test your microbalance. How accurate is it?

When evaluating any device ask yourself the following questions.

- Does the device do the job it was designed for?

- How effective or accurate is the device?

- Are there any improvements that could be made to the design or how it is constructed?

- What are the environmental impacts of the device when it is disposed of?

Activity A

List the features of the microbalance that you would use when you evaluate it. Imagine that you had purchased the microbalance and then list the features in order of importance.

BTEC Assessment activity 16.2 P4 M2 D2

1 As a science technician, you have been asked to use your design to produce a microbalance to save money rather than buying one. How could you produce a microbalance quickly on a low budget? **P4**

2 Discuss how you can calibrate your microbalance and research how other scientists do this. **M2**

3 Evaluate the effectiveness of your design and give suggestions for future improvements. **D2**

Grading tip

For the grading criteria for this unit you will create a portfolio of evidence.

For **P2** , **P4** , **P6** and **P8** you need to produce various devices using your designs and instructions. For **M2** you will need to calibrate your weighing scale, comment on how accurate it is and weigh a number of pre-weighed objects to evidence this.

PLTS

Independent enquirers

You will analyse and evaluate information relating to the effectiveness of your end product following construction of the device.

WorkSpace

Mike Bowen
Engineering Manager, JST(UK) Ltd.

STEM AMBASSADORS ILLUMINATING FUTURES

Nationally coordinated by STEMNET

I manage a team of design engineers, laboratory technicians and technical support staff working on the development of electronic connectors. Our work covers design, technical analysis, project management, testing and problem solving. I oversee all engineering work carried out by my team and have to make sure that it's completed professionally and on time. It's my job to ensure that we provide the best technical solution within budget.

When we design a new connector we have to make many decisions about materials and processing methods as well as considering component shape and size. Calculations are made to help predict how the connector will perform when in use.

Before we are ready to start production it's important to test the new connector to ensure that it performs as expected and meets all necessary specifications. Testing can take many weeks and usually includes a combination of electrical, mechanical and thermal tests. When the results are ready we can decide if the connector is safe to use.

Think about it!

1 Why is it important to think about the use of the connector when selecting materials?

2 Why must a connector be safe to use?

Just checking

1 Explain why is it important to think about the properties of the materials you use to make different devices.
2 Write down three important points to consider when planning a design.
3 List three things you must think about when designing a pinhole camera.
4 Name two things you need to do when you are making a periscope.
5 List two ways to provide evidence for the effectiveness of your device.

Assignment tips

To get the grade you deserve in your assignments remember the following.

- In your portfolio provide details for each device of: your initial design, evidence of your completed product, an evaluation of the effectiveness of your end product.

- Explain the science behind how your device works and include suggestions for future improvements to your design.

- Show the effectiveness and accuracy of your microbalance and calibration method.

Some of the key information you'll need to remember includes the following.

- Your materials must be fit for purpose so your end product works effectively and safety.

- Health and safety is important when planning, constructing or evaluating devices.

- An evaluation is very important to provide evidence for the effectiveness of your end product. This evidence needs to be added to your portfolio for future reference.

- Calibration is essential when building a device to ensure it is effective in its function including precision and accuracy.

- Batteries are not good for the environment. Renewable sources of energy are much more environmentally friendly and don't run out.

You may find the following resources useful as you work through this unit.

For information on...	Visit...
designing and making a periscope	Producing a periscope
designing and calibrating a microbalance	Making a microbalance
making your own batteries	Making batteries 1 Making batteries 2 Making batteries 3

17 Chemical analysis and detection

Some chemists will be looking at samples taken from the environment – what hazardous substances are reaching rivers, lakes and the soil? Other chemists will be taking samples to check that the products of a chemical plant are pure enough to sell – they will look for any signs of contaminating substances.

Some analyses will use complex and expensive equipment. In this unit you will learn that many of the techniques used are very similar to practical methods which you can use in your school or college laboratory.

Simple test-tube reactions can show the presence of common inorganic ions while the use of indicators or a pH meter allows acids or alkalis to be detected.

Flame tests and paper chromatography are easy to use in your laboratory. The more complicated methods used in analytical laboratories are based on the same chemical ideas.

Learning outcomes

After completing this unit you should:

1 know the reagents and techniques used to analyse a variety of chemical compounds

2 be able to classify compounds according to their pH

3 be able to show how chromatography is used to analyse materials

4 be able to detect different chemicals in unknown compounds.

Assessment and grading criteria

This table shows you what you must do in order to achieve a **pass**, **merit** or **distinction** grade, and where you can find activities in this book to help you.

To achieve a **pass** grade, the evidence must show that the learner is able to:	To achieve a **merit** grade, the evidence must show that, in addition to the pass criteria, the learner is able to:	To achieve a **distinction** grade, the evidence must show that, in addition to the pass and merit criteria, the learner is able to:
P1 Identify the reagents needed to analyse inorganic chemicals **See Assessment activity 17.1**	**M1** Describe the hazards associated with the reagents needed to analyse inorganic chemicals **See Assessment activity 17.4**	**D1** Explain how to avoid the risks associated with analysing inorganic chemicals **See Assessment activity 17.4**
P2 Describe the techniques needed to analyse inorganic chemicals **See Assessment activity 17.1**	**M2** Explain the results of using these techniques by providing the formula of an unidentified compound **See Assessment activity 17.1**	**D2** Evaluate the accuracy of techniques used to analyse inorganic chemicals and explain how they could be improved **See Assessment activity 17.1**
P3 Carry out experiments to classify compounds according to their pH **See Assessment activity 17.2**	**M3** Explain the uses of the classified compounds in the laboratory and home **See Assessment activity 17.2**	**D3** Explain, with examples, the difference between an acid, a base and an alkali **See Assessment activity 17.2**
P4 Carry out experiments to show how chromatography is used to analyse materials **See Assessment activity 17.3**	**M4** Demonstrate how chromatography works to separate materials **See Assessment activity 17.3**	**D4** Evaluate the advantages and disadvantages of using chromatography to analyse materials **See Assessment activity 17.3**
P5 Carry out experiments to identify chemicals in unknown compounds. **See Assessment activity 17.1**	**M5** Explain the scientific principles behind the tests used to identify the chemicals in unknown compounds. **See Assessment activity 17.1**	**D5** Evaluate the results of the analysis, considering how to improve subsequent experiments. **See Assessment activity 17.1**

How you will be assessed

Your assessment could be in the form of:

- a practical write-up e.g. of your analysis of inorganic compounds, acids and alkalis or unknown chemicals and reagents

- a practical write-up, including explanation of results e.g. of a chromatography investigation

- a risk assessment for your practical work

- a presentation on your results.

Kalil, 19 years old

It has been great in this unit to have so much practical work to do. A lot of it looks really easy but it actually takes a lot of practice to get things like chromatography right. When you use all the tests and other experiments to work out what chemical you've got you almost feel like a police detective.

We found out a lot about the hazards of chemicals – it's scary when you find out how dangerous they can really be. I understand now why our teacher tells us so much about safety when we do experiments.

I was really surprised when we tested some of things you find at home like oven cleaner and even cola and found that they contained really strong alkalis and acids.

What was also really good was our visit to a local chemical company, where we saw some of the modern equipment which they use to test the chemicals they have made.

Catalyst

What is it? Is it dangerous?

Imagine you are looking at a white powder and a colourless liquid. They could just be harmless salt and water but what if they were dangerous substances? In what way might they be dangerous? How could you safely try and find out what they are? In small groups think about these questions.

- What are the different ways in which a chemical could be dangerous?
- What simple and safe tests could you use to find out whether the substances are really water and salt or something nastier?

17.1 Chemical analysis

In this section:

Key terms

Reagent – a chemical which is added to another substance to cause a reaction.

Precipitate – a solid substance formed when solutions are mixed together.

Ionic compounds

Inorganic chemicals usually come from things that were not once alive, like rocks. Many are ionic compounds. Solutions – including liquids like mineral water – contain a mixture of ions dissolved in water.

Ionic compounds contain positive ions (cations) and negative ions (anions). The charges in an ionic compound must balance out, so that the compound is neutral. Some common charges on ions are shown in the tables on these pages.

Unit 4: See page 102 to review ionic compounds.

Precipitation tests

We can test for ions in solutions using **reagents** in test-tube reactions. Many of the tests involve a reaction called precipitation. If a precipitation reaction occurs a solid (called a **precipitate**) is formed from a solution. The solution may become cloudy or you may see a solid settling out. A precipitate is a positive result and shows you that a particular ion is present.

Unit 17: See pages 308–309 for information about the hazards of these reagents and how to assess the risks.

There are other, more advanced, tests for ions like sulfate (SO_4^{2-}), nitrate (NO_3^-) and ammonium (NH_4^+). More information can be found on the Internet. A good site is **Doc Brown's chemical tests**.

The precipitation method isn't reliable if the ions are very diluted – even compounds that aren't very soluble stay dissolved if there is a lot of water. You can try boiling some of the water off to improve accuracy.

Sometimes it can be hard to tell the colours of precipitates apart, for example, precipitates with chloride and bromide ions (see the table). Adding ammonia solution can help you tell apart chloride and bromide ions. The white precipitate formed from chloride ions will dissolve in ammonia solution (also called ammonium hydroxide) whilst bromide precipitates don't dissolve.

Table: Common ions and how to test for them

To test for...	Reagent added	Positive result
Copper Cu^{2+}	Sodium hydroxide	Pale blue precipitate
Iron Fe^{2+}	Sodium hydroxide	Green precipitate
Iron Fe^{3+}	Sodium hydroxide	Brown precipitate
Chloride (Cl^-)	Silver nitrate	White precipitate
Bromide (Br^-)	Silver nitrate	Cream precipitate
Iodide (I^-)	Silver nitrate	Pale yellow precipitate

Activity A

A green precipitate is formed when sodium hydroxide is added to a solution. What ion is present in the solution?

Flame tests

Many metal ions give out brightly coloured light when they are put into a flame. Different metals give out different colours so the colour can help you work out what metal ions are in a substance. This test works best on solids, and is very useful for ions of the elements in groups 1 and 2 of the periodic table.

Impurities in the solid can mask the flame colour of the metal ions you are testing. For example, even tiny amounts of sodium impurity produce a strong yellow colour. To improve the accuracy of flame tests, we use loops of nichrome wire cleaned before use in concentrated hydrochloric acid.

Other tests

There are many other tests used to identify ions. For example, to identify carbonate ions (CO_3^{2-}) you may add an acid to the substance you are testing. If it contains a carbonate ion you will see bubbles of carbon dioxide gas being given off. You can check that the gas is carbon dioxide by bubbling it into limewater – the limewater will go cloudy.

Gases

Some gases are easy to identify if a test tube of the gas is collected.

- A glowing splint placed in a test tube of oxygen gas will relight.
- A test tube of hydrogen gas will burn with a squeaky 'pop' if a lighted splint is placed at the mouth of the tube.

The bright colours seen in fireworks come from different metal ions.

Table: Group 1 and 2 ions and their flame colours

Metal ion	Colour of flame
Lithium (Li^+)	Red
Sodium (Na^+)	Yellow
Potassium (K^+)	Lilac
Calcium (Ca^{2+})	Brick red
Strontium (Sr^{2+})	Red
Barium (Ba^{2+})	Green

Worked example

What is the formula of a compound containing potassium ions and carbonate ions?

First find out the charges on the ions: potassium = K^+ so the charge is 1+, carbonate = CO_3^{2-} so the charge is 2–.

The 2– on the carbonate needs to be balanced by 2+. So two potassium ions are needed for each carbonate.

The formula is written as K_2CO_3.

Activity B

Use the charges on the ions to work out the formula of potassium chloride.

Did you know?

Brackets sometimes need to be used if you need more than one sulfate, hydroxide or carbonate in the formula.

$Cu(OH)_2$ means two hydroxide ions (OH^-) are needed for every one copper ion (Cu^{2+}).

BTEC Assessment activity 17.1 **P1 P2 M2 D2 P5 M5 D5**

1 You are in charge of a team which analyses the waste from a mining operation. You think a stream is contaminated by copper ions and bromide ions, and that the solid waste dumped in a nearby field contains barium ions and carbonate ions.

(a) Write a set of instructions for other team members explaining how the water and solid waste should be analysed. **P1 P2**

(b) Explain how the formula of the compound in the stream and in the solid waste can be worked out. **M2**

2 A student has analysed some water from a river. He thinks that precipitation reactions show chloride ions in the water. He thinks that his flame test results show sodium ions in the water.

Write a short comment on his work to explain why his conclusions might not be completely reliable and what he should do to improve the reliability and accuracy of his analysis. **D2**

Grading tip

For **P1** make sure that you clearly name which reagent your team members should use in which tests.

For **P2** give accurate details of the tests that should be used to analyse the different chemicals.

For **M2** remember to think about the charge on each ion.

For **D2** think about the accuracy of the techniques used.

17.2 Acids and alkalis

In this section: P M D

Key terms

Acid – a substance with a pH of less than 7.

Alkali – a base dissolved in water; alkalis have a pH of greater than 7.

Base – a substance which neutralises an acid.

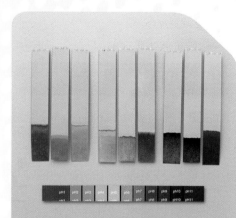

Mixing toilet cleaners can release poisonous chlorine gas.

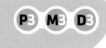

The pH of a liquid can be measured using universal indicator paper.

The hazards of cleaning products

Cleaning toilets and drains is an unpopular but very important job in the home or the workplace. It can also be dangerous – if you don't read the labels. Some toilet cleaners are acids and others are alkalis. They can react together if they are mixed and in some cases release toxic gases like chlorine.

Identifying acids and alkalis

You can do this easily using universal indicator paper or solution. The colour of the solution or the paper is compared with a chart. This will tell you the pH of the solution. A pH below 7 means the liquid is an **acid**, a pH above 7 means it is an **alkali**. If the pH is exactly 7 then the liquid is neutral.

Uses of acids and alkalis

Acids and alkalis are used in many different ways in the home and the laboratory.

In the laboratory

Important uses of acids include acting as catalysts in some reactions, like the breakdown of proteins and sugars. Alkalis, like sodium hydroxide, can also sometimes act as catalysts in these reactions and are also used in chemical analysis when testing for ions.

Unit 17: See pages 302–303 for more information on testing for ions.

In the home

Acid name and formula	Use
Hydrochloric acid (HCl)	Removing limescale
Phosphoric acid (H_3PO_4)	In cola drinks
Ethanoic acid (acetic acid) CH_3COOH	In vinegar (to preserve and flavour food)

Alkali name and formula	Use
Sodium hydroxide (NaOH)	As an oven cleaner (reacts with grease)
	In bleach
Ammonia (NH_3)	In window-cleaning fluids

Activity A

Two solutions used in home cleaning have lost their labels. Some students measured the pH of the solutions to try and find out what they are. Solution A has a pH of 2, solution B has a pH of 10. Which solution was the oven cleaner and which solution was the limescale remover?

The reactions of acids, alkalis and bases

When acids react with alkalis or **bases** they neutralise each other to produce substances called salts. A salt contains an ion from a base (usually a metal) combined with an ion from an acid. Some water is also produced in the reaction. A lot of heat can also be produced when acids and alkalis react and this is one reason why these neutralisation reactions can be dangerous.

You can see how this works in the following table:

Acid	Base	Salt
Hydrochloric acid	Sodium hydroxide	Sodium chloride
Sulfuric acid	Calcium hydroxide	Calcium sulfate
Nitric acid	Copper oxide	Copper nitrate

Sodium hydroxide and calcium hydroxide can dissolve in water. When they are dissolved in water they are called alkalis. Copper oxide doesn't dissolve in water at all. It can only be called a base.

Writing chemical equations

The word equation for a neutralisation reaction is:

Acid + base (or alkali) → salt + water

If you put in the formulas of the substances you can write a balanced equation.

Word equation:
hydrochloric acid + sodium hydroxide → sodium chloride + water

Balanced chemical equation: HCl + NaOH → NaCl + H$_2$O

Activity B

(i) Name the salts produced when these acids and bases react:

(a) sulfuric acid + potassium hydroxide

(b) hydrochloric acid and iron oxide.

(ii) Write a word equation and a balanced chemical equation for these reactions:

(a) hydrochloric acid + potassium hydroxide

(b) sulfuric acid + calcium oxide.

Assessment activity 17.2 P3 M3 D3

Write a short guide for customers of a supermarket to explain why household cleaning products contain acids and alkalis or bases.

Explain how the substances can be tested to check what they contain and what the names of the contents mean, and describe what will happen if acids and alkalis react together. P3 M3 D3

Grading tip

For P3 and M3 make sure you find examples of both acids and alkalis. Use research from the Internet to try and find out more information to explain why they are used in your chosen examples.

For D3 you will need to explain what is meant by the words acid, base and alkali; you should also try and include equations (word and balanced) to explain what happens when they react.

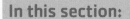
17.3 Spot the difference – chromatography

Spots of ink samples are dropped onto a piece of paper, which is then placed in a solvent. The pattern of spots produced is called a chromatogram. Chromatography can show whether a substance is pure, and can separate mixtures and identify their compounds.

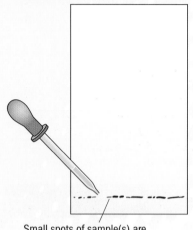

Small spots of sample(s) are dotted onto plate or paper

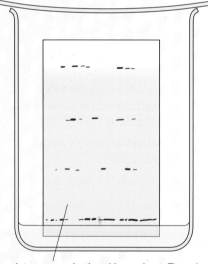

The plate or paper is placed in a solvent. The solvent rises up, taking the substances in the samples with it.

The stages in chromatography.

Paper chromatography

Chemists often need to find out whether a sample is pure. Medicines and food products must be tested very carefully before they are sold. Forensic scientists often need to find out whether a substance (e.g. ink) found at a crime scene and at a suspect's home are the same. Paper chromatograpy separates out the different dyes in the ink. By testing the two sources of ink at the same time, forensic scientists can see whether the inks contain the same dyes.

Thin–layer chromatography (TLC)

Very often, chemists prefer to use thin-layer chromatography (TLC) to test samples. Instead of paper, the sample is spotted onto a glass or plastic plate covered in a white solid (called silica gel).

How does chromatography work?

When the paper or TLC plate is placed in the solvent, you will see the solvent start to rise up. As it passes the sample, the different substances dissolve in the solvent and are carried up with it.

Some substances are attracted more than others to the paper or the silica. These move more slowly upwards, and so the different substances reach different heights by the time the plate is removed from the solvent. If two spots are at the same height, this will help chemists to decide that they are the same substance.

Did you know?

Although the word 'chromatography' comes from a Greek word meaning coloured, the process can be carried out on colourless substances. For colourless substances the spots need to be made visible; ultraviolet light is sometimes used.

Case study: Safe to eat?

Josh works for the environmental health department of his local council. They are worried that some of the fruit on sale in the local market may contain some pesticide which could make people ill if too much of it gets into their body.

He mashes up the skin of a fruit and filters it to make a solution, then spots it onto a TLC plate.

Several spots show up on the plate. How can he tell whether any of them are the pesticide?

Evaluating chromatography

In modern chemical laboratories there are a whole range of advanced chromatography methods used to overcome some of the problems with simple chromatography.

Problems with simple methods	How they are overcome using advanced methods
If the sample is not coloured then the spots cannot be seen easily	Ultraviolet light is shone on the chromatogram – the spots show up by glowing a different colour to the plate
Sometimes the different substances do not separate very well	A method called high-performance liquid chromatography uses high pressure to help separate the substances
It cannot be used for gases and liquids which evaporate quickly (they will evaporate from the plate before they can be detected)	A method called gas–liquid chromatography is used for liquids which evaporate easily
Although it will separate the different substances from each other, it is then difficult to get them off the chromatogram to collect them	A method called column chromatography makes it easy to collect the separated substances at the bottom of a glass column

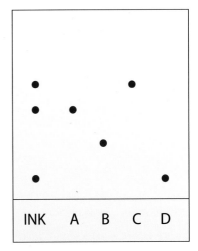

INK   A   B   C   D

The chromatogram of a sample of ink. Pure samples of dyes are also put onto the plate to compare with the ink.

Activity A

(i) Study the chromatogram of a sample of ink. Give the letters of the dyes which are probably present in the ink sample.

(ii) Which dye is most strongly attracted to the silica gel of the plate? Explain your answer.

 Assessment activity 17.3   P4 M4 D4

Imagine that you work for the Forensic Science Service. Prepare a demonstration and presentation which you could give to a group of science students to describe how you test the ink on documents to find out whether it matches the ink found in a suspect's pen.

Include practical details of a simple chromatography technique and also explain why more advanced techniques are now used.
P4 M4 D4

Activity B

Research advanced methods on the Internet. A useful website to start from is **Advanced chromatography**

Grading tip

For P4 you need to describe how you carried out chromatography. Remember to add important details like exactly where the spots of solution should be dropped.
For M4 you will need to demonstrate how chromatography works. You should also explain what sort of results you would expect if the two inks matched up.

You should be able to suggest some problems with the simple chromatograpy method using the information on these pages but for D4 you will need to research a method which might give more reliable and accurate results.

17.4 Hazards and risk assessment

In this section: M D

Key terms

Hazard – a chemical is a hazard if it is able to cause harm in some way.

Risk assessment – a document which lists the risks in an activity and the ways of avoiding or reducing these risks.

Did you know?

It is a legal requirement that any bottle of a hazardous substance must be labelled with the appropriate symbol.

Practical activity: *testing for metal ions*

Procedure: *add 0.5 cm³ of test reagents to 2 cm³ of unknown solution*

Substance being used	Nature of hazard	Steps taken to minimise risk
Sodium hydroxide solution	Irritant – particularly to the eyes	Wear eye protection Wipe up spills immediately Wash off skin immediately

A full **risk assessment** needs to be carried out before a chemist carries out any practical work.

Chemical risk assessments

Before anyone is allowed to handle chemicals at work, it is a legal requirement that an employer carries out a risk assessment on the chemicals and the way in which they will be handled. Of course, as well as these precautions, you must follow all the usual laboratory rules, such as not eating and drinking to avoid any danger of swallowing the chemicals, and wiping up any spills immediately.

Hazard symbols

Different chemicals pose different **hazards**. Some hazard symbols are shown below. There are also symbols for toxic, flammable, explosive and oxidising, which you find on chemical containers.

Symbol	Stands for	Meaning	Example	Safety precautions
	Corrosive	Destroys living tissues on contact	Sodium hydroxide	Wear eye and skin protection Wash off skin immediately
	Harmful	May have health risks if breathed in, swallowed or absorbed through the skin	Copper sulfate	Wear eye protection, wash off skin immediately Handle in a fume cupboard or well-ventilated lab
	Irritant	Not corrosive but can cause reddening or blistering of the skin	Iron sulfate	Wear eye protection Wash off skin immediately

More information about chemical hazards is, of course, widely available on the Internet. A good place to start is **About chemical hazards**. A full list of the hazards of chemicals can be found at **Chemical safety database**.

Assessment activity 17.4

Choose three of the chemical analyses given on pages 302–303. For each one write a full risk assessment that identifies the hazards and describes what safety precautions will be necessary. **M1** **D1**

Grading tip

For **M1** a good place to start would be the hazard symbols on the bottles of chemicals from your school or college prep room.

For **D1** you will need to provide more detail about the hazards and how to avoid them. You may want to find out more about each chemical from the Internet – but be careful only to include hazards and risks that are relevant to the experiment you are carrying out – for example, if you are using a substance in solution then there is no need to include information about the hazards of the solid.

If corrosive substances like sodium hydroxide are not handled in the correct way they can cause serious burns.

WorkSpace

Adam Hughes

Chemist, BAE Systems Submarine Solutions

STEM
AMBASSADORS
ILLUMINATING
FUTURES

Nationally coordinated by STEMNET

I work within a small team of chemists. We use a variety of techniques to analyse gases, liquids and solids that go into building nuclear submarines.

One of the good things about the job is that there is no such thing as a typical day. There is some routine testing, but most tasks are non-routine.

I go on board the submarines to take samples, including water samples. One important test we carry out on water samples looks for fluoride, chloride and sulfate ions, as they can damage stainless steel. We do this using a technique called ion chromatography, which uses a resin column to separate the negative ions according to their size and electrical charge. The equipment shows the results as peaks on a graph. Smaller fluoride ions appear from the column first. The bigger sulfate ions appear last, with chloride in the middle. The time at which the ions appear identifies the type of ion, and the size of the peak the amount.

Think about it!

1 If you passed a negatively charged ion and a neutral molecule of the same size through a column containing positively charged resin, which would pass through quicker? Why?

Just checking

1 A solution that contains an unknown substance produces a white precipitate when silver nitrate solution is added and produces an orange flame colour.
   (a) What ions are present?   **(b)** What is the name of the substance?
   (c) What is the formula of the substance?

2 Name **(a)** an acid that you might find in the home and **(b)** an alkali that you might find in the home. Describe one use of each substance in the home.

3 **(a)** Name the two simple chromatography methods used in school or college laboratories.
   (b) List one problem with the use of these methods to identify substances and explain how the problem could be solved.

Assignment tips

To get the grade you deserve in your assignments remember to do the following.

- Know that a proper chemical risk assessment describes hazards and explains how to avoid the risks.

- To achieve criteria **P5**, **M5** and **D5** you will need to look back at all of the work you have done in this unit and use it in writing your report. The assignments will test ideas from across the whole unit.

- When you are analysing chemicals, as for **P5**, make sure that you know how to identify the ions in a solution and how to use indicator paper to find pH.

- When explaining the scientific principles in these experiments, as for **M5**, remember that different ions produce different-coloured precipitates, and acids and alkalis react with the dyes in the indicator.

- When you are evaluating the experiments you have done, as for **D5**, remember that some tests do not always work reliably. You should be able to suggest ways of making them more reliable.

Some of the key information you'll need to remember includes the following.

- Ionic compounds contain both negative and positive ions. If you get a positive test for each one then you can easily combine them to work out the name and formula of the compound.

- When an acid is neutralised by a base, a new compound called a salt is produced, as well as some water.

- In chromatography some substances move more slowly than others because they are more strongly attracted to the plate or the paper.

You may find the following websites useful as you work through this unit.

For information on...	Visit...
hazards of chemicals	About chemical hazards
testing for the ions in inorganic chemicals	Inorganic ion tests
acids, bases and salts	Acids, bases and salts
chromatography	Paper and advanced chromatography

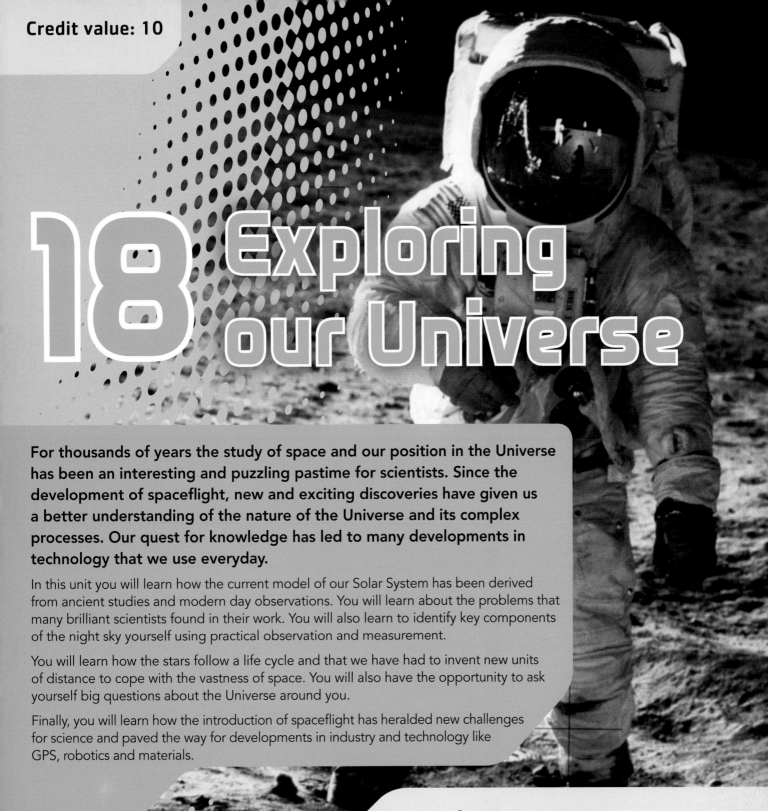

18 Exploring our Universe

For thousands of years the study of space and our position in the Universe has been an interesting and puzzling pastime for scientists. Since the development of spaceflight, new and exciting discoveries have given us a better understanding of the nature of the Universe and its complex processes. Our quest for knowledge has led to many developments in technology that we use everyday.

In this unit you will learn how the current model of our Solar System has been derived from ancient studies and modern day observations. You will learn about the problems that many brilliant scientists found in their work. You will also learn to identify key components of the night sky yourself using practical observation and measurement.

You will learn how the stars follow a life cycle and that we have had to invent new units of distance to cope with the vastness of space. You will also have the opportunity to ask yourself big questions about the Universe around you.

Finally, you will learn how the introduction of spaceflight has heralded new challenges for science and paved the way for developments in industry and technology like GPS, robotics and materials.

Learning outcomes

After completing this unit you should:

1 know how ideas about the Universe have developed
2 understand the processes involved in astronomical theories
3 know the methods used for space exploration
4 be able to investigate astronomical objects.

Assessment and grading criteria

This table shows you what you must do in order to achieve a **pass**, **merit** or **distinction** grade, and where you can find activities in this book to help you.

To achieve a **pass** grade, the evidence must show that the learner is able to:	To achieve a **merit** grade, the evidence must show that, in addition to the pass criteria, the learner is able to:	To achieve a **distinction** grade, the evidence must show that, in addition to the pass and merit criteria, the learner is able to:
P1 List the main historical developments and discoveries in astronomy **See Assessment activity 18.1**	**M1** Describe some major historical changes which have occurred in astronomical concepts **See Assessment activity 18.1**	**D1** Give examples of how the inadequacies of early instruments and/or mistaken preconceptions influenced our understanding of the Universe
P2 List the main characteristic components of the present day model of the Solar System **See Assessment activity 18.1**	**M2** Explain the main features of the Earth–Moon system	**D2** Explain, non-mathematically, the main gravitational influences within the Solar System
P3 List the principal stages in the lives of most stars **See Assessment activity 18.2**	**M3** Describe the principal stages in the lives of most stars **See Assessment activity 18.2**	**D3** Explain the differences between the lives of those stars with initial masses significantly smaller and those with initial masses greater than that of the Sun **See Assessment activity 18.2**
P4 List the main methods of measuring astronomical distances from within the Solar System to the edge of the observable Universe **See Assessment activity 18.2**	**M4** Explain the theory and practice of two methods of measuring astronomical distances that are based upon different physical principles **See Assessment activity 18.2**	**D4** Explain the astronomical distance ladder showing how uncertainties accumulate as the distance becomes larger **See Assessment activity 18.2**
P5 Review the theories for the origin and fate of the Universe **See Assessment activity 18.3**	**M5** Explain the factors which determine the fate of the Universe **See Assessment activity 18.3**	**D5** Explain, with examples, which of the current Big Bang variants is best supported by the evidence **See Assessment activity 18.3**
P6 Name the principal methods used in the exploration of space **See Assessment activity 18.4**	**M6** Explain why unmanned exploration is preferred to manned exploration at the moment **See Assessment activity 18.4**	**D6** Explain the developments likely to influence future space missions. **See Assessment activity 18.4**
P7 Carry out practical investigations collecting primary data on some of the characteristics of the night sky **See Assessment activity 18.5**	**M7** Analyse your investigations, drawing any possible conclusions. **See Assessment activity 18.5**	
P8 Investigate using secondary sources some of the characteristics of our Universe. **See Assessment activity 18.5**		

How you will be assessed

Your assessment could be in the form of:

- a written account, timelines or video presentation on the history of astronomy

- models, posters, data tables or computer presentations e.g. on the Solar System

- a report on e.g. the measurement of astronomical distances

- a report or presentation e.g. on the life of the Universe

- a log-book e.g. for your observations of the night-sky.

Louis, 18 years old

I really enjoyed learning all about space and how technology has changed. Some of the pictures taken by the Hubble space telescope are amazing and many facts that we learned in the section are almost unbelievable. I found that I needed practice looking at the night sky without a telescope before I was able to complete any real observations.

I have always been interested in spaceflight and hope that we will all have the chance to go into space in the future. My group in class discussed how the Universe might end and we came up with our own theory about other possible universes.

By studying this unit I am now able to name many of the objects in the night sky quite easily and I can remember a lot of facts about the biggest and brightest ones.

Catalyst

Is the Earth flat?

In some parts of the world many people still believe the world is flat. We have the benefit of modern technology and education, but how would you convince others that the world is round? Is there any evidence that you can use to argue the case? Apart from satellite photographs, is there any other way of proving that the world cannot be flat? Discuss this in groups and try to support your points. Once you have finished, make a PowerPoint presentation to your class.

18.1 Models of the Solar System

 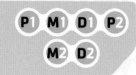
Our understanding of the Earth and its position in the Universe has developed over thousands of years. It was not until the invention of telescopes and satellites that scientists' ideas and theories could be properly tested or proved. As a result, many old ideas about our place in the Universe have changed.

Key terms

Moon – a large object having its own gravity which orbits a planet.

Planet – a larger object which orbits the Sun shining by reflected light.

Activity A

Describe the main differences between the two Solar System models.

The Universe according to Ptolemy, A.D. 120–180 from an established print A.D. 1600. Compare this with the current model of the Solar System shown on page 63.

Date	Development/discovery
A.D. 150	Ptolemy produces his model of the Universe
1543	Copernicus' new Solar System model
1608	Lippershey invents telescope
1610	Galileo observes Sun and planets
1781	Herschel discovers Uranus
1923	Hubble discovers galaxies are accelerating
1957	The first satellite, Sputnik 1, launched
1961	First manned space flight
1969	First manned moon landing
1990	Hubble space telescope launched
1998	In-orbit construction of International Space Station (ISS) begins

Table: key dates in the history of astronomy.

Activity B

Using the Internet and books, investigate Copernicus' new model of the Solar System. How does it differ from Ptolemy's model of the Universe?

Features of the Solar System

The Solar System is made up of:

- nine recognised **planets** (actually eight planets and one dwarf planet) and their **moons**

- **asteroids**

- **comets**

- **meteors**.

All of these features orbit the Sun as a result of the Sun's gravitational influence. Any object which has a mass also has a gravitational field. This is noticeable to us only when the object is very large, like a moon or planet, and we can see the effects. This means that all the objects in the Solar System are affected by the gravity of each other as well as the Sun.

Table: Solar System data.

Planets	Average distance from Sun (million km)	Diameter (Earth = 1)	Relative mass (Earth = 1)	Surface gravity (N/kg)
Mercury	58	0.4	0.1	3.8
Venus	108	0.95	0.8	9
Earth	150	1.0	1.0	10
Mars	228	0.5	0.12	3.8
Jupiter	778	11	320	26
Saturn	1427	9	95	11
Uranus	2870	4.01	15	11
Neptune	4497	4	17	12
Pluto	5900	0.2	0.002	0.6

The Earth–Moon system

The Earth is 12 756 km in diameter, spins on its axis once in 24 hours and takes just over 365 days to complete one orbit of the Sun. The Moon, in comparison, is 3479 km in diameter and orbits the Earth at an average distance of 376 284 km. It spins on its axis slowly, once in just over 27 days, and also takes the same length of time to go once around the Earth. This is why we can only ever see one side.

Did you know?

The gravitational forces between the Earth and the Moon are having a serious effect on both of them. The Earth's rotation around the Sun is slowing down and the Moon is getting further away from the Earth at a rate of 3 m every 100 years.

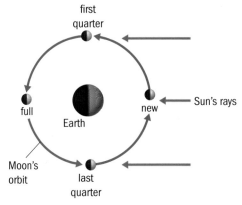

The phases of the Moon at various points around a central Earth, with the Sun shining on one side.

BTEC **Assessment activity 18.1** **P1** **M1** **P2**

As a junior writer for a science journal, you must present an article showing the developments in astronomy and our current knowledge of the Solar System. In your article, answer the following questions.

1. Why did more accurate and reliable observational astronomy develop after 1600? Give examples. **P1** **M1**

2. Investigate the main differences between the planets inside the asteroid belt and those outside. There is one planet which does not fit this pattern. Why is this? **P2**

Grading tip

You will need to list all the main historical developments and discoveries shown in the timeline to achieve **P1** . For **M1** you will need to describe at least three changes in astronomical concepts. For **P2** you should include the differences in the sizes of the planets and what they are made of. Look at the orbital path and the fact that Pluto is essentially made of rock and ice to help you.

Science focus

The Moon has a strong influence on the Earth because of its mass. As it orbits the Earth, it pulls on the oceans, causing the water to overlap the land. The Sun has a smaller tidal effect on the oceans but when the Sun and Moon are lined up the result is a large Spring Tide.

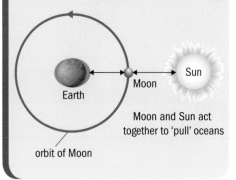

Moon and Sun act together to 'pull' oceans

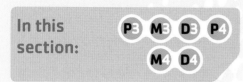
18.2 Understanding processes in the Universe

In this section: P3 M3 D3 P4 M4 D4

Key term

Nuclear fusion – the process of joining the nuclei of atoms which is carried out in stars and is responsible for a star's light.

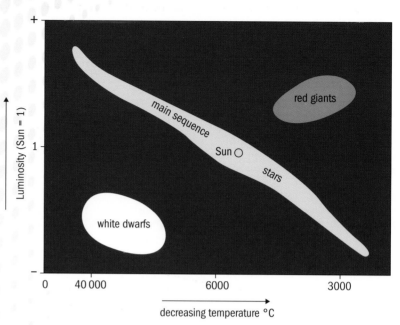

Simplified Hertzsprung–Russell diagram. This diagram shows the different stages of life for a star.

The exciting life of stars

Stars begin their lives when clouds of gases are pulled together by gravity. The star begins to shine brightly as a **nuclear fusion** reaction giving out heat and light starts. From this moment, the future life of a star depends on its mass. If it has a small mass, like our Sun, it will shine for many millions of years then expand to become a **red giant** and end its life as a small **white dwarf**.

If it has a large mass, it will use up its fuel quite quickly and may explode as a **supernova** to finally end its life as a **neutron star** or even a **black hole**.

Measuring distances

Measuring the distance to even the closest objects in space is difficult. Most methods rely on observing the light from the object and some maths. The distances are so great that a whole new set of units has been developed.

Did you know?

A black hole is not a hole – it is a star that has come to the end of its life but is incredibly dense. Its gravitational pull is so strong that not even light can escape from it.

Activity A

What process causes stars to shine?

Activity B

Work out how many astronomical units there are in one light year.

Table: Units used for measuring distances in space

Name of unit	Meaning
Astronomical Unit	Average distance of the Earth from the Sun, approximately 150 million km
Light year	The distance light travels in 1 year, approximately 9 million million km
Parsec	3.26 light years or about 31 million million km

When we look at a distant object with one eye closed and look again with the other eye closed, we see a slightly different view against the background. Astronomers use this principle to view a nearby star against distant background stars at two points in the Earth's orbit around the Sun. The angle made can then be used in trigonometry to work out the distance to the star. This is known as the **trigonometric parallax method** of distance measuring.

Astronomers use many different ways of measuring distances. When the light of a distant object is analysed (stellar spectroscopy), lines which appear in its electromagnetic spectrum can be shifted to the red or blue end, compared with a laboratory spectrum. This can tell scientists how far away the object is and also whether the object is accelerating away from our Solar System or towards it.

Unit 2: See pages 64–65 for more information on red shift as well as the life of the Universe.

Astronomical distance ladders

It is very difficult for astronomers to measure all distances using the same method. It is quite easy, for example, for you to estimate the length of your classroom in metres but a lot more difficult to estimate the length of your school unless you have other objects in the area of known distance.

Astronomers use a variety of methods for different distances. Each significant change in distance can be seen as another rung up the **distance ladder**. It is much more likely that distances measured to nearby objects are more accurate than those measured to objects in deep space.

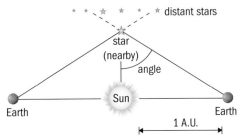

Trigonometric parallax method of distance measuring.

Table: some methods used to measure distances in space

Object	Methods used to measure distance
Planets in our Solar System	Planetary probe data and orbital observations
Nearest stars	Parallax
Other objects in our galaxy	Cepheid variables, star spectroscopy
Other galaxies	Cepheid variables, supernovae, spectroscopy
Deep space	Bright galaxies, galaxy spectroscopy

BTEC Assessment activity 18.2 P3 M3 D3 P4 M4 D4

Your work in a planetarium includes organising a visit and presenting a talk on processes in the Universe and distance measurement for local schools. Write a presentation which includes the following:

1 Draw a diagram which shows the main stages in the lives of most stars. **P3** Show how astronomical processes can influence a star's life on your diagram. **M3 D3**

2 Produce a PowerPoint presentation which outlines and explains a method to measure the distance to a nearby star and a method to measure the distance to a galaxy. You may find the following websites useful: **Astronomical distances** and **Measuring the distance to a galaxy**. **P4 M4**

3 Using the Internet, provide an explanation of what the astronomical distance ladder is and why many other methods are used for larger distances. **D4**

Grading tip

Remember to include details from the Hertzsprung–Russell diagram to achieve **P3**. Include the stages of all types of stars and details of gravitation for **M3**. For **D3** think about mass differences and what these can lead to. To achieve **P4**, a clear diagram of the trigonometric parallax method will be useful. **D4** requires some description of the limitations for parallax methods at greater distances. You should also include uncertainties in using brightness of stars, variable stars and supernovae.

18.3 The beginning and the end of the Universe

In this section:

Key terms

Big Bang – theory which claims that the Universe began from an explosion.

Red shift – when light from an object which is moving away from us is analysed, the spectral lines show a slight movement in the red end of the spectrum.

Critical density – the amount of mass and energy needed in a given volume of space to 'close the Universe'.

How the Universe may have begun.

How the Universe began and how it will end are questions that may never be answered. When accurate observations of the stars and galaxies were made at the start of the 20th century, it was discovered that the galaxies were moving away from each other. This means that the Universe is expanding. Two theories about the nature of the Universe developed: the Steady State theory and the **Big Bang**.

1   The Steady State theory – the Universe will expand into itself, gaps will continue to be filled and there is no beginning and no end. Try to imagine yourself in a swimming pool with no boundaries – up, down or side to side. As you move through water more water fills the space that you moved from.

2   The Big Bang – everything began from one single point in space and time. This theory has considerable observational evidence. For example, distant galaxies are moving faster as if they are still moving apart after the initial explosion (measured using **red shift** techniques discussed in Unit 2).

The most accepted theory is the Big Bang, although there are many problems with it. One problem is that there isn't as much mass or energy in the Universe as would be expected. It is thought that material we can't see or detect in space (dark matter) might make up some of this extra mass and that dark energy accounts for the energy difference.

Activity A

Working with a partner make a poster explaining the Big Bang theory. Use the Internet to research supporting evidence and problems with the theory and add these to your poster.

Other theories have also been put forward which are mainly variations of the Big Bang. For example, according to **inflationary theory**, time had already begun and the Big Bang happened within it. This is different from the original theory in that time is thought to have started at the same time as the Big Bang.

Inflationary theory is a relatively new addition to the Big Bang concept and it may help to explain why observations show that the Universe is expanding continually.

Cosmologists think that the end of the Universe will depend on how dense it is. They have given three possible outcomes:

• Universe density is less than **critical density** – it will expand forever and fade away (open)

- Universe density is equal to critical density – it will begin to slow down but never stop (flat)

- Universe density is greater than critical density – it will slow down and be pulled back to a pin point by gravity (closed).

However, until the extra mass and energy are accounted for, accurate measurements of the real density of the Universe cannot be achieved.

BTEC Assessment activity 18.3

1 Gather together information about other variants of the Big Bang theory on the origin of the Universe. List the differences and similarities between them and decide which theory appears to have the most supporting evidence. You may find the following website useful: **Big bang theory**. **D5**

2 Discuss as a group and make notes on how the Universe may have begun and how it may end. **P5 M5**

Grading tip

Review the different theories clearly for **P5**. Most will be very similar to the main theory of the Big Bang. Use information on these pages to help you to decide how the Universe may end for **M5** but you will also need some extra information about density measurements. Provide a detailed table of differences and similarities for **D5** and use as many research sources as you can.

WorkSpace

Dr John McCormick

Lead Antenna Engineer, SELEX Galileo

STEM AMBASSADORS ILLUMINATING FUTURES

Nationally coordinated by STEMNET

I am in charge of Research and Development of active electronically scanned array (AESA) antennas. I assign tasks to other members of staff, review and report progress and develop research plans.

In addition I advise each of the research programmes that I manage.

Aircraft, satellites and spacecraft need sophisticated antenna sensor arrays to sense their environment. Spacecraft must also be able to communicate with their mission controls, so they need to send signals across space.

For remote sensing, spacecraft in orbit send radiofrequency signals into the space around them. They analyse the reflections of these signals to sense what is around them or on planets below them.

The same sensor array technology is used on a larger scale for ground-based radioastronomy, where large arrays search far out into space. Most efforts to contact extraterrestrial civilisations use antennas to send radiofrequency signals in the hope that aliens will have the technology to sense these signals and reply using their own antennas.

Think about it!

If humans are to explore the Universe, why do you think it is important to be able to:

1 communicate across vast distances?

2 sense the environment of planets and areas of space that are very remote from Earth?

18.4 Journey into space

In this section: P6 M6 D6

Key terms

Electromagnetic spectrum – the full range of different wavelengths usually from radio waves to gamma rays (see page 50).

James Webb Space Telescope (artist impression) due for launch, 2014.

Science focus

Charge Coupled Devices or CCDs are light-sensitive integrated circuits. They are made of many pixels which convert light from an object into a small electrical charge. The image is improved and becomes much clearer.

Activity A

Using the Internet, find out which wavelengths are used by the following instruments: Arecibo, COBE, IRAS, Hubble space telescope, ROSAT, Chandra space telescope, Compton space telescope.

Satellites and instruments

Our knowledge of space really began using ground-based instruments. Modern space exploration is now carried out using advanced telescopes and instruments located both on the ground and in space. We can now study astronomical objects close up using space probes and without the interference of the Earth's atmosphere.

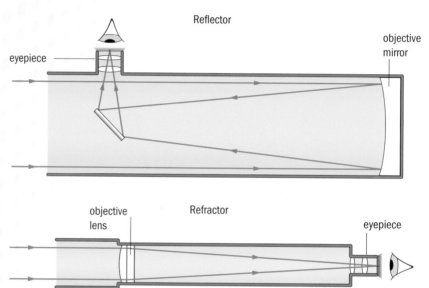

Both these telescopes use optical methods (visible light, mirrors and lenses) for observing astronomical objects.

New methods for discovering new objects

Many ground-based and orbiting telescopes are advanced versions of the basic designs shown above, but new methods of observation have also been developed over the last century.

Many instruments in use or planned for the future are designed to study the Universe using other wavelengths (not visible light) from the **electromagnetic spectrum**. This allows us to 'see' more detail.

Information Computer Technology (ICT) has made a huge difference in astronomy and space sciences. Some of its uses include: processing images from space telescopes; providing 3-D displays and carrying out complicated calculations; handling large amounts of data from projects such as collective joining of radio telescopes and classifying and cataloguing new objects.

Search for Extraterrestrial Intelligence (SETI)

You have probably asked the question 'do aliens exist?'. At the moment, there is no scientific evidence to answer the question, but the SETI projects have been set up to 'listen' for signals from space which may show signs of intelligence.

Planetary and observational satellites and probes

A number of different satellites and probes have been launched over the past 20 years or so. One example is 'Galileo', which travelled to Jupiter in 2003, flew past the four Galilean moons and successfully entered Jupiter's atmosphere. Others include Pathfinder, the Hubble space telescope, New Horizons and the Mars Reconnaissance Orbiter.

Activity B

Choose two of the satellites/probes listed and use the Internet to investigate their missions.

Useful products from spaceflight

The development of spaceflight, the technology used and the need to solve extremely difficult problems have all given rise to advances in science and products which are used on Earth everyday. Some examples are:

1. Materials – swimsuit design and manufacture, golf ball aerodynamics and structure, scratch-resistant lenses, lubricants, shoe materials, fire-resistant materials and heat insulation products.

2. Robotics – multi-control systems, advanced computer programming, finely tuned sensors and hydraulic systems for use in space have led to the widespread use of robotic systems in industry.

3. Prosthetic limbs – the development of lightweight, strong and resistant materials has helped to advance the design of prosthetic limbs.

Other areas where space exploration science has influenced technology on Earth include solar panels and computing.

Table: The history and development of manned spaceflight

Organisation/ spaceflight programme	Details
NASA (National Aeronautics and Space Administration)	US organisation set up in 1958 to develop manned and unmanned space exploration
Apollo	NASA programme which took humans to the Moon for the first time in 1969
Skylab	Large (100 tonne) space station, orbited the Earth from 1973 to 1979
ESA (European Space Agency)	Set up in 1975 with members from many European countries
Space Shuttle	Launched by NASA with its first orbital test flight in 1981
ISS (International Space Station)	Orbiting laboratory and observation platform, manned by astronauts from different countries
World Space Programmes	There are now many space agencies taking part in missions in Earth's orbit. These include China, Japan, India and Russia
Commercial spaceflight	Some well-known companies are designing space ships in the hope of offering people the 'trip of a lifetime'

BTEC Assessment activity 18.4   P6 M6 D6

1. List as many different telescopes and instruments as you can which help us to study space. **P6**

2. Work in groups and make a presentation to the class which details the problems faced by the spaceflight industry. **M6** Include in your presentation how you think space exploration will develop in the future. **D6**

Grading tip

To achieve **P6** link the instrument or telescopes listed to the part of the electromagnetic spectrum that is used, e.g. the Chandra X-ray telescope, and include examples of planetary probes and instruments on board the satellites named. For **M6**, remember to give all points of view in your presentation and to explain why manned Space exploration is particularly problematic.

18.5 Observing the night sky

Key terms

Constellations – groups of stars forming shapes named by the ancient Greeks.

Celestial coordinates – a system used to pinpoint the position of an object in the night sky.

Naked eye – looking at objects without any help from instruments or telescopes.

Did you know?

The galaxy M31 in Andromeda is the furthest object which can be seen with the naked eye on a clear night. It is 2 million light years away!

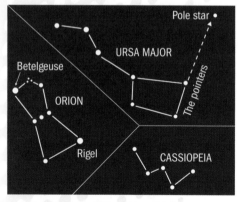

Constellations – Orion, Ursa Major, Cassiopeia.

Activity A

Why do we need celestial coordinates?

Star maps and patterns

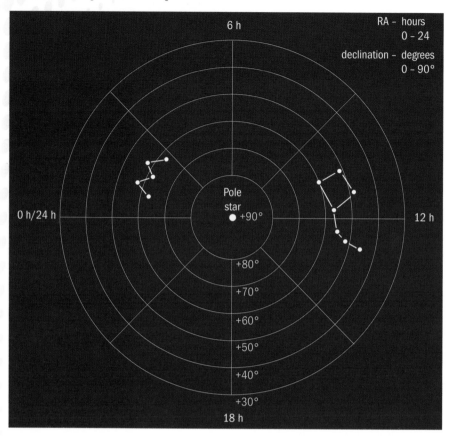

Celestial map of the northern hemisphere showing the lines of right ascension and declination.

To most of us, the night sky appears to be full of randomly positioned stars. Some seem brighter than others and some seem to twinkle more than others.

On closer inspection, however, it is possible to identify patterns which appear to link many stars together – these are **constellations**. Within these constellations exist many remarkable astronomical objects, most of which need the use of a telescope to make them visible. Here are some examples:

- globular star cluster in the constellation *Hercules*
- trifid nebula in the constellation *Sagittarius*
- the Andromeda galaxy near the constellation *Pegasus*.

When we observe the constellations, it is important to be able to provide an accurate position of the objects we are observing, just as it is when giving information about places on a map of the Earth. To do this astronomers have divided the sky using lines of **right ascension** (similar to longitude) and **declination** (similar to latitude). These are called **celestial coordinates**. Right ascension is measured in hours and declination in degrees.

Practical observing

Large telescopes provide accurate and colourful images and are essential when making observations of difficult or very small and distant objects. The brilliant flares and coronal mass ejections from the surface of the Sun have been well filmed and photographed. Asteroids, too, have been observed as their orbits may bring some of them closer than normal to the Earth as they travel around the Sun. These particular close encounters are called Near Earth Objects (NEOs).

A lot of useful observations of the objects in space can be achieved without telescopes or other instruments. This is called **naked eye** observation. With practice, we can easily identify the most noticeable objects before beginning to use a telescope or binoculars. Suitable objects for this method include: the Sun and Moon; inner planets (Mercury and Venus); Mars; outer planets (Jupiter and Saturn); bright stars; main constellations; regular meteor events and some comets.

When making observations of the night sky, it is useful to identify the Pole Star (Polaris). This lies at a point in the sky which almost exactly matches with the Earth's magnetic North Pole if it was projected into space. Navigators have been able to use this star to help them find their latitude on the Earth – provided they were north of the equator.

Transits, eclipses and occultations

The movement of one object in front of another as seen from Earth happens quite often. The three different types are shown in the pictures on the right.

Safety and hazards

When observing the Sun **never look directly at it or through binoculars or a telescope!** Use the projection method.

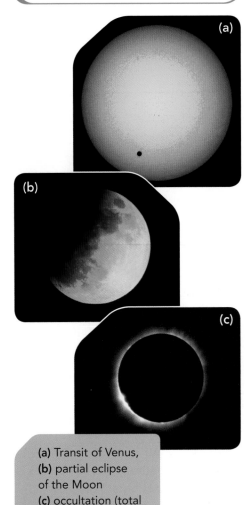

(a) Transit of Venus,
(b) partial eclipse of the Moon
(c) occultation (total eclipse) of the Sun.

BTEC **Assessment activity 18.5**   P7 M7 P8

Many discoveries in astronomy have been made by amateur astronomers who study the night sky as a hobby. Your work as a general assistant for a local amateur astronomical society involves making observations ready for discussion at the next monthly meeting.

Using safe procedures, make an observational log over a suitable time period of the phases of the Moon and movement of a planet across the night sky. **P7**

Predict the next phase change of the Moon over the next 24 hours by drawing this in your log. Plot your predicted path of a planet by looking carefully at your observation records. **M7**

Research details of your chosen objects from activity **P7** and **M7** on the Internet and compare professional data found with your own observations. List the differences and similarities. **P8**

Grading tip

For **P7** you will need to draw a star map showing constellations and coordinates. Compare your predictions with actual recorded observations to achieve **M7** . To achieve **P8** you should draw or print maps from research and compare them for the same dates and times as your own observations were carried out.

Just checking

1   List and draw the planets in the Solar System, starting with the smallest.
2   Why do we only see one side of the Moon?
3   Draw a simple block chart to show the probable sequence of events in the life of our Sun.
4   Outline two methods to measure astronomical distances.
5   Why do we use instruments to view the night sky in different wavelengths?
6   Detail three problems associated with spaceflight.
7   Draw and label 10 of the most easily identified constellations.
8   Why is the Pole Star important?

Assignment tips

To get the grade you deserve in your assignments remember to do the following.

- Summarise the most important developments in astronomy from a long list which you can put together from Internet and book research. Timelines are most useful if set out in a clear table.

- Begin by completing all the required tasks for the **P** criterion. For most of these in this unit a suitable list will be enough, but make sure that the list covers all the key points discussed in your tutor sessions.

- To achieve **M** and **D** criteria you should include descriptions within your lists and explanations where asked. Use a variety of websites and text books for the topic and compare the information to be sure that your sources are accurate.

Some of the key information you'll need to remember includes the following.

- The life and death of stars is dependent on their mass. The more massive, the more violent is their 'death'.

- Shorter astronomical distances can be measured using maths. Longer distances need the analysis of light from other distant objects.

- The Big Bang is the preferred theory of the origin of the Universe at present. The end of the Universe is thought to be dependent on the density of the material from which it is made.

- Many satellites and space probes have been used to explore our Solar System and beyond. Manned spaceflight has more problems than unmanned and costs of spaceflight in general are extremely high.

You may find the following websites useful as you work through this unit.

For information on...	Visit...
telescopes and instruments	Hubble Site and NASA's Space telescopes
life and death of stars	The BBC on stars
origin of the Universe	The origin of the Universe
spaceflight	Human Space flight, The Space review and Virgin galactic

19 Electronics in action

Imagine the world without smart mobile phones, MP3 players and computers. The world is full of electronic gadgets and every month new ones go on sale. Engineers and other scientists are working hard to develop smaller and smarter devices. As well as the typical high-street gadgets, medical applications that use electronics are becoming increasingly popular, for example electronic pacemakers and artificial limbs that use computer chips.

All these useful devices depend on the understanding of some basic electronic components that you will study in this unit. You will be introduced to important electronic components that are used to build up electronic circuits. The unit presents a variety of applications of these components including a system that can sense temperature and one that will turn on an alarm if a car door is opened with the headlights still on. You will learn how to design circuits to solve particular problems, how to build these circuits, and what sort of measurements you should take to test them.

Learning outcomes

After completing this unit you should:

1 know the components used in electronic systems

2 be able to conduct experiments on electronic circuits safely

3 be able to safely construct an electronic system to help solve an everyday need

4 be able to assess the constructed electronic system safely.

Assessment and grading criteria

This table shows you what you must do in order to achieve a **pass**, **merit** or **distinction** grade, and where you can find activities in this book to help you.

To achieve a **pass** grade, the evidence must show that the learner is able to:	To achieve a **merit** grade, the evidence must show that, in addition to the pass criteria, the learner is able to:	To achieve a **distinction** grade, the evidence must show that, in addition to the pass and merit criteria, the learner is able to:
P1 Identify symbols that represent electronic components **See Assessment activities 19.1, 19.2 and 19.3**	**M1** Describe the operation of electronic components **See Assessment activities 19.1, 19.2 and 19.3**	**D1** Compare the properties of analogue and digital components **See Assessment activity 19.2**
P2 Perform an electrical measurement on an electronic circuit safely **See Assessment activity 19.6**	**M2** Describe what factors may limit any measurements made **See Assessment activity 19.6**	**D2** Compare the results of measurements on the electronic circuit with that predicted by theory **See Assessment activity 19.6**
P3 Design an electronic system, which contains an active device, to help solve an everyday need **See Assessment activities 19.3, 19.4, 19.5 and 19.7**	**M3** Design an electronic system, which contains an active device, to include calculating a numerical parameter **See Assessment activities 19.3, 19.4 and 19.7**	**D3** Justify components selected for the electronic system **See Assessment activities 19.3, 19.4, 19.5 and 19.7**
P4 Produce an electronic system, which contains an active device, that could be used to help solve an everyday need	**M4** Describe the operation of the constructed electronic system that could solve an everyday need **See Assessment activities 19.3, 19.4, 19.5 and 19.7**	**D4** Describe the limitations of the working electronic system **See Assessment activity 19.3**
P5 Perform an electrical test on the constructed electronic system safely.	**M5** Explain the outcomes of the electrical test.	**D5** Explain how your working electronic system can be further improved.

How you will be assessed

Your assessment could be in the form of:

- a table e.g. of circuit symbols

- results from your measurements on an electrical circuit

- a leaflet or report including circuit and block diagrams

- computer simulations of your planned electronic system

- photographs of your completed circuit.

Joshua, 16 years old

I like this unit because it has lots of experiments which are related to how real gadgets work. It is definitely an interesting unit that will be useful to anyone who uses electronics in their lives and also for those who would like to follow this subject at a higher level.

There are many ways that teachers can assess us for this unit, which is good because you learn different skills and there will be an assignment to fit your learning style.

The best thing about this unit is that you don't need to know anything about electronics – you learn from scratch and come out making a circuit that can be used in the real world!

Catalyst

Name that gadget!

In groups of four discuss two electronic gadgets that you have used.

- Discuss the good and bad things about the gadgets.
- Decide what the input and output of these devices are.
- See if you can work out how the device works.

19.1 Electronic systems and passive components

Key terms

Input transducer – converts a physical change into an electrical change.

Processor – actively changes an electronic signal.

Output transducer – converts an electrical signal to a useful physical change.

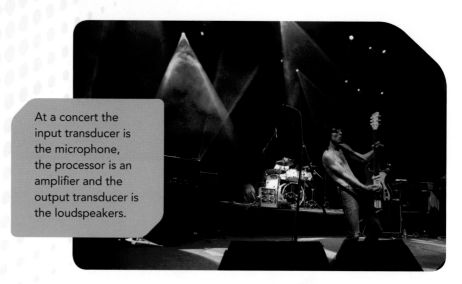

At a concert the input transducer is the microphone, the processor is an amplifier and the output transducer is the loudspeakers.

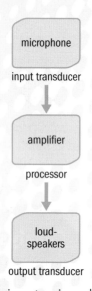

An electronic system is made up of an input transducer, a processor and an output transducer.

Activity A

Draw a block diagram showing the input transducer, processor and output transducer for a system which lights up a garden when it gets dark.

Every electronic device is either an **input**, **processor** or **output** device. An input device (also called an input transducer) converts a physical change in the environment to an electrical change. An example is a microphone which converts sound (pressure changes in the air) into an electrical change. An output device (or output transducer) causes some change to the environment by converting an electrical signal into another form of energy. An example is a loudspeaker which converts an electrical signal into sound energy. A processor is a device that changes an electrical signal. An example is an amplifier which takes the input electrical signal from a microphone and amplifies it before it is output by loudspeakers.

Passive components

Passive electrical components do not actively change an electronic signal; often they simply reduce a signal. Examples of passive components include **resistors**, **diodes** and **capacitors**. More examples are given in the table, along with their symbols.

Diodes

Diodes are components that allow current to pass in one direction only. They are often used to direct current the right way or stop current damaging a component. A diode has a positive end (anode) and a negative end (cathode). When a voltage tries to push a current through the diode from anode to cathode, current flows quite easily. If a reverse voltage is applied, almost no current will flow. Silicon diodes need the anode to be 0.7 V higher than the cathode for current to start flowing, whereas germanium diodes start conducting with a voltage of 0.3 V.

If too much current flows through a diode, it will overheat and could be destroyed.

Some passive components and their symbols and uses.

Passive component	Use
Fixed resistor	Restricts the current by a fixed amount within a circuit
Variable resistor	Similar to fixed resistors but the resistance can be mechanically changed
Signal diode	Used to rectify circuits (change ac to dc). Are also used to protect transistors from high voltages
Light emitting diode	Used in displays such as alarms and digital clocks
Photodiode	Used to detect light levels in sensing circuits
Polarised capacitor	Stores charge. Used, for example, in timing circuits and power supplies. These need to be connected the correct way round in a circuit

Passive component	Use
Non-polarised capacitors	These can be connected either way round in a circuit
Filament lamp	Used to provide light
Buzzer	Provides a sound output such as an alarm
Toggle switch	On/off switch. Used, for example, for lighting in homes
Tilt switch (no standard symbol)	Used to detect changes in angle of an object
Reed switch	Uses a magnetic field. Used in burglar alarms

Resistor markings

The British Standard code BS1852 identifies the value of the resistor in ohms (Ω), for example 4K7 means 4700 Ω. Resistors usually have coloured bands printed on them that represent the value of the resistor and the tolerance.

Activity B

Using the coloured bands, describe the fixed resistor shown below in as much detail as possible. You will need to research colour markings of resistors using the Internet, and find out what 'tolerance' means.

 Assessment activity 19.1 P1 M1

You are working as a junior technician in a local electricity company. You have been assigned to produce a leaflet for the public that explains the basic components that make up circuits. Your leaflet should include the following points.

1 Identify the types of switches that can be used **(a)** to detect the position of objects and **(b)** in burglar alarms. **P1**

2 Describe how a diode works and give examples of how they can be used. **M1**

 Activity C

(i) An electrician calculates that he needs a resistor with a value of 170 Ω. But you can only buy resistors with certain preferred values. Using the Internet find out what the closest preferred value for the resistor is.

(ii) The resistor that the electrician buys has a tolerance of ± 2%. Write down the closest preferred value and tolerance of the resistor using the BS1852 code, and produce a coloured drawing to show the colour code for the resistor.

Grading tip

Being able to draw the symbol for a component, or name a component from its symbol, is a skill that will help you attain pass grade **P1**. To gain merit **M1** you need to be able to describe how a component works in a circuit.

19.2 Active devices

In this section: **P1** **M1** **D1**

Key terms

Amplify – make larger or more powerful.

Noise – electronic signals that are not part of the signal you want.

The computer that was on board the first lunar landing module used a lot of active components.

Did you know?

Television programmes used to be transmitted by terrestrial analogue signals. The signals would often suffer from noise, making the picture fuzzy. The digital signals that are now used by televisions are much sharper because the noise is eliminated. In the past you could get a very fuzzy picture if the signal was weak and the noise was strong. These days when the signal is weak and the noise is strong, you get no picture at all!

Activity A

(i) Explain what an active device is.

(ii) Give one example of a digital device and one example of an analogue device.

An active device is one that **amplifies** either the current or voltage signal in an electric circuit. They need their own power supply. Active devices used in analogue electronics include transistors and operational amplifiers. In digital electronics logic gates are common active devices.

Digital vs. analogue

Digital electronics (for example digital clocks and computers) usually use two values (denoted 1 or 0): in voltage these values will be the supply voltage (VS) and 0 V. You can think of something digital as being able to take one of two values, such as on and off, but nothing in between. **Analogue** electronics (for example audio amplifiers and microphones) can take a whole range of values. **Noise** can be a problem in analogue electronics, for example the 'hum' in some microphones and analogue televisions. In digital electronics noise can be eliminated.

Examples of active devices

Some active devices and their symbols and uses.

Active component	Use
Bipolar junction transistor (BJT) base — collector / emitter	Used in amplifying or switching circuits, such as temperature sensors
Metal-oxide-semiconductor field-effect transistor (MOSFET) S / G / D	Used in high-speed amplifying or switching circuits, such as computers
Operational amplifier V_{S+} / V_+ / V_- / V_{out} / V_{S-}	Used to amplify signals in analogue electronics, e.g. guitar amplifiers
Logic gate A / B — Y	Used in digital electronic systems such as traffic lights and alarm systems

npn bipolar junction transistors (npn BJTs)

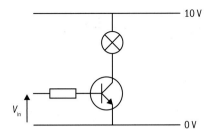

An npn BJT can be made to work either as an 'electronic' switch or as an amplifier.

In the circuit shown, if the input voltage (V_{in}) is less than 0.5 V then the transistor switch is off and no current flows through the transistor, so the bulb is off. If V_{in} is above 0.7 V then the transistor switch is on, current flows through the transistor and the bulb comes on. A very small voltage change can switch the bulb on or off. The small voltage change might come from a sensor circuit that detects light levels or temperature changes.

Circuit diagram showing a BJT being used to switch a bulb on and off. The resistor protects the transistor from being damaged by too high a current.

Operational amplifiers

Operational amplifiers (Op Amps) are used in many areas of electronics, for example in audio recordings and detectors.

An Op Amp can be thought of as a 'black box' containing transistors and other components, such as resistors. You don't need to worry about what is going on inside the Op Amp, you just need to know what the 'black box' does: an Op Amp amplifies the difference between two input voltage signals giving an output voltage. The gain of an Op Amp (ratio of output voltage to input voltage) can be enormous – up to 200 000: if the input to the amplifier changes by 10 microvolts (10 millionths of a volt), the output can change by 2 volts. Engineers can calculate voltage gain using the equation:

$$\text{voltage gain} = \frac{\text{output voltage}}{\text{input voltage}}$$

Science focus

npn BJT as an amplifier

The same type of transistor can also be wired slightly differently to act as an amplifier. Small changes at the input produce larger changes at the output. This could be used to help amplify the output of a microphone so that it can drive a loud speaker.

BTEC Assessment activity 19.2 **P1** **M1** **D1**

You have just started a trial period of employment with a multinational electronics company. You are required to show your knowledge of electronics by answering some questions. A BFY51 transistor is a common type of npn BJT. You will need to find further information to help you answer the questions.

1 Draw the circuit symbol for a BFY51 transistor and label the connections. **P1**

2 Describe how a BFY51 transistor can be used to switch a bulb on and off. **M1**

3 Explain why mobile phones use digital signals rather than analogue ones, and why the quality of the digital signal is so much better than the old analogue signals. **D1**

Grading tip

To get the **M1** grade you need to describe how passive and active components work in a circuit and how the circuit works as a whole.

Activity B

The voltage gain of a particular Op Amp is about 100 000. If the input to the amplifier is 0.000 01 V, what is the output voltage?

Functional skills

Mathematics

Correctly obtaining the output voltage requires the use of appropriate mathematical operations.

19.3 Logic gates and timers

In this section: 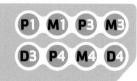 P1 M1 P3 M3 D3 P4 M4 D4

Key terms

Logic gate – electronic component that makes a decision.

Monostable timer – circuit that gives out a single timed pulse, used to switch something on and off.

Astable timer – circuit that gives out a continuous stream of timed pulses, used to switch something on and off repeatedly.

Logic gates

Logic gates are active components that respond to voltages at their inputs and give a single output. There are two logic states, 1 and 0. These represent voltage levels, high or low. One type of logic gate is an AND gate which will give a high voltage output when one input *and* the other input are high. For example, an AND logic gate could be used to turn on a light when someone walks past a door *and* it is dark.

There are three main types of logic gate: AND, OR and NOT. The way they operate can be described using **truth tables**. The truth table for a logic gate shows what happens to the output when different input signals are provided.

Additional logic gates can be formed from the three basic gates described above. A NAND gate is formed by connecting a NOT gate to the output of an AND gate. A NOR gate is formed by connecting a NOT gate to the output of an OR gate.

Combining logic gates

In practice an individual logic gate is not very useful on its own. In most applications several logic gates are connected together.

AND

Input A	Input B	Output Q
0	0	0
0	1	0
1	0	0
1	1	1

NOT

Input A	Output Q
0	1
1	0

OR

Input A	Input B	Output Q
0	0	0
0	1	1
1	0	1
1	1	1

NOR

Input A	Input B	Output Q
0	0	1
0	1	0
1	0	0
1	1	0

NAND

Input A	Input B	Output Q
0	0	1
0	1	1
1	0	1
1	1	0

Some different logic gates and their truth tables.

Worked example

An electronic system that uses logic gates will open a window in a greenhouse when it is hot and daylight, or if the grower overrides it with a manual switch.

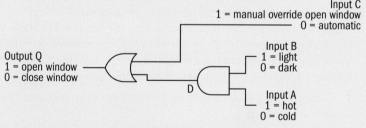

Input C
1 = manual override open window
0 = automatic

Input B
1 = light
0 = dark

Input A
1 = hot
0 = cold

Output Q
1 = open window
0 = close window

Construct the truth table for the logic circuit.

To work out the output follow each input through each stage of the circuit until you get to the output.

Truth table for AND gate:

A	B	D
0	0	0
0	1	0
1	0	0
1	1	1

Truth table for OR gate:

C	D	Q
0	0	0
0	1	1
1	0	1
1	1	1

Truth table for the whole circuit:

A	B	C	D	Q
0	0	0	0	0
0	0	1	0	1
0	1	0	0	0
0	1	1	1	1
1	0	0	0	0
1	0	1	0	1
1	1	0	1	1
1	1	1	1	1

The 555 timer

Sometimes we want to switch something on and off repeatedly for a certain amount time. For example, a cooker alarm might be designed to beep once a second for 2 minutes and then stop. Or we may want the alarm to go on continuously for a certain time and then stop, for example a school bell. Both of these functions can be achieved using a **555 timer**. The 555 timer can be set either as **monostable** or **astable**. You can find circuit diagrams for the 555 timer as an astable and as a monostable at **Doctronics: 555 timer**.

Monostable timers

When a monostable timer is switched on it will stay on for a specific time and then switch off until it is switched on again. The output will be a pulse, as shown in the diagram. The length of time that it is switched on for can be chosen by selecting values of the resistor (in ohms) and the capacitor (in farads). The equation below shows how the time, resistor value and capacitor value are related.

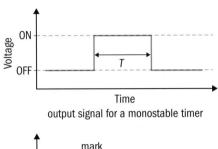

output signal for a monostable timer

$$T = 1.1 \times R \times C$$

where T is the time in seconds, R is resistance and C is the capacitance.

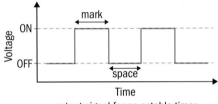

output signal for an astable timer

Output signals from monostable and astable timers.

Astable timers

The 555 timer also allows you to design a repeating on/off output signal. When an LED is connected to the output, the LED will flash on and off at a constant rate. At very high frequencies it makes the LED look like it is on all the time. The amount of time that the output of the timer is on is called the mark. The amount of time when the output is off is called the space. The time for the mark and space can be chosen by selecting values of the two resistors and the capacitor.

> The mark (time on): $T_M = 0.7 \times (R_1 + R_2) \times C$
> The space (time off): $T_S = 0.7 \times R_2 \times C$

BTEC Assessment activity 19.3 P1 M2 P3 M3 D3 M4 D4

You are doing some work experience at a local electronics company. You have been asked to design an alarm system to alert kitchen staff if a freezer door has been left open. You have decided to use a 555 timer.

1 The alarm needs to go on for 20 seconds and then switch off. What kind of 555 timer circuit should you use? **P1 P3 D3**

2 You have a capacitor with a value of 100 mF. What resistor value will you need? **M3**

3 Describe how the alarm system works. Are there any limitations to the system? **M1 M4 D4**

Grading tip

To get the **P3** grade you will need to be able to design a system containing an active device. To get **M3** you will need to calculate a numerical parameter when designing your electronic system – including a 555 timer is a good way to do this.

19.4 Circuits

In this section: **P3** **M3** **D3**
P4 **M4**

Potential divider circuits are used in electric guitars.

Useful circuit calculations

Quantity	Calculation
Total resistance of a series circuit e.g.  8 kΩ 2 kΩ	Sum the values of the individual resistors, e.g. for a circuit containing two resistors: $R_T = R_1 + R_2$
Total resistance of a parallel circuit e.g. 8 kΩ 2 kΩ	e.g. for a circuit containing two resistors: $\frac{1}{R_T} = \frac{1}{R_1} + \frac{1}{R_2}$ e.g. for a circuit containing three resistors: $\frac{1}{R_T} = \frac{1}{R_1} + \frac{1}{R_2} + \frac{1}{R_3}$
Voltage	Ohm's law: $V = I \times R$ where V is voltage (in volts), I is current (in amps) and R is the resistance (in ohms)
Power dissipated in a resistor (amount of energy transferred into heat each second)	$P = V \times I$ where P is power (in watts), V is voltage (in volts), and I is current (in amps)

Science focus

For the potential divider circuit shown below the voltage across the thermistor changes with temperature level. The variable resistor controls the sensitivity of the circuit to the temperature level.

The voltage across the thermistor and variable resistor is shared between them; the bigger the resistance the bigger the share of the voltage. This means that if the thermistor's resistance increases, the voltage across the thermistor increases so V_2 gets bigger and V_1 gets smaller. This voltage change could turn on a transistor which could control a heater.

The voltages can be calculated using these formulas:

$$V_1 = \frac{(V_s \times R_1)}{R_1 + R_2}$$

$$V_2 = \frac{(V_s \times R_2)}{R_1 + R_2}$$

Activity A

Two resistors with values 8 kΩ and 2 kΩ are connected **(i)** in series and **(ii)** in parallel (see diagrams in table above). Calculate the total resistance in both cases.

Potential divider circuits

Potential dividers are circuits that divide voltage across two or more components that are connected in series.

Worked example

What is the output voltage, V_2, of the potential divider circuit shown? The input voltage is 5 V. The variable resistor (R_1) is set to 8 kΩ and the thermistor (R_2) has a resistance of 2 kΩ. Remember, 1 k is 1000.

$$V_2 = \frac{(V_s \times R_2)}{R_1 + R_2} = \frac{5 \times 2000}{8000 + 2000} = 1\,V$$

Note: the rest of the voltage is across R_1. So $V_1 = (5 - 1) = 4\,V$

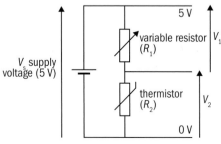

A potential divider circuit containing a variable resistor and a thermistor.

Activity B

For the temperature sensing circuit shown (bottom of previous page), find the voltages across the variable resistor and the thermistor if the value for the thermistor is now 12 kΩ. The input voltage is still 5 V and the variable resistor is still at 8 kΩ.

Light detector circuit

The light detector circuit shown is a potential divider circuit that can be used to turn a light on automatically when it gets dark. As the light level drops, the voltage across the LDR rises and turns the transistor on. The transistor turns the bulb on.

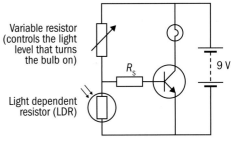

A light detector circuit that turns a light on as it gets dark.

Case study: Garden lighting

Jeslyn is a technician and has been assigned to design a system for switching a garden light on at sunset. She has measured the dark resistance of the LDR she is using to be 2 MΩ. Jeslyn will use a 9 V battery in the circuit. The other resistor in the potential divider circuit has a value of 1 kΩ. She wants to calculate the voltage of the LDR in the dark. She substitutes the values into the potential divider equation.

In the dark: $V_{LDR} = \dfrac{(V_s \times R_{LDR})}{(R_{LDR} + R)}$

$= \dfrac{(9 \times 2 \times 10^6)}{(2 \times 10^6 + 1 \times 10^3)} = 9\,V$

What is the voltage at the output during the day (when light falls on the LDR)? Assume that the resistance of the LDR falls to 1 kΩ.

Science focus

Sensing temperature

A temperature sensing circuit can be made using a thermistor and resistor together, in the same way that the LDR and resistor were combined in the light detector circuit. Remember, when it is cold the thermistor has a high resistance so the voltage across the thermistor will be greater. When it is hot the thermistor's resistance is low so it has a small voltage across it.

BTEC Assessment activity 19.4

You are a junior science technician, working for a government environmental department. You are required to design and build a circuit that could detect temperature changes in a greenhouse.

1 Draw a thermistor circuit showing what components should be used and how they should be connected. **P3 D3**

2 Calculate what value resistor you will need in order for the voltage across the thermistor to be 5 V, assuming that its resistance is 10 kΩ and the power supply is 9 V. **P3 M3 M4**

Grading tip

To meet **P4** you need to include a transistor switching an output device in your circuit.

To achieve **D3** you will need to explain why you've chosen the components you've used in your circuit. To do this you will need to explain what the component does to help solve the problem you've been given.

19.5 More circuits

In this section: P3 M3 D3 P4 M4

Key terms

Comparator circuit – a circuit that compares two input voltages and gives an output voltage.

Reference voltage – a voltage that stays the same all of the time (used in a comparator circuit).

A comparator circuit can be used to switch on a heater in a greenhouse if it gets too cold.

Did you know?

The first operational amplifiers were made using glass bulbs and needed hundreds of volts to make them work. These days the Op Amp is so small it is nearly invisible and can work with voltages below 10 volts.

Comparator circuits

A **comparator circuit** compares two voltage inputs. The output depends on which of the two input voltages is the highest. Comparator circuits are used throughout electronic systems, for example in converting analogue signals (such as sound) to digital signals. They are also used in circuits that sense temperature and light.

Operational amplifier as a comparator circuit

An Op Amp is used in a comparator circuit to compare the two signals at its inputs, V_1 and V_2, and gives an output, V_{out}.

- An Op Amp produces a positive output, $V_{out} = V_S$ (the value of voltage used to power the Op Amp), when V_1 is greater than V_2 ($V_1 > V_2$).

- An Op Amp produces a negative output, $V_{out} = -V_s$, when $V_2 > V_1$.

Temperature control using a comparator circuit

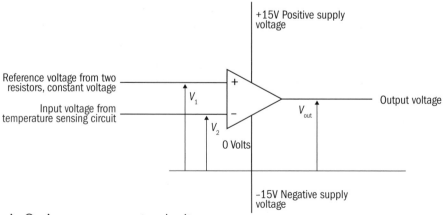

An Op Amp as a comparator circuit.

A comparator circuit can be used in a temperature control system. The circuit shown compares the voltage from a temperature sensing circuit (V_2) with a **reference voltage** (V_1). This could be used to control the temperature of a greenhouse.

When it is too cold V_2 is high, giving a negative output which can be used to turn a heater on. The opposite happens when the temperature is too high.

The voltage V_1 is provided using a potential divider circuit composed of two resistors. This voltage stays the same all the time because the resistors are fixed. This is called a 'reference voltage'. If the resistors are changed, the reference voltage will change. If the reference voltage is changed, the Op Amp will switch at a different temperature. The voltage V_2 is provided by a potential divider circuit composed of a variable resistor and a thermistor. The comparator circuit is similar to the circuit on page 334. The comparator circuit here is far more sensitive than the transistor circuit, so the temperature can be controlled much more precisely.

Logic circuits

Logic gates can be used to build circuits that respond to certain criteria. For example, in most cars a warning buzzer needs to come on when the headlamps are on and the driver's door opens. The logic circuit shown can be used to control the warning buzzer.

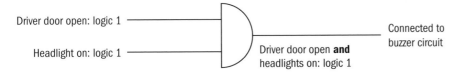

Driver door open: logic 1

Headlight on: logic 1

Driver door open **and** headlights on: logic 1

Connected to buzzer circuit

Logic circuit that could be used to turn on a buzzer when a car's headlamps are on and the driver's door opens.

Activity A

Describe two other situations where a very sensitive circuit might be needed to control a device.

If a driver leaves the headlights on when they leave their car, the battery will run down and they won't be able to start the car next time they want to go out. A logic circuit can be used to control a warning system.

BTEC Assessment activity 19.5 **P3** **D3** **M4**

You are working for a company that manufactures motor accessories such as electronic car door mirrors and door locks. You are working on a warning buzzer system like the one described above.

1 Draw a truth table for the logic circuit, shown above, that is used to warn that the headlights are on when the driver gets out of the car. **P3**

2 Explain, using your truth table, how the system works. **D3** **M4**

Grading tip

To get the **D3** grade you need to explain why you use certain components. You should support your explanation using evidence, such as a truth table or a simulation, and by explaining what each component does.

19.6 Working with electricity

In this section: P2 M2 D2 P5 M5

Key term

Residual current device – a sensitive safety device that monitors current and will disconnect the supply if there is a fault in the circuit.

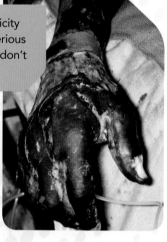

Mains electricity can cause serious burns if you don't work safely.

This sign is displayed in areas where there is a risk of electrocution.

Activity A

Describe three risks that you need to think about before you work with electricity, along with what you should do to reduce these risks.

Working safely with electricity

When working with circuits and electronics equipment there is a risk that somebody could get harmed. For example, a shock from the mains electricity could cause severe burning or even stop the heart from working.

Safety considerations

Some things you should do to keep safe are listed below.

- Use a **residual current device** (RCD) – this is a sensitive circuit breaker that is designed to prevent electrocution. For example, if you run over a cable with an electric lawn mower, and cut through it, the RCD will disconnect the supply.

- Make sure your wiring is safe and appropriate: Are there any bare wires visible? Is the cable cover gripped securely?

- Make sure you understand cable colour-coding (for example the colours of live and earth wires). Knowing the colours of the insulation surrounding the conductors in electrical wiring helps anybody working on the cables to connect them correctly.

- Take care when cutting wires with a wire cutter – there is a risk that you may cut yourself, and also that a piece of wire may fly off and hit someone in the eye.

- Take care when using electrical components such as resistors or diodes, which may get hot and burn skin if touched.

The Health and Safety at Work Act, 1974, requires that a risk assessment is carried out, to reduce the likelihood of injury.

First aid

If someone suffers from an electric shock, the first thing you should do is remove the source of electricity, by disconnecting the power supply or unplugging the equipment. You should then telephone for an ambulance immediately.

If the casualty has suffered electrical burns these should be cooled with water. Again, an ambulance should be called immediately.

Electrical measurement instruments

There are three main measuring devices that you will need to use to help you assess your electrical circuits and find reasons for faults in them. These are shown in the table (top of next page).

Measuring instrument	Use
Multimeter	Used to measure voltage, current and resistance. Some multimeters can also measure capacitance
Cathode ray oscilloscope	Used to measure, and display, voltage of ac and dc, with changing time
Logic probe	Used to test logic signals, i.e. the probe can tell you whether the signal is 1 or 0 at the part of the circuit you are testing

Some common reasons for circuits not working are given below.

Component	Problem	Possible reason
Timer	The pulse time is too long	The capacitor is faulty
Amplifier	No gain is observed	One of the connections has not been soldered correctly to the circuit board
Transistor	No switching effect is seen	One of the legs is broken
LED	The LED is not switching on	The anode and cathode are connected the wrong way round

Activity B

(i) What instrument would you use to measure the base current of a BJT?

(ii) How would you check if a 741 operational amplifier has +15V and −15V connected to its terminals?

Case study

Keeping the plants warm

Ben is an engineer out on call at a garden centre. He is working on a comparator circuit that is part of a system that makes sure that a greenhouse is kept at the right temperature to keep the plants alive. Ben is trying to determine why the system isn't working. The output of the comparator circuit is connected to an astable 555 timer, and this is connected to an LED. When the system is working the LED flashes on and off when it is too hot or too cold in the greenhouse. The problem is that the LED is not turning on at all. Ben has a multimeter and a cathode ray oscilloscope. He measures the voltage across the LED using the multimeter and finds that it reads 0.4V and not the 1.5V needed for the LED to light up. When he measures the output of the timer, using the oscilloscope, he sees a flat line instead of the pulse that he expected to see.

What else could Ben measure, and with what instrument, in order to get more clues as to why the system is not working?

Why might the system not be working?

Assessment activity 19.6 P2 M2 D2

You are employed as an electrician by a security systems company. You need to test that a security alarm system is working correctly after it has been installed.

1 What measuring device would you use to check that the correct 12V voltage supply was powering the system? **P2**

2 There are test points built into the logic section of the circuit. One of these is supposed to give a stream of pulses, 100 per second. Explain why a logic probe would not be the most useful instrument to test this logic output. What would you use instead? What would you expect to observe. **M2**

3 You find that one of the voltages you measure is 5.2V rather than the 5.0V you expect. Explain why this is not a fault and is quite normal in electronic circuits. **D2**

Grading tip

You can meet the **P2** and **P5** criteria by combining your testing with the measurements on your complete electronic system. Similarly, **M5** and **M2** can be combined, but notice that **M5** requires you to explain the result of your testing: what did your testing tell you about your circuit?

19.7 Designing and producing an electronic system

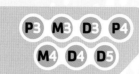

sensor.yka

Play Content

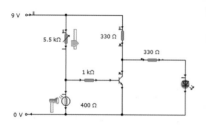

You can use computer software to simulate your circuit.

Image supplied by Yenka.

Grading tip

Just guessing values of *R* and *C* and substituting them into the equation for the time until *T* is almost equal to 20 seconds is called a reverse calculation and won't get you the merit grades.

Activity A

For an astable 555 timer rearrange the equation so you could calculate the value of the resistor R_1 for a given mark time.

Hint: Mark time: $T = 0.7 \times (R_1 + R_2) \times C$

Design

Before you start actually making a circuit you need to plan your electronic system. Think about the three points below.

- **Need:** What do you want the system to do? Do you want it to alert someone or something, control a moving object, or turn something on or off? Where will it be used?

- **Design:** You will need to draw a system diagram of your proposed system. That is, you need to show the input, processor and output of your complete circuit.

- **Circuit type:** Identify the components you will need at each stage – input, processor and output. For example, if you want an alarm to sound as the output your output transducer will be a buzzer.

Calculations

You will need to calculate a value for a parameter when designing your circuit. For example, if you wanted to calculate the capacitance to make a monostable 555 timer give a 20 second pulse you would use the equation:

$$T = 1.1 \times R \times C$$

But first you would need to rearrange the equation to give the capacitance as the subject:

$$C = \frac{T}{1.1 \times R}$$

Using the value of an available resistor and $T = 20$ you would then calculate the capacitance.

Simulations

Before building circuits, engineers usually simulate their circuits on a computer. This helps them to find out whether their circuits will behave the way they want them to before they start to construct them. There are various free software programs available online that you can use.

Evaluation

When you have completed your working system, you need to evaluate it. Ask yourself the following questions.

- **Does your circuit behave the way you expected it to?** For example, imagine you constructed a system that will turn a fan on when it gets hot. At what temperature did the fan actually switch on? What temperature did you expect your fan to be turned on?

- **What are the limitations of your system?** For example, if you used a potential divider, did you get the required voltage you needed? Is the Op Amp giving you a high enough output voltage?

- **What can you do to improve your system?** For example, you could extend your system by including a warning device that will inform someone that it is getting hot before the fan goes on.

BTEC Assessment activity 19.7 P3 M3 D3 M4

You are working for a multinational engineering company. You have been assigned to produce an electronic system that will alert farmers when their crops need watering.

1. Use a system diagram to plan an appropriate system. **P3**
2. Plan the details of your design, including calculating a numerical parameter that could be used in your circuit, and drawing a circuit diagram. **M3**
3. Describe your electronic system and explain how it will work. **M4**
4. For each component you have included in your design explain why you have chosen it. **D3**

Science focus

It is likely that you will use a protoboard to build your electronic system. The components will have wires that you can push into plastic holes on the board. A protoboard is often known as a 'breadboard'.

PLTS

Creative thinkers

When you design your electrical circuit to help farmers you will have to think creatively to solve the problem.

 **WorkSpace** # Rebecca De Souza
Electronics Technician, McCaskill Ltd

I work as a technician in the service department of a large consumer electronics firm. Faulty equipment is returned to us – TVs, computers, digital cameras and so on.

Very often a TV or something similar isn't working because just a single electronic component has failed. My job is to identify which component has failed.

One way of doing this is to apply a signal to the circuit. I then use an oscilloscope to check if the signal is what it should be at various points in the circuit. If the signal isn't right at one point then I can usually trace the fault and work out which electronic component has failed. With luck we can replace it and get the equipment working again.

Modern digital oscilloscopes are so much better to work with than the old ones; we can select settings that reduce the noise on the oscilloscope and so get much more reliable results.

Think about it!

1. What is meant by a signal that can be displayed on an oscilloscope?

2. When faulty electronic equipment is returned to the manufacturers, it is often cheaper to replace it rather than repair it. What reason could you give for trying to repair equipment rather than throwing it away?

Just checking

1 Draw the circuit symbol for **(a)** a resistor, **(b)** an AND gate and **(c)** a thermistor.
2 Draw a circuit diagram for three resistors connected **(a)** in series and **(b)** in parallel.
3 What does a potential divider circuit do? Give an example.
4 Construct a truth table for **(a)** a NOT gate and **(b)** a NAND gate.
5 What is an Op Amp? Give one example of its use.
6 Draw and explain the output signal you get when a 555 timer is connected as **(a)** monostable and **(b)** astable.
7 Draw a system diagram of an electronic system used to monitor temperature change. The input is a thermistor, the processor is a comparator and the output is an LED.

edexcel

Assignment tips

- In order to produce a well-designed and fully working electronic system you need to plan well. To have a good plan you need to know what problem you are trying to solve.

- Make sure that your design contains an active device, otherwise you will not meet the grading criteria.

- Beware of 'reverse calculations' (when you calculate your numerical parameter, rearrange the equation if necessary rather than guessing values that will give you the right outcome).

- Carry out a computer simulation of your circuit before you build it. Remember, a simulation does not necessarily show how your circuit will work in real life, but it's a good indication.

- Here are two websites that allow you to simulate electronic circuits:

  - **Yenka electronics**

  - **Circuit wizard**

- If your constructed circuit does not work the first time you try it, don't panic! This unit is not just about getting the circuit working first time, it's about investigating why things don't work. You will need a multimeter, an oscilloscope and possibly a logic probe to examine your circuit.

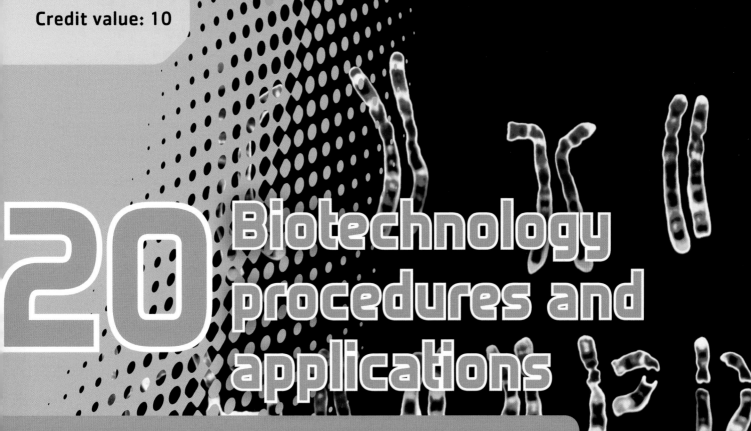

20 Biotechnology procedures and applications

Biotechnology is the use of scientific and engineering principles to modify and improve biological materials. These materials cover a wide range including crops, animals, energy sources and medicine.

In this unit you will take the role of biotechnology science technicians. This will help you decide whether you want to be part of a UK industry that:

- has 450 biotechnology companies
- produces 45% of biotechnology drugs trialled in Europe
- is one of Britain's fastest growing industries.

Modern biotechnology involves genetic engineering but biotechnology has been around for thousands of years. In the past, it was used to develop the best characteristics in crops or animals. These would then be produced in large numbers. Biotechnology is also used in the making of alcoholic drinks, bread, cheese and other well-known foods. In recent years, biotechnology has been able to do much more, including producing renewable sources of energy from waste and making new medicines to treat diseases. It could even help us in the future to cure diseases like multiple sclerosis.

Learning outcomes

After completing this unit you should:

1 know how the biotechnology industry has developed
2 know how biotechnology is used in our everyday lives
3 be able to perform simple biotechnology procedures
4 know how biotechnology may help treat or cure disease.

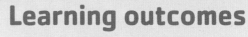

Assessment and grading criteria

This table shows you what you must do in order to achieve a **pass**, **merit** or **distinction** grade, and where you can find activities in this book to help you.

To achieve a **pass** grade, the evidence must show that the learner is able to:	To achieve a **merit** grade, the evidence must show that, in addition to the pass criteria, the learner is able to:	To achieve a **distinction** grade, the evidence must show that, in addition to the pass and merit criteria, the learner is able to:
P1 Identify technologies used in the past to change biological characteristics in plant and animal systems **See Assessment activity 20.1**	**M1** Describe the differences in the ways in which genes have been selected in the past and present **See Assessment activity 20.1**	**D1** Evaluate the use of biotechnology procedures in plant and animal systems **See Assessment activity 20.1**
P2 Describe the processes currently used in the biotechnology industry **See Assessment activity 20.1**		
P3 Identify the influence of current uses for biotechnology in our everyday lives **See Assessment activities 20.1 and 20.2**	**M2** Describe the influences of these biotechnology uses upon our everyday lives **See Assessment activity 20.1**	**D2** Explain the issues surrounding the uses of biotechnology in our everyday lives **See Assessment activity 20.1**
P4 Carry out practical experiments to demonstrate some biotechnology procedures	**M3** Describe how one of these procedures would be undertaken on an industrial scale **See Assessment activity 20.2**	**D3** Explain the scientific principles behind one biotechnology procedure **See Assessment activity 20.2**
P5 Identify some diseases or illnesses where biotechnology is used in their treatment or cure. **See Assessment activity 20.2**	**M4** Describe how one disease is being treated using biotechnology. **See Assessment activity 20.2**	**D4** Discuss the issues surrounding the use of biotechnology to cure diseases. **See Assessment activity 20.2**

How you will be assessed

Your assessment could be in the form of:

- a leaflet e.g. on the development and uses of biotechnology

- an article

- an experiment e.g. demonstrating some biotechnology procedures

- a presentation (ICT, video or poster) e.g. about biotechnology in the treatment of illness.

Leanne, 17 years old

I think that this unit is easy to understand with clear assignments. The unit gives you some insight and information for the assignment topics, meaning that people will understand what is expected of them in their assignments. The pages contain grading criteria which are clearly linked to each assignment and there are tips on how to get the best grade. This is also an interesting unit to do and I myself find the subject of biotechnology and its uses very interesting. The unit is also useful as it relates to different jobs and work-related activities, meaning it may help me to get a job in this field.

Catalyst

Biotechnology

What do you know about biotechnology?

Discuss with a partner what you think the following terms mean.

- Genetically modified food.
- Recombinant DNA technology.

20.1 Biotechnology past and present

In this section: P1 M1 D1 P2 M2 D2 P3

Key terms

Agriculture – the production of food and goods through farming.

Trait – a feature or characteristic of an organism.

Hybrid – a living thing that contains genes from two different organisms, e.g. a mule is a cross between a donkey and a horse.

Transgenic – an organism whose genetic material has been changed.

Yield – the production of a certain amount.

Biological products – products that are made using living things.

Biodegradable – something in biology that can be broken down by microorganisms.

Biotechnology is involved in the manufacture of both soft-centred chocolates and washing liquid.

Have you eaten Quorn™ or yoghurt recently or do you know someone who has diabetes and needs to take insulin? If so, you will be aware of some of the useful products that are made using biotechnology.

Biotechnology uses living things, especially microorganisms but also plants and animals, to make useful products. Biotechnology is used in important areas such as medicine, the environment, food sciences and **agriculture**.

Biotechnology has been around for thousands of years but in the past few decades it has become an exciting, multi-million pound industry which joins bioscience with engineering.

Biotechnology developed in the past

Fermentation using microbes to make beer, cheese, wine and yoghurt and using yeast to make bread

Selective breeding of plants and animals with good **characteristics** by farmers, e.g. plants with bigger or sweeter fruit or animals with more meat and less fat

Biotechnology developed in the past

Vaccines Edward Jenner produced the first vaccine for smallpox

Gathering and processing of herbs to make useful medicines, e.g. opium poppies were used as a tranquilliser in Victorian times

Biotechnology in the present

The early applications of biotechnology are still in use today, along with many more new processes and technologies. Bacteria and enzymes are now used to make useful products. A process called industrial fermentation is also used in many biotechnology processes to produce a large amount of product. For example, yeast is used in industrial fermentation to make millions of vitamin tablets very quickly.

Activity A

What advantages are there to using the current biotechnology processes?

Biotechnology using genetic engineering

Processes used in biotechnology include artificial selection. This involves selecting and breeding from offspring that have the most desirable **traits**. Recombinant DNA technology is a type of genetic manipulation. This involves moving one or more genes which code for the best characteristic from one organism to another. This changes the DNA molecule in that organism.

Here are some ways in which these processes are used.

- Producing crops which are resistant to pests and diseases.

- Producing **hybrid** plants and animals, e.g. to produce crops such as corn that will grow well in certain areas.

- Producing **transgenic** animals with the best characteristics.

- Giving animals genes that cause certain diseases so they can be used for research.

- Increasing the **yield** of crops to prevent waste and make more money.

- Improving a crop so that it looks better, lasts longer or tastes better.

- Producing fuel as a source of energy for vehicles and factories using **biological products** – this replaces the need to use non-renewable energy sources such as coal, oil and gas.

- Making **biodegradable** plastics from plants that make plastic in their tissues – this helps to reduce pollution.

Table: Current uses of biotechnology

Using bacteria	Treating raw sewage
	Making useful medicines, including antibiotics and insulin, on a large scale using fermentation
Using enzymes	Producing some foods including yoghurt and cheese
	Making detergents

Did you know?

Sugar cane can be used to produce ethanol for fuel and the waste **biomass** can also be turned into fuel using biotechnology.

A biotechnologist at work.

BTEC Assessment activity 20.1 P1 M1 D1 P2 M2 D2 P3

1 In groups, research and discuss how biotechnology processes and techniques differ from the past to the present day. Produce a poster using your findings and then present this to the rest of your class. **P1 M1 D1 P2**

2 Describe the ways in which biotechnology affects you, your health and the environment. **M2 D2 P3**

Grading tip

For all criteria make sure you know what the command words are for each task. The study guide will help you with this.

Science focus

Uses of transgenic animals

Cows are given the gene to make insulin so they produce milk containing insulin. This can then be used to treat people suffering from diabetes.

Sheep can be made to produce a protein in their milk to treat emphysema, a disease of the lungs.

Examples of genetically modified crops

Tomatoes which have a long shelf life; Golden Rice which has been genetically altered to contain the gene for production of vitamin A, to improve the quality of rice available to people in less economically developed countries where there is little food.

20.2 Biotechnology techniques

In this section: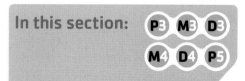

Key terms

Polymerase chain reaction (PCR) – a technique to make lots of copies of DNA from a sample.

Human genome – the genetic map of the human body's DNA including chromosomes and genes.

Transplant – to place the organ of one person into another person.

Ethical – concerning people's beliefs, values and rules about a situation.

Many biotechnology processes used in industry today depend on knowledge of genetics. This allows biotechnologists to find exactly where and on which chromosome different genes are located. By having this knowledge the biotechnologists are able to play around with genes. They can move them from one organism to another to end up with the product they want. One procedure involves using *Agrobacterium tumefaciens*. This bacterium acts as a **vector** in the process of moving different genes into plant cells to change the plant in some way.

Special **microbes** and vectors are used to move the genes from one place to another. These can be yeasts, bacteria, plasmids and viruses. Other biotechnology procedures include the **polymerase chain reaction** (PCR), which is used in forensic science.

Unit 13: See page 251 for an application of PCR.

Activity A

How can biotechnology help to treat or cure disease?

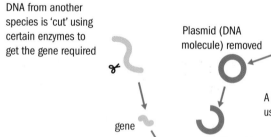

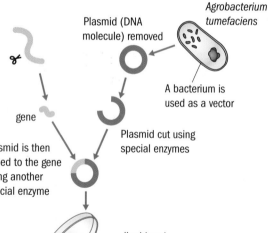

DNA from another species is 'cut' using certain enzymes to get the gene required

Plasmid (DNA molecule) removed

Agrobacterium tumefaciens

A bacterium is used as a vector

gene

Plasmid is then joined to the gene using another special enzyme

Plasmid cut using special enzymes

Leaf put into a liquid containing the plasmids and nutrients. Some of the leaf discs will take up the plasmid

liquid and plasmids

The discs that take up the plasmid containing the new gene are grown. Cells of the new plants now contain a new gene so will be different from before

leaf discs from plant to be changed

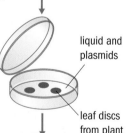

Genetically modifying plants using bacteria.

Biotechnology can help to treat or cure disease

Biotechnology requires knowledge of the **human genome**. It is useful for the following applications.

- Treating or curing some diseases – some conditions currently being treated using biotechnology include diabetes, cystic fibrosis, multiple sclerosis, lymphoma, tissue loss and heart disease.

- Fertility treatment for couples who can't get pregnant naturally.

- Making human enzymes and proteins.

- Making new cells or tissues, starting with stem cells, that can be used for **transplants**.

Ethical implications of biotechnology

After reading about biotechnology you may think it is very useful and that it has many advantages. One disadvantage of biotechnology is that there are **ethical** implications when the DNA of organisms is changed. People have different attitudes towards the use of biotechnology and the risks to themselves and the environment. Many scientific trials have taken place. As there is not yet any long-term evidence on the possible side effects of using **genetic engineering**, people do have the right to be concerned.

PLTS

Self-managers

You will use this skill to carry out your practical experiments when isolating DNA and manipulating an organism using biotechnology.

BTEC Assessment activity 20.2 P3 M3 D3 M4 D4 P5

1 Write a paragraph using bullet points to describe how three useful products can be produced on a large scale using biotechnology and how they affect our lives. P3 M3 D3

2 Make a list of six diseases or conditions that are treated or cured using biotechnology. Explain how this is done for each disease and discuss any issues raised. M4 D4 P5

Grading tip

Search for the term 'using biotechnology to treat disease' on the Internet. This will help you understand the information needed for this unit and complete the grading criteria for merit and distinction.

 **WorkSpace**

Aoife Sullivan
Microbiologist, MicroBio Safety

I run a laboratory testing food products for microorganisms such as *Salmonella*, *E. coli* and *Listeria*. It is my job to let food manufacturers know whether their food products are safe to go out to supermarkets.

I manage a team of laboratory technicians who carry out a variety of roles including:

- registering food samples coming in to the laboratory using a specialist computer data system
- preparing food samples ready for analysis
- analysing samples for microorganisms after an appropriate period of incubation.

Listeria is a bacterium that can cause pregnant women to miscarry. Polymerase chain reaction (PCR) technology can check samples for potentially harmful bacteria fast. It is a biotechnology that amplifies genetic material, allowing microbiologists to look for particular genes from pathogenic bacteria.

After microbiologists have checked their products, food manufacturers can transport the products to supermarkets quickly after production in the knowledge that they are safe for sale.

Think about it!

1 What would you do if you found *Listeria* in a sample?

2 Would it matter if you found small amounts of *Listeria* in a food sample that required heating by the consumer?

Just checking

1 Define 'biotechnology'.
2 How have biotechnology processes changed from the past to present day?
3 Describe the process of gene transfer as a biotechnological technique.
4 Which organisms can be used as vectors in the transfer of DNA?
5 What are the advantages and disadvantages of using biotechnology?
6 Discuss how people feel about genes being moved between organisms.

Assignment tips

To get the grade you deserve in your assignments remember the following.

- You can use information from Units 3 and 6 to help you complete this unit successfully. For example, your knowledge of DNA and genes (pages 72–75) and of gene therapy (pages 158–161) will help you to understand the processes involved in biotechnology and how it is used.

- When doing practical investigations remember to make sure that you follow all health and safety guidelines and follow instructions carefully.

- When writing up a report for the investigation make sure you include everything. This could include a plan, results, graphical evidence, a conclusion and an evaluation.

- Finally, check your work for errors before handing it in and make sure it contains all the necessary information presented in the correct way.

Some of the key information you'll need to remember includes the following.

- Biotechnology processes and the potential for its uses in medicine, agriculture, food sciences and the environment have greatly improved from in the past to the present day. Biotechnology has a dramatic impact on our everyday lives.

- Different microorganisms and other living things can be used to produce different useful products using biotechnology processes.

- Biotechnology can be used to treat and even cure some illnesses and conditions.

- Biotechnology produces a wide variety of useful products but there are areas that can cause concern, for example, genetically modified foods where genes are transferred from one organism to another.

You may find the following websites helpful to you as you work through this unit.

For information on...	Visit...
biotechnology processes and their uses	Biotechnology Online
	National Center for Biotechnology Information

Flu Shot

21 Science in the world

The media love a good story and science provides some of the best. Miracle cures, robot surgery, cloned animals – the list is endless and sounds good in newspapers, radio and television news and magazines.

Can you believe everything you see and read about science? Who are the people writing the news, and are they scientists themselves? Are scientists under pressure to report discoveries to the media before they are sure of their results?

In this unit you should think about how the media influence what you think and believe about science. How many times has a headline caught your attention, but when you've read the article it hasn't been about what you thought? *'Robots take over the world'* could be about a car manufacturer using robots to spray paint.

You should also think about questions like:

- How fast can news be spread?
- How quickly do people find out about new scientific discoveries, and is it a good thing to know quickly?
- How do you decide if scientific discoveries are 'good' or 'bad'?

Learning outcomes

After completing this unit you should:

1 understand some of the factors that can influence scientific progress

2 understand how science can be represented in the media

3 know how some scientific discoveries have been used in society

4 know some of the consequences associated with scientific discoveries/advances.

Assessment and grading criteria

This table shows you what you must do in order to achieve a **pass**, **merit** or **distinction** grade, and where you can find activities in this book to help you.

To achieve a **pass** grade, the evidence must show that the learner is able to:	To achieve a **merit** grade, the evidence must show that, in addition to the pass criteria, the learner is able to:	To achieve a **distinction** grade, the evidence must show that, in addition to the pass and merit criteria, the learner is able to:
P1 Discuss how external factors can influence scientific progress **See Assessment activity 21.1**	**M1** Explain why some developments have not had the impact expected of them **See Assessment activity 21.1**	**D1** Assess which influences have had the greatest impact on scientific progress **See Assessment activity 21.1**
P2 Discuss a recent controversial scientific topic that was in the media spotlight **See Assessment activity 21.1**	**M2** Explain why this scientific topic was treated as controversial by the media **See Assessment activity 21.1**	**D2** Assess whether the media is making a positive or negative contribution to understanding science **See Assessment activity 21.1**
P3 Describe the advantages and disadvantages of how some scientific discoveries are used in society **See Assessment activity 21.2**	**M3** Explain how some of the consequences that caused problems were overcome. **See Assessment activity 21.3**	**D3** Assess the potential impact of these consequences on quality of life or standard of living. **See Assessment activity 21.3**
P4 Describe some consequences associated with recent scientific discoveries/advances. **See Assessment activity 21.3**		

How you will be assessed

Your assessment could be in the form of:

- a report for a science journal

- an article for a student magazine

- a website e.g. explaining scientific discoveries and advancement.

Karl, 18 years old

I found this unit helped me focus on whether I wanted to work in a biological science or whether I should look at working in a job that covered all the sciences.

I liked finding out about discoveries in science especially where nanoparticles were used in sports clothes, medicine and cosmetics.

It was great seeing how robots could be used in areas where it was too dangerous for humans, showing how science is used in the world. We saw publicity shots from the army showing how their robots could blow up unexploded bombs. We also saw robots being used to produce microchips in computers.

Catalyst

New technology

What was the last piece of new technology you heard about, or the latest new medicine?

What was this new development designed to do? What other things could it be used for? Might any of these be dangerous?

21.1 Influencing and representing scientific progress

In this section: P1 M1 D1 P2 M2 D2

Key terms

Animal rights – the idea that animals should have basic rights like humans.

Civil rights – the rights of an individual human being.

Without your voice the torture will continue

Animals in Labs Need Your Help

www.idausa.org

People can feel very strongly about some scientific research, e.g. research on animals. They try to influence scientists.

Did you know?

It cost about £2.5 billion (€3 billion) to build the Large Hadron Collider in Switzerland. A new hospital might cost about £300 million, so about eight new hospitals could be built for the same amount of money.

Influencing science

There are many factors which influence scientific progress.

The effect that these influences can have varies.

Influence	Effect
Public concern	If the general public, and the media, are worried that the science being researched is not safe or could be misused then they may try and stop it being carried out. An example of this is the public's concern over genetically modified (GM) crops. Public concern can also speed up research. In the case of the swine flu pandemic of 2009, public concern meant that more money was spent on making a vaccine quickly
Financial matters	Research in some branches of pure science is very expensive. The Large Hadron Collider at CERN cost 3 billion euros and its research results may not benefit society for perhaps 50 or more years. It has relied on funding from several nations. Finance usually only comes from rich countries, so they control what research is carried out
Economic pressures	In times of recession, for example, companies have less money to spend. This can mean that money for research is hard to find. Prosperous countries can still support research, so some scientists may move countries to continue their research
Animal rights/civil rights groups	Protest groups may campaign to stop research happening if they believe it harms animals or people. Some protests are peaceful, but some very extreme groups use terrorism to frighten scientists and pressurise governments into stopping research work
Scientific opinion	Experienced scientists choose research topics and supervise students doing them. In this way scientific opinion is influenced. The senior scientists themselves may not be able to support a project because there is insufficient funding or parliament may have decided not to allow some research, e.g. that involving embryos or the cloning of genetic material
Overseas pressures	Other countries may have a monopoly on the raw materials needed for a new technology and they may use this as a way of influencing new developments. Objections to exploitation of a country's resources which may not benefit that country immediately can mean materials are difficult to find. Objections to developments may also be made because of cultural and religious beliefs
Funding for research	Quick returns on investments are required by companies funding research. This influences the type of research carried out. The research needs to have immediate and long-term benefits to recoup the money invested

Science and the media

Coverage of scientific and medical topics in the media shows our natural interest in these subjects. Some examples are shown in the table below.

Specialist reporting	Scientists publish their research results and exciting discoveries in specialist scientific journals. These are usually very technical and are not easy for the public to access. Examples are the science journal *Nature* and the medical journal *The Lancet*.
General reporting	Newspapers have science reporters whose job is to read the scientific and medical journals, and then to write about the results and discoveries in a way that the public can easily understand.
Films	Science is often represented in films. For example, the film *The Day After Tomorrow* shows devastating weather events caused by global warming. Films like this can be based on science, but can you trust that what they show could really happen?
Soaps	Television and radio 'soap operas' often tackle health or science issues in their storylines; for example, an elderly person might suffer from Alzheimer's disease.
Serials	An episode of the popular science fiction series *Torchwood* was broadcast on BBC Radio 4. It was set in the Large Hadron Collider in Switzerland and contained a mixture of scientific fact and fiction.
Internet	The Internet contains a vast amount of scientific information. A lot of it is of good quality and a lot of it is not. How does a reader searching for information on a scientific topic know whether to trust their source of information on the Internet?

Science focus

Genetically modified (GM) crops were developed to help food production in developing countries. The media published summaries of the scientific papers. From this the public became aware of drawbacks to using GM crops. Pressure groups became active and in this country the result has been that public opinion has prevented the widespread use of GM crops. In the USA and France, GM crops are more widely used. This shows that the influence of scientists, lobby groups, etc. in some countries has resulted in different outcomes.

Activity A

(i) Name two television programmes, radio programmes or films (not already mentioned) that represent science or medicine.

(ii) For each programme or film you have listed, say whether or not it covers science facts, or science fiction. How do you know?

BTEC Assessment activity 21.1 P1 M1 D1 P2 M2 D2

Pick a controversial scientific topic which was in the news recently. You must produce a presentation to show a group of non-scientists a balanced view of the science. In your presentation you must answer the following questions.

1 What factors have influenced the way the science has been researched? Which of these influences have had the greatest effect? **P1 D1**

2 What effect was expected from the scientific development? **M1** Explain whether it had this effect and why.

3 Why it was in the headlines? **P2 M2**

4 How did the media deal with the science? Think about how accurately they reported the facts and whether more focus was given to some parts of the research than others. **M2**

5 What effect do sensational headlines have on public perception of the science involved? There may be positive and negative effects. **D2**

Grading tip

Make sure you have discussed the topics asked for. You need to explain the information you have researched. Give all sides to an argument but give your own opinion and back it up with facts and figures. Don't forget to mention any resources you have used.

21.2 Using scientific discoveries

In this section: P3 P4

Wind turbines – just one of the ways in which a scientific discovery has been developed.

Key terms

Nanotechnology – a technology that deals with structures of 100 nm or less in size and involves developing materials or devices of that size.

Chemotherapy – the use of drugs to treat disease; the term is usually used in the treatment of cancer.

Have you seen any films made in the 1990s where someone is talking on a mobile phone as big as a brick? Modern mobile phones have benefited from scientific discoveries over the last 20 years. They are now much smaller and can give you instant images which you can print out on your computer. A similar technology is used in speed cameras. Mobile phones rely on satellites for good communications, but satellites can also be used for spying purposes.

How does society use scientific discoveries?

The table below lists some discoveries in science and the ways society has used them.

Activity A

Take one or two of the discoveries listed in the table that interests you and add some more examples to those already given.

Did you know?

A nanoparticle is very tiny, between 1 and 100 nm in diameter. A nanometre is one-billionth of a metre and the width of a human hair is 50 000 nm.

Discovery/development	Examples
Alternative energy sources	Nuclear power, wind and water turbines, solar, eco fuels (ecologically friendly fuels)
Computer-controlled developments	Robots, CAD/CAM (computer-aided design/computer-aided manufacturing) systems
Nanotechnology	Use of polymers in cosmetics, nanofibres in tyres, nanoparticles in clothing, medicine
Materials	Stainless steel, carbon fibre, plastics, foam
Medical apparatus	Laser instruments, **ECG**, **MRI**, **endoscopes**
Materials used in space	Non-stick substances, heat-resistant tiles
Communication systems	Satellites for **GPS** and TV, **microwaves**
Pharmaceuticals	Vaccinations

Advantages and disadvantages

For many scientific discoveries, there will be some negative as well as positive effects. Sometimes the public will decide that the advantages outweigh the disadvantages. For example, most people believe that, although **chemotherapy** causes sickness and hair loss, it is worth the discomfort if the chemotherapy cures their cancer.

Advantages and disadvantages of scientific discoveries.

Discovery/advance	Example	Advantages	Disadvantages
Medical advances that save lives	New drugs to help target cancer	Improved **life expectancy** and chance of recovery for patients	Very expensive; quality of life may be poor
Alternatives to **fossil fuels** to generate electricity	Nuclear power	Electricity generated without producing any CO_2 that can affect the climate	Nuclear waste is dangerous and needs to be stored for years
New technology for faster/better communication	Satellite phones	Used by people in remote places to call for help or coordinate aid to stranded people	Satellites can also be used to spy on people without their knowledge
New materials with better properties than others	Heat-resistant tile	Used on the space shuttle to allow it to re-enter the atmosphere without being destroyed	Costs of developing the material can be high and can use limited natural resources
Computer-controlled devices	Robots working in factories to perform repetitive or unpleasant jobs	Robots can often work faster or more effectively in areas not suitable for humans	Robot aircraft can be used by the military to attack the enemy without risking their own soldiers and this may make the military value life less
Nanotechnology	Nanoparticles are used in a wide range of products, e.g. drug delivery systems	Using nanoparticles to deliver drugs into the body means that the drug can get to exactly the right place, so less of the drug may be needed and it can work more quickly and effectively	There is concern that, as nanoparticles are small enough to cross cell membranes, they could cause biochemical damage in cells and tissues

Activity B

Write down two alternative forms of fuel not mentioned in the table above. List at least one advantage and one disadvantage for each one.

BTEC Assessment activity 21.2 P3

1 Find examples of scientific discoveries not already mentioned on these pages.
2 Explain in detail the nature of the scientific discoveries.
3 Describe some advantages and disadvantages for society of this discovery. P3

Grading tip

For P3, try to think about advantages and disadvantages that might not be immediately obvious, as well as the more obvious ones. They might be to do with people's enjoyment of life, rather than global issues that may affect the whole of society.

Use books, newspapers, scientific publications (e.g. *New Scientist*) and the Internet to help you with your research.

Robots on a car production line, an example of computer-controlled devices.

21.3 Consequences of scientific discoveries

In this section:  M₃ D₃ P₄

Home testing for diseases

Modern medicine has made it much easier and quicker to test and monitor the progress of a whole range of diseases from cancer to Alzheimer's.

Some tests, like the testing of blood sugar levels in patients with **diabetes**, are simple enough to do at home. Other tests normally need patients to consult a doctor or are done in hospitals. But some drugs companies have developed kits which allow people to test for diseases such as cancer or **AIDS** at home. Other companies are working on ways to allow us to test our **DNA** to find out whether our **genes** will make it more likely we will develop diseases like Alzheimer's in the future. They claim that these kits will be very popular as they will help people diagnose diseases earlier. They say this will save lives and improve people's quality of life.

> **Key term**
>
> **Stem cell** – a cell taken from a living organism which is able to develop into a range of different types of tissue.

What could the consequences be?	How are these possibilities being controlled?
Many doctors oppose the use of self-testing kits. • The tests might not be as reliable as the tests used in hospitals. Some patients might be told they are clear of the disease when they are not. Others may be falsely told that they have the disease and be caused unnecessary worry. • If tests are done by a doctor, patients may be able to get immediate treatment or counselling. This wouldn't happen straightaway for a home test.	Doctors think it is very important that laws or regulations be brought in to make sure that the testing kits should only be used under the instruction of a medical professional. Because some of these kits can now be ordered on the Internet it may be very difficult to control this.

> **Activity A**
>
> Use the Internet to find out about miniaturisation, such as the production of microscopic motors or cameras. Produce your own table like the ones on this page.

Genetic modification and cloning

Scientists are now able to produce genetically modified (GM) animals. They can insert genes from other animals into, say, a sheep. The new GM sheep may then be able to produce better milk or wool or even produce useful medicines in its milk. Scientists could then clone the sheep to produce large numbers of GM animals.

What could the consequences be?	How are these possibilities being controlled?
• Cloned animals may be less healthy than normal animals. Dolly died at the age of six; sheep normally live to 11 or 12 years old • If large numbers of cloned GM animals are released into the wild they may affect the ecosystem in an unpredictable way • The technology now exists to clone human beings. Most people are very opposed to this on ethical grounds	• Animal cloning is allowed in most countries in the world, but there are strict regulations about the use of GM technology; in some countries, such as the UK, they can be used in research but not used commercially • Cloning gets a lot of media attention and because of this research into the cloning of a complete human being has been made illegal in most countries. Even research using human **stem cells** is banned in some countries, often because of objections from religious groups

Dolly the sheep was the first cloned mammal.

BTEC Assessment activity 21.3 M3 D3 P4

Mobile phone technology has developed enormously since it was first used in the 1970s. How will this technology develop in the future? Already some mobile phones can be used like mini computers.

1 Prepare a presentation to describe how the advances in mobile phones and computers have been useful to society.

2 Have there been any problems caused as a consequence of the way in which they are used? Have these been overcome or controlled – for example have there been any laws or regulations necessary as a result of this use? M3 P4

3 Sum up your presentation by explaining whether you think mobile phones have a positive or negative impact on the quality of life and whether this will be true in the future as well. D3

Grading tip

For P4 you need to describe some positive and negative consequences of the use of mobile phones. You could think about consequences for individuals and for society.

For M3 you will need to find out about things that have been done to overcome or control any of the problems which are a consequence of the use of mobile phones.

For D3 you will need to think about both sides of an argument – and you must show that you have the imagination to think about what could happen in the future as well.

WorkSpace

Claire Edwards
Science Editor for a national newspaper

In my job I decide which scientific developments and stories would make good copy for our pages, and then work with our writing team to produce an article for the next edition.

Sometimes the news comes to us – I read a lot of press releases written by scientists who want publicity for their research. But we also have to work quickly to write articles on the scientific background to news stories, like demonstrations about climate change and the use of animals in research. We all bring our ideas about the big stories to editorial meetings. Once we've decided which stories to cover, one of the writing team needs to be briefed to write the story.

It is so important to get the facts right – it's easy to make something like a new drug for treating cancer sound really exciting – but we need to remember that we are often only reporting early trials. In stories that are controversial and when the facts aren't clear we need to make sure that we give both sides of the argument.

Think about it!

1 Why might editors be under pressure to make a scientific discovery sound more important than it is?

2 What could the danger be in reporting a new treatment in this way?

3 Can you think of other subjects that might be controversial?

Just checking

1 Give an example of an area of science where public protests have made it more difficult to carry out research.
2 Where do the general public get their information about science from? Are these sources always reliable?
3 What are GM organisms? Give one example of a GM organism and explain why it was developed.
4 What is nanotechnology? Describe disadvantages or undesirable consequences which could occur as a result of using nanotechnology.
5 Give one example of a scientific advance or discovery which has been controlled in order to prevent possible undesirable consequences.

edexcel

Assignment tips

To get the grade you deserve in your assignments remember to do the following.

- You will need to do quite a lot of reading and research to help you describe and explain some of the impacts which science has on the world. This is very time-consuming so it's important to spread your research over a sensible period of time.

- When you are using a search engine to help you find information about a particular application, you may have to type in different sets of words to get what you want. For example, 'carbon fibre' will bring up many hits, most of which you do not want. Try and refine your search by adding other words such as 'carbon fibre uses'.

- Think very carefully about whether material you find on the Internet or in other sources is relevant. Is it up-to-date? Is it from a reliable source? Does it actually link in with the grading criteria? Is it biased? If it is, you could try and find information from a source that takes the other point of view.

Some of the key information you'll need to remember includes the following.

- The speed and success of scientific research is affected and influenced by a number of different factors. These include public opinion, availability of funding and the opinions of other scientists.

- The general public often form their opinion from the way in which science is presented in the media. These include television programmes, films and the Internet.

- Scientific discoveries and advances have helped to improve life for individuals and society. Still, some of these discoveries have been used in ways which have created problems, disadvantages or ethical issues which need to be solved or controlled.

- Examples of some of these discoveries and advances include GM organisms, cloning, nanotechnology and new methods of communication such as mobile phones.

You may find the following websites useful as you work through this unit.

For information on...	Visit...
controversial advances in biology (cloning, GM etc.)	Bioethics Education Project
controversial advances in physics (mobile phones, robotics etc.)	Physics and Ethics Education
nanotechnology	Nanotechnology Now

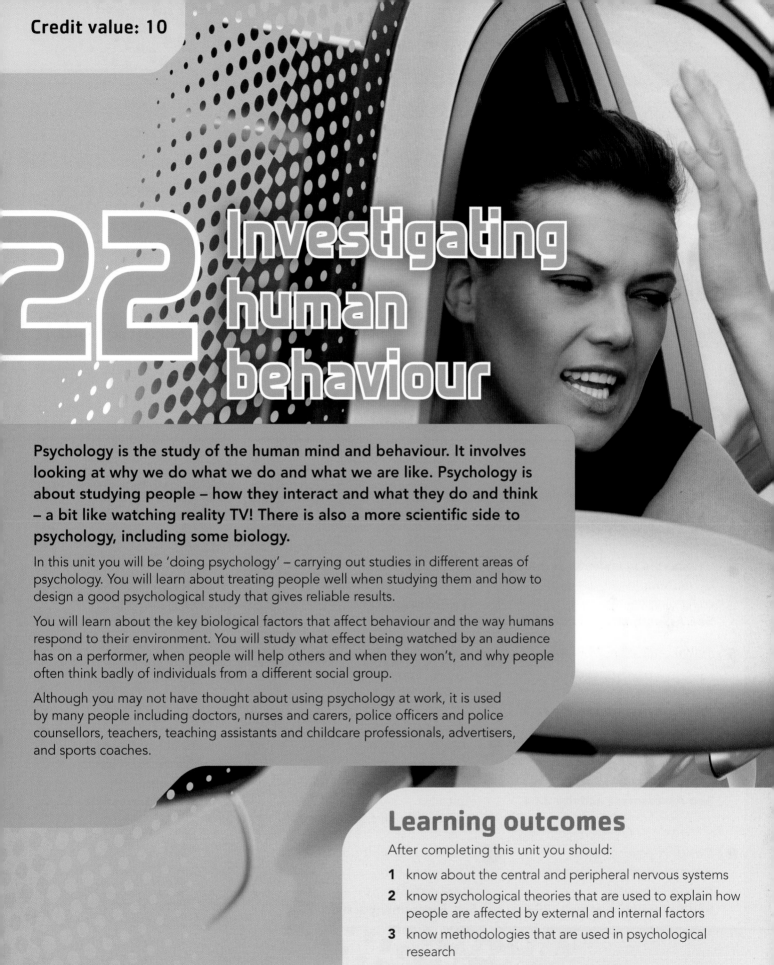

Credit value: 10

22 Investigating human behaviour

Psychology is the study of the human mind and behaviour. It involves looking at why we do what we do and what we are like. Psychology is about studying people – how they interact and what they do and think – a bit like watching reality TV! There is also a more scientific side to psychology, including some biology.

In this unit you will be 'doing psychology' – carrying out studies in different areas of psychology. You will learn about treating people well when studying them and how to design a good psychological study that gives reliable results.

You will learn about the key biological factors that affect behaviour and the way humans respond to their environment. You will study what effect being watched by an audience has on a performer, when people will help others and when they won't, and why people often think badly of individuals from a different social group.

Although you may not have thought about using psychology at work, it is used by many people including doctors, nurses and carers, police officers and police counsellors, teachers, teaching assistants and childcare professionals, advertisers, and sports coaches.

Learning outcomes

After completing this unit you should:

1 know about the central and peripheral nervous systems

2 know psychological theories that are used to explain how people are affected by external and internal factors

3 know methodologies that are used in psychological research

4 be able to design a psychological study.

Assessment and grading criteria

This table shows you what you must do in order to achieve a **pass**, **merit** or **distinction** grade, and where you can find activities in this book to help you.

To achieve a **pass** grade, the evidence must show that the learner is able to:	To achieve a **merit** grade, the evidence must show that, in addition to the pass criteria, the learner is able to:	To achieve a **distinction** grade, the evidence must show that, in addition to the pass and merit criteria, the learner is able to:
P1 Outline the structure and function of the central nervous system **See Assessment activity 22.1**	**M1** Describe how the central nervous system works to process information **See Assessment activity 22.1**	**D1** Explain how the central and/or peripheral nervous system operate for a particular purpose **See Assessment activity 22.1**
P2 Outline the structure and function of the peripheral nervous system **See Assessment activity 22.1**	**M2** Describe how the peripheral nervous system is involved at times of stress **See Assessment activity 22.1**	
P3 Outline the biological and social explanations of how various factors affect human behaviour **See Assessment activity 22.2**	**M3** Describe the research which has been associated with these explanations **See Assessment activity 22.2**	**D2** Assess how the psychologists involved in the research reached their conclusions **See Assessment activity 22.2**
P4 Outline the main features of methodologies used in psychological research **See Assessment activity 22.3**	**M4** Discuss the strengths and weaknesses of each psychological research methodology **See Assessment activity 22.3**	**D3** Explain how a suitable methodology is chosen for a specific study **See Assessment activity 22.3**
P5 Outline the methodologies used in a psychological research case study **See Assessment activity 22.3**		
P6 Identify the main ethical guidelines related to psychological research **See Assessment activity 22.4**	**M5** Describe the consequences of not carrying out a psychological investigation ethically **See Assessment activity 22.4**	**D4** Evaluate the suitability of the research study for its intended purpose. **See Assessment activity 22.4**
P7 Design a psychological research study to investigate human behaviour. **See Assessment activity 22.4**	**M6** Describe the methodologies selected to carry out the investigation. **See Assessment activity 22.4**	

How you will be assessed

Your assessment could be in the form of:

- a diagram e.g. of the peripheral and central nervous system

- written material or a presentation e.g. on fight, flight and arousal

- a discussion paper e.g. on factors affecting human behaviour

- a write-up for a study investigating an aspect of human behaviour.

Steven, 16 years old

I really enjoyed this unit – it was fascinating to learn why we behave the way we do.

My favourite practical was the experiment we did with a computer game. Two of us had played the game before and two had not. For the study we each played the game twice and recorded our scores. Once we had the class watching and the other time we played on our own (without an audience). We found that the two who had played before did better with an audience than without and those who had never played before did not do very well at all, but did even worse with an audience. I was surprised that I did better with people watching – maybe I'm a show-off! My results agreed with what we learned, which I was happy about.

Catalyst

Helping out

- List four situations where someone would need help (e.g. an older person opening a heavy door, or someone in a library who has just dropped a load of books).

- Draw up a questionnaire to find out how many people would help in the situations you listed.

22.1 The central and peripheral nervous systems

In this section:

Key terms

Central nervous system (CNS) – the brain and spinal cord.

Peripheral nervous system (PNS) – the nerves that connect the CNS to the rest of the body.

Arousal – a state of excitement or anxiety.

All living things respond to stimuli (changes in their environment). These responses help them survive. Humans have some simple (instinctive) behaviour patterns where we don't think about what we do. For example, we duck if we see a ball coming towards our head. These reactions involve reflexes. We also show very complex behaviour, because we have large, well-developed brains. Our behaviour can be influenced by other people and how they make us feel. How we feel depends also on our brains and how neurons communicate with each other by electrical signals and by chemicals.

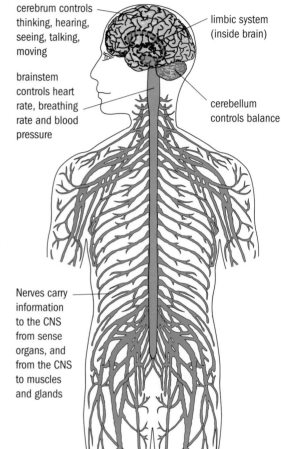

cerebrum controls thinking, hearing, seeing, talking, moving

limbic system (inside brain)

brainstem controls heart rate, breathing rate and blood pressure

cerebellum controls balance

Nerves carry information to the CNS from sense organs, and from the CNS to muscles and glands

☐ Central nervous system (CNS)

☐ Peripheral nervous system (PNS)

The central nervous system and peripheral nervous system.

Did you know?

The brain cannot function properly without information about the rest of the body or the outside world and when people suffer sensory deprivation (kept in the dark with no sounds and nothing to touch) they begin to hallucinate.

The central nervous system (CNS)

Structure

The **central nervous system** (CNS) is made up of the brain and the spinal cord. Because the front part of the spinal cord receives a lot of information from sense organs in the head, the brain has developed over time to have complex parts and functions.

Function

The CNS receives information from various parts of the body, processes it and sends information back to muscles and glands (**effectors**) so that the body can make suitable responses. Sense organs send information to the CNS about the outside world. Some of the functions of the CNS are given below.

- The folded outer part of the brain, called the **cerebrum**, controls movement, memory, hearing, talking, reading and writing, learning, thinking and solving problems.

- The **cerebellum**, at the back of the brain, controls balance.
- Deep inside the brain is the **limbic system**. This plays a key role in emotions and memory.
- The **brainstem** controls heart rate, breathing rate and blood pressure.

The peripheral nervous system

The **peripheral nervous system** (PNS) is made up of the nerves that carry information between the body (organs and limbs) and the CNS. The PNS connects the CNS to the rest of the body so the brain can get information about the body and the outside world. For example, if you see a lion, your eyes (sense organs) sense it and send information along the nerves of the PNS to your brain. Your brain (CNS) has a memory that lions are dangerous and it sends impulses along nerves (PNS) to your leg muscles (effectors) so you can run away.

The stress response

Running away from a lion makes sense. If you avoid getting eaten you survive. If you survive long enough you may reproduce and pass on your genes to the next generation. This is what evolution is all about – those best adapted (in this case seeing the lion, recognising danger and escaping) survive and breed. They pass the genes for those features, such as running fast, to their offspring. Over very long periods of time, if some animals in a population adapt to different conditions they become different enough from the others in the initial population to be a new species. Modern humans all belong to one species and are different from their extinct ancestor species.

Activity A

Name the five main sense organs. Discuss with your partner what sort of information each of the sense organs conveys to the central nervous system.

Many ordinary situations are potentially dangerous. Even seeing this picture can cause us to have a fear arousal response.

Activity B

The hormone **adrenaline** is involved in the stress response. When you see a dangerous situation, adrenaline is released from adrenal glands and from some nerve endings. The adrenaline makes your heart rate, breathing rate and blood pressure increase and you become **aroused**. How do these responses help you escape danger?

PLTS

Independent enquirers

You will develop your skills as an independent enquirer when you research to find out more detail about the structure and functions of the CNS and PNS and to find out about the nerve pathways and responses involved.

BTEC Assessment activity 22.1 P1 M1 D1 P2 M2

You are a psychology student at university taking part in an open day for school pupils.

1. Make a poster with an annotated diagram that shows the structure and functions of the central and peripheral nervous systems. **P1 M1 P2**

2. On your poster, indicate the nerve pathways and responses involved when someone is swimming and sees a shark's fin nearby. **M1 D1 M2**

Grading tip

To achieve **P1** and **P2** you will need to show the main structures and functions of the CNS and PNS on your diagram. To get **M1** and **M2** describe how the CNS processes information and describe which organs and nerves are involved in responding to a stressful situation. To get the distinction grade **D1** explain how the CNS and PNS work to enable you to see the shark, sense danger and swim away as fast as you can.

22.2 Responding to external and internal factors

In this section: D2 P3 M3

Key terms

Biological theory – used in psychology to explain human behaviour in terms of physiology (physical and chemical processes in the body).

Altruistic behaviour – helping someone else without expecting to get a reward.

Social identity theory – a theory developed by Tajfel in 1978 that says human behaviour is determined by the social groups we belong to.

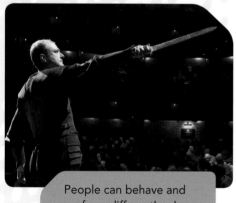

People can behave and perform differently when they are being watched.

The inverted U demonstrates that arousal increases at first performance but then it falls. Catastrophe theory shows a more sudden drop in performance after an optimum level of arousal is reached.

Activity A

Think of something you are good at. Are you better or worse at it when being watched? Does it make a difference who is watching you?

Some human behaviour is driven by instincts, such as avoiding danger and finding food and a mate. Our physiology and genes play a key role (internal factors). But we have complex brains and we live in groups so our behaviour is also affected by other people (external factors).

Biological theory

Biological theory is used to explain our behaviour in terms of our physiology. For example, you have seen on pages 364–365 that the nervous system and hormones, such as adrenaline, get us ready for fighting or fleeing when faced with danger.

Audience effect

Many actors say they get scared before a performance and this arouses or excites them and they perform well, due to the effects of adrenaline. However, some people perform badly when they are nervous and anxious. Many young drivers fail their first test because they are aware they are being assessed and do not drive as well as when having a normal lesson. Their fear and arousal has had a negative effect.

Some stress is good for us but too much can be harmful. In the long term it can weaken our immune system and cause illness.

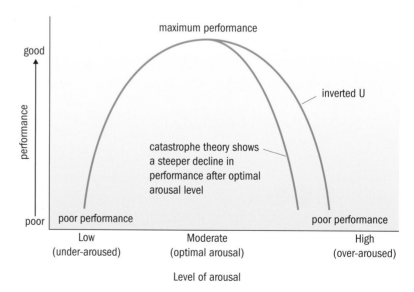

The psychologist Zajonc (pronounced Zy-unce) produced **drive theory** to explain the audience effect. Zajonc said that the presence of an audience causes a performer to become aroused. This usually causes the performer to perform better if they are already skilled, but causes the performer to perform worse if they are new to the skill.

Altruistic behaviour

Case study: The story of Kitty Genovese – a real life story

In 1964 in New York a 28-year-old woman called Kitty Genovese was returning home when she was attacked. She screamed and neighbours heard her, but no one did anything. She was murdered. A reporter later said that 3 people knew what was happening and did nothing.

Why do you think no one did anything to help?

Two psychology researchers, Latané and Darley, said it was bystander apathy. There might be a diffusion of responsibility, which means that if there are other people nearby we think someone else will act, or if others do not act then there is nothing wrong. Latané and Darley set up an experiment to show this. Participants were in a room, waiting to be interviewed. Also in the waiting room were confederates (part of the experiment). When smoke poured under the door, the participants watched the others, who were (deliberately) doing nothing. Even though it appeared to be a very dangerous situation, the participants also did nothing.

Social identity theory

Some people have irrational dislikes (prejudices) of certain groups of people. Individuals identify with certain groups (in-groups) and feel hostile to other groups (out-groups). Psychologists studying this phenomenon have called it **social identity theory** and have used it to help explain human behaviour such as violence between fans of opposing football teams.

Activity B

Use the Internet to find out why people do or don't behave altruistically in different situations. Search for studies by Piliavin and Darley and Batson.

Science focus

The researchers Reicher and Haslam conducted the BBC prison experiment which looked at group behaviour. They randomly assigned participants to roles as either guards or prisoners, and found that the 'prisoners' formed a strong social identity group.

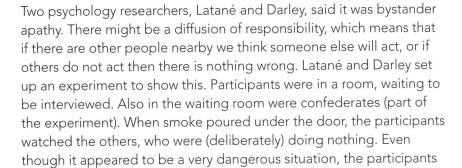

BTEC Assessment activity 22.2 D2 P3 M3

You are a cognitive behavioural therapist. You have to outline the various factors that affect human behaviour for clients who have anger management problems.

Produce a leaflet that you can use to help explain these factors to your clients.

Include the following factors in your leaflet.

1 Describe the biological and social explanations that affect human behaviour. P3
2 Describe some of the research that supports these explanations, and assess it.  D2 M3

Grading tip

To get M3 you need to describe the research – you need to include: who did it, when it was done, who the participants were and how many there were, and what the results were. You can use psychology textbooks and the Internet to find out about these studies.

For D2 you need to assess the research: Was it ethical? Did it involve deception? Were all variables controlled? Was it reliable and valid?

22.3 Methodologies used in psychological research

Key terms

Laboratory-based experiment – an experiment that is carried out in a laboratory.

Field-based experiment – an investigation similar to a laboratory-based experiment, but carried out in a natural setting.

Naturalistic observation – watching people in a natural setting with no manipulation of independent variables.

Questionnaire – a self-report method of investigation, in which subjects directly answer questions about their actions, thoughts and behaviour.

Questionnaires may be used to find out about people's attitudes – but we can't be sure if people are telling the truth.

 Activity A

List three field-based investigations on behaviour that you could carry out in the classroom.

Carrying out a psychological study

Psychologists use a variety of methods to carry out their studies, depending on the area of psychology they are investigating. They use scientific method.

- The researchers may start with an observation of behaviour in a few people – at this stage evidence is anecdotal. It has not been tested so is not valid.

- They then make a hypothesis – a simple, testable statement.

- The researchers then design and carry out studies to test the hypothesis. They set up studies to measure how a certain behaviour (the **dependent variable**) changes according to a particular circumstance (the **independent variable**). They keep all other variables the same so that the dependent variable is changing *only because* of the independent variable. This makes the experiment valid – which means it measures what it says it measures.

- They and other researchers repeat their experiments and if they get the same results the experiment is reliable.

Types of psychological studies

Experimental method

Laboratory-based experiments are best for studying how biological factors, such as the brain, nerves and hormones, affect behaviour. Researchers can change the independent variables and keep other variables, such as time of day, background noise or temperature, the same. They measure how the dependent variable changes. In laboratory-based experiments, for example finding out if caffeine affects people's reaction times, people are not in a natural setting and they know they are being watched (audience effect) – these factors may affect their behaviour. Another disadvantage is that people have to agree to participate and it may be that certain types of people are more likely to participate and be studied.

Field-based experiments are like laboratory-based experiments but are done in more realistic settings. An example would be watching to see if people will sit on a chair, in a crowded airport lounge, if it has a coat on it.

Naturalistic observations

Like field experiments, these are carried out in realistic settings. However, the researcher simply observes the participants and has no control over the variables – the independent variables change naturally. Some examples of different types of **naturalistic observation** are given in the table.

Type of observation		Example
Covert	Participant	Researcher pretends to be a football supporter and joins a group of fans to observe whether they are more hostile to the opposing team when playing away or at home. The group of fans don't know they are being watched so their behaviour won't change, but it is difficult to control the variables.
	Non-participant	Researcher attends football matches and secretly watches the behaviour of a group of supporters but he has not joined the group. Again, the group of fans don't know they are being watched so their behaviour won't change.
Overt	Participant	Researcher joins a group of homeless people. He says he wants to find out how they help each other. The presence of the researcher could change the behaviour of the people he is watching.
	Non-participant	Researcher openly watches children at play to investigate levels of aggression in boys and girls. The presence of the researcher could change the behaviour of the children.

Questionnaires

There are two main types of question that can be used to create a questionnaire.

- **Closed questions** – these are questions with only a restricted set of possible answers, such as a rating scale (e.g. 1–10) or yes/no options. Closed questions are easy to analyse but don't really investigate people's feelings or attitudes in depth.

- **Open questions** – these are questions without a restricted set of answers. For example, participants might be asked to describe how they feel about something, or they might be given a word and asked what other word comes into their mind first. Open questions can provide researchers with more information, but the analysis is usually more difficult and time consuming.

Science focus

Psychologists sometimes use **case studies**. This is where one particular person or group is studied in depth. For example, a researcher may study how a patient with damage to a particular part of the brain behaves in order to find out what that part of the brain does. The individual may not behave the way some other people would, because of some other factor (such as how he/she was brought up), so the findings from the case study cannot be generalised to the whole population.

BTEC Assessment activity 22.3 D3 P4 M4 P5

You are a psychologist and have been asked to design some studies for psychology pupils to carry out.

1 Prepare a table outlining the main features and strengths and weaknesses of the methodologies that the pupils will be using, including case studies. **P4 M4 P5**

2 For each of the studies below choose a suitable methodology, explain how it would be used and explain why that methodology is most suitable:

(i) Why do people shop at supermarkets? (ii) Do boys in primary schools play more aggressively than girls? (iii) Does background noise affect pupils' ability to solve maths problems? (iv) How does alcohol affect people's reaction times? **D3**

Grading tip

For **P5** you will need to use the Internet to find out more about case studies and give a particular example. A useful website is **Wikipedia: case studies in psychology**: follow the links to famous case studies in psychology.

22.4 Designing a psychological research study

In this section: D4 M5 P6 M6 P7

Key terms

Ethical guidelines – a set of rules laid down by the British Psychological Society that researchers have to follow.

Quantitative data – data in the form of numbers.

Qualitative data – data not in the form of numbers, but in the form of written notes or pictures, for example.

Descriptive statistics – ways of analysing quantitative data, such as graphs and calculations.

When psychologists covertly observe people in a natural setting they must follow ethical guidelines, such as not distressing anyone and not intruding.

Unit 15: To find out more about analysing data see pages 288–289).

Activity A

Imagine you are covertly observing how homeless people help each other. How would you go about it? What would you have to do, after the study, to comply with ethical guidelines?

Ethical guidelines

Psychologists must follow the **ethical guidelines** laid down by the British Psychological Society (BPS) in order to protect participants. These guidelines are based on what society deems to be acceptable. Any psychologist who does not comply with the ethical guidelines will face a reprimand from the BPS and may have their name removed from the register, so that they will no longer be allowed to carry out research.

Some of the responsibilities of a researcher under these guidelines are listed below.

- Explain to participants what the study is about and ask them to give **informed consent** if they are happy to take part.

- Not publish the names of participants and assure participants that the study is **confidential**.

- Allow participants to stop taking part, at any time during the study, if they wish – i.e. they have the **right to withdraw** from the study.

- The researcher must **debrief** participants afterwards – the researcher must tell the participants exactly what the study is about and ask them if they are still happy to have their data used.

Planning a psychological research study

When a team of psychologists has developed a hypothesis to test, they will design their study. They decide: which methodology is most suitable to use; who the participants will be; what they will tell the participants to get consent (debrief to follow); what the independent, dependent and control variables are; whether they will gather **qualitative** data or **quantitative** data; and how they will analyse the data.

Analysing data

If the results of a study are in the form of quantitative data, **descriptive statistics** can be used – researchers may find the mean, median, mode and range and use graphs such as line graphs, bar graphs and pie charts to visually represent their findings.

Report writing

Psychologists write reports of their findings. These are then peer-reviewed (checked by other psychologists) and published in journals to inform other people. Psychologists have to write their reports in a certain way.

BTEC Assessment activity 22.4 · D4 M5 P6 M6 P7

You are a psychologist and you want to test the hypothesis 'People in smaller communities are more altruistic than people in larger communities'. You decide to leave some self-addressed, stamped letters in different settings. You expect altruistic people to post them.

1 Write a plan for carrying out this study. **M6 P7**

2 How could you tell, when letters came to you by post, which came from which area? **M6 P7**

3 What type of investigation is this? **M6 P7**

4 What are the main ethical issues involved in this study and what are the relevant guidelines? What will happen if these guidelines aren't followed? **M5 P6**

5 Is this a suitable study for testing the hypothesis? **D4**

Science focus

A report should include:

- an informative title
- the aim of the research, the background and any ethical issues
- the hypothesis
- the method
- the results, plus graphs and statistics
- conclusions
- an evaluation of the study.

Grading tip

For **D4** you need to evaluate how suitable the study is to test the stated hypothesis. You should ask yourself whether all control variables are kept the same and whether the study is valid.

WorkSpace

Carol Dombrowska
Accredited CBT Counsellor, Glynton Associates

I work as a cognitive behavioural therapy (CBT) counsellor, helping people to recognise and control thoughts that affect their behaviour negatively.

Each counselling session lasts fifty minutes, then I write up notes.

My main responsibilities are:

- assessing clients to find out their goals and what might benefit them, explaining what CBT is and how we work and agreeing a contract with them
- working with regard to ethics and good practice
- working under supervision and keeping my manager updated.

One example of my work is helping clients with anxiety. A vicious cycle might start with the thought 'I will have a panic attack if I go shopping', leading to feelings of anxiety and fear, which cause physical symptoms such as breathing difficulties and increased heart rate. Therefore the person avoids shopping, and in consequence never 'unlearns' the fear. CBT helps by tackling the thoughts that cause anxiety and thus inappropriate behaviour, and reducing the physical response, leading to a cycle of more healthy feelings, physical reactions and behaviour.

Think about it!

1 What do you do if something in a session is worrying for you?

2 Are all problems suitable for CBT?

3 Is CBT better than medication for mental health issues?

Just checking

1 Is the brain part of the central or the peripheral nervous system?
2 Who was Kitty Genovese and what is her story?
3 Does the audience effect make someone perform better or worse if they are already good at a task?
4 Explain what is meant by a laboratory experiment. What are the advantages and disadvantages of using a laboratory experiment for psychological research?
5 Explain two ethical guidelines that should be followed when carrying out psychological research.

Assignment tips

To get the grade you deserve in your assignments, remember the following.

- Always include evaluation comments, for example, say what is good or bad about a study or theory.

- When asked for two features, such as the 'structure' and 'function' of the CNS and PNS, make sure you address both.

Some of the key information you'll need to remember includes the following.

- Biological information – the structure and function of the central and peripheral nervous systems and how arousal works.

- Social information – who helps (altruism), the effects of the audience and arousal, and the effects of being in a group (social identity theory).

- The three main methodologies used in psychological research – experiment, questionnaire and observation.

You may find the following websites useful as you work through this unit.

For information on...	Visit...
central and peripheral nervous system	The purchon biology website BBC Science
arousal	Brian Mac sports coaching
audience effect	Psychlotron
social identity theory	PsychExchange resources
altruism	S-cool psychology
methodology	Gerard Keegan's psychology site

Glossary

555 timer – device used in timing circuits that allows a signal to go on and off at pre-set times

abiotic factors – the non-living components of an ecosystem

active site – the part of an enzyme where a chemical reaction happens

adenine – a base that pairs with thymine to make a base pair in DNA

adjuvant – a substance that increases the effects of another drug

adrenaline – a hormone secreted from the adrenal glands; involved in the 'fight or flight' response

aerobic respiration – the process of making energy in the body that requires oxygen

agglutination – the clumping together of red blood cells when different blood types are mixed

AIDS – Acquired Immune Deficiency Syndrome; caused by the HIV virus that attacks the immune cells making a person vulnerable to infections

allele – a gene that exists in one of two or more alternative forms and that codes for a certain characteristic

alternating current – a current that regularly changes direction with time

Alzheimer's disease – a physical disease affecting a person's mental processes such as memory, movement and ability to think clearly

amino acids – the building blocks that make up a protein

ammeter – instrument that measures current

amplitude – the maximum displacement of a wave from its equilibrium position

amylase – an enzyme that catalyses the breakdown of starch

anaemia – condition caused by a lack of iron

analogue – signal that can take a range of values

anode – contact where there is a surplus of positive charge

anorexia nervosa – a psychological eating disorder in which the sufferer does not eat or eats very little

antenatal screening – testing for diseases or conditions in a fetus or embryo before birth

antibody – protein produced in the blood that attacks and kills antigens such as bacteria or viruses

anti-diuretic hormone (ADH) – a hormone released from the pituitary gland in response to low water levels in the body

antigen – a foreign substance that stimulates the production of antibodies

artery – blood vessel that carries oxygenated blood to the body

asteroid – a small, rocky object that orbits the Sun

atom economy – a measurement of how efficient an industrial process is; a reaction has a high atom economy if most of the atoms in the reactants end up in useful products

autonomic nervous system – nerves that automatically control some normal bodily functions

B lymphocyte – a cell that produces antibodies

bacteria – single-celled microorganisms; one of the kingdoms of organisms

biomass – the total mass of living material *or* plant or animal material that can be used as a fuel

biosphere – the part of the Earth that is habitable

biotic factors – the living components of an ecosystem

bipolar junction transistor (BJT) – type of transistor; often used in switching circuits

black hole – an area in Space in which the gravitational field is so strong that nothing, including light, can escape

booster injection – an additional dose of an antigen given to strengthen immunity

brainstem – the lower part of the brain that continues into the spinal cord

bulimia nervosa – a psychological eating disorder in which the sufferer eats large amounts and then vomits intentionally

buzzer – device that provides a sound output

cancer – uncontrolled cell division leading to formation of a tumour or growth

capacitor – device that stores charge

capillaries – very thin blood vessels that connect arteries to veins

carcinogen – a substance that can cause cancer

cardiac arrest – when the heart stops beating temporarily or permanently

cardiovascular – relating to the heart and blood vessels

catalase – an enzyme that catalyses the breakdown of hydrogen peroxide

cathode – contact where there is a surplus of negative charge

cell (battery) – a device used to supply electrical energy from chemical energy

cell (biological) – the smallest structural unit of an organism

cellulose – a polysaccharide that makes up plant cell walls

central nervous system (CNS) – the brain and spinal cord

cerebellum – part of the brain; responsible for balance

cerebrum – the largest part of the brain; responsible for movement, memory, hearing, talking, reading and writing, learning, thinking and solving problems

chlorophyll – green pigment required for photosynthesis; found in the chloroplasts of plant cells

closed question – a question with a restricted set of answers

comet – object made of frozen gas and ice that has come from outside our Solar System

community – a group of organisms that live together and are dependent on each other

complementary base pairing – the pairing of adenine with thymine and cytosine with guanine in DNA

compressive force – a force that acts to 'squash' a material

conduction – the transfer of thermal energy by the vibrations of atoms, and by delocalised electrons in metals

confidential – kept secret

conservation – the protection and preservation of something, usually natural resources such as plants and animals

consumer – an organism that feeds on other organisms

convection – the transfer of thermal energy by the movement of a liquid or gas

covert – secret; a covert observation is one where the participants do not know they are being observed

current – the flow of charge in an electrical circuit

cytosine – a base that pairs with guanine to made a base pair in DNA

debrief – when, at the end of a psychological study, participants are told all about the study and their part in it

deceleration – negative acceleration; to slow down

declination – the angular distance 'north' or 'south' of the celestial equator; similar to latitude on Earth

denature – to change the structure of a protein or enzyme causing it to become inactive; caused, for example, by high temperatures or a change in pH

deoxyribonucleic acid (DNA) – a large molecule carrying genetic information found in the cells of living organisms

dependent variable – the variable that is measured in an experiment

diabetes – a disease caused by too little insulin being produced or the inability of the body to properly use insulin resulting in poor regulation of blood glucose levels

differentiation – the development of cells to perform specific functions

digital – signal that has discrete values (e.g. high or low, 1 or 0)

diode – device that allows electricity to flow in only one direction

direct current – a current that only flows in one direction

dissolved oxygen meter – a device for measuring the amount of oxygen dissolved in a solution

distance ladder – the succession of methods used to determine distances to astronomical objects

dominant gene – a gene that is always expressed; represented by a capital letter

Doppler effect – a change in frequency of a wave seen when an observer is moving relative to the source of the wave

drive theory – the theory that a person will perform their dominant response – the response most likely given the individual's skill – in the presence of an audience

eating disorder – psychological disorder involving abnormal eating habits

ecological niche – the role an organism plays in its community, including what it eats

effector – a muscle or organ that can respond to a stimulus from a nerve

electrical conductivity – a measurement of how well a substance conducts an electric current

electrical generator – a machine that produces electricity

electrocardiogram (ECG) – a measure of the electrical activity of the heart

electrolyte – chemical substance that is used in a battery

embryo – a developing organism; either an animal in the womb or egg, or a plant in a seed

endoscope – a long thin flexible tube with a light source and camera that provides images of the inside of the body

energy efficiency – the fraction of energy that is usefully used

energy transfer – the change of energy from one type to another

enzyme – a protein made by cells that acts as a catalyst (speeds up chemical reactions)

equilibrium position – the point where there is no disturbance

ethical issue – a situation in which a decision has to be taken on whether a particular action is right or wrong

excretion – the elimination of the waste products of metabolism from the body

extinction – when a species no longer exists

extracellular – outside a cell

fetus – a developing mammal after the fertilisation stage and before birth

filament lamp – device that provides a light output; contains a filament, usually made from tungsten

fixed resistor – a fixed value of resistance that limits current flow

follicle-stimulating hormone (FSH) – female hormone that helps maturation of the follicle before ovulation

fossil fuels – fuels such as coal, gas and oil formed from the remains of prehistoric animals and plants

frequency – the number of cycles in one second

fungi – one of the kingdoms of organisms

galaxy – a cluster of billions of stars held together by gravity

gene – the basic unit of genetic material in an organism; a particular section of DNA

genetic engineering – manipulating genes in an organism, usually to produce a desired characteristic

genetically modified crop – a crop plant with genes altered to give characteristics such as disease resistance

genotype – the particular type and order of genes an organism has

global positioning system (GPS) – a system in which radio signals from satellites are used to determine the receiver's location on Earth

glucagon – a hormone produced in the pancreas when blood glucose levels get too low; converts glycogen to glucose

growth hormone – hormone produced by the pituitary gland that stimulates growth

guanine – a base that pairs with cytosine to make a base pair in DNA

habitat – the place where an organism lives

haematologist – a medical specialist who studies the blood and bone marrow

histamine – a chemical released in the body during an allergic reaction

homeostasis – the maintenance of a constant internal environment in the body

hormone – a chemical released from a gland and transported in blood in response to a stimulus

host – an organism on which a parasite lives or feeds

humus – decaying animal and plant material found in soil

hydrogen bond – a weak bond between a hydrogen atom and another atom

hypothesis – a statement that can be tested

identification key – a tool used to identify different biological organisms

immune system – the tissues and cells that defend the body against disease or allergies

incinerate – to burn something completely in a controlled way

independent variable – a variable that is purposely changed in an experiment

inflammatory response – an immune response triggered by a cut or damage to the skin

inflationary theory – theory that the Universe expanded exponentially and faster than light early in its evolution

informed consent – agreement by a participant to take part in a study which they have been given full information about

insulin – a hormone secreted by the pancreas to help body cells take up glucose from the blood

interdependence – how organisms in an ecosystem depend on each other to survive

intracellular – a term referring to the inside of a cell

in-vitro fertilisation (IVF) – artificial joining of the nuclei of a sperm and egg in a test tube

ion – an atom or group of atoms with a positive or negative electrical charge

kinetic energy – energy an object has because it is moving

latent prints – fingerprints which are not visible to the naked eye and need a suitable development technique to make them visible

legislation – law or set of laws passed by parliament

life expectancy – the average length of life of individuals within a population

light-emitting diode – a diode that emits light when a current passes through it; used in displays

limbic system – part of the brain; produces emotions and is involved in memory

line transect – a line along which measurements or samples are taken

lipase – an enzyme that catalyses the breakdown of lipids

liposome – a very small 'sac' that can be used to deliver drugs, vaccines, genes etc. into a cell

liver – a large organ in the body with many functions

logic probe – instrument that detects logic ouputs

luteinising hormone (LH) – hormone involved in regulation of the menstrual cycle; produced by the pituitary gland

macrophage – a large white blood cell that surrounds and digests foreign substances and micoorganisms

magnetic resonance imaging (MRI) – a method used in medical physics to provide detailed pictures of the inside of the human body

metabolism – all of the chemical changes in the body, especially those used to provide energy

metal-oxide-semiconductor field-effect transistor (MOSFET) – type of transistor; used in computers etc.

meteor – a piece of debris from Space that glows as it enters the Earth's atmosphere (known as a shooting star)

meteorite – a rock fragment from Space that has survived atmospheric burn-up and landed on Earth

meteoroid – a rock fragment in Space

microbe – a microscopic living organism that can cause disease

microbial activity – the decay of dead material by microorganisms

microclimate – the climate of a small area that differs from its surroundings

microorganism – an organism that can only be seen with the aid of a microscope

microphage – a small white blood cell that surrounds and digests foreign substances and micoorganisms

microwave – an electromagnetic wave with wavelength between 1 mm and 1 m

Milky Way – the name of the galaxy that contains our Solar System

monomer – a molecule that can join together with others to make a long-chain polymer molecule

motor neuron – a neuron that carries a nerve impuse from the central nervous system to an effector

multimeter – an instrument that can be used to measure voltage, current resistance and sometimes frequency

multiple sclerosis – a progressive disease that damages the myelin surrounding neurons

nerve impulse – an electrical signal that travels along a neuron in response to a stimulus

neurotic illness – mental illness linked to social or emotional factors

neutron star – a star at the end of its life, which is very dense and hot and is composed of neutrons

non-polarised capacitor – a capacitor that can be connected either way round in a circuit

non-renewable energy – energy from a source that can't be replaced once it is used

non-specific immune response – immune response which is not pathogen-specific

noradrenaline – hormone secreted from the adrenal glands when the body is stressed

nuclear energy – potential energy that is stored in the nucleus of an atom

oestrogen – female hormone involved in building and maintaining the lining of the uterus

open question – a question without a restricted answer

operational amplifier – device that can make signals bigger at the output

optimum conditions – the most favourable conditions to work in, e.g. relating to pH, temperature etc.

oscilloscope – measuring instrument that measures voltage as a function of time

overt – not secret; an overt observation is one where the participants know they are being observed

pancreas – an organ in the body that produces hormones

parasite – an organism that lives in or on another living organism

pathogen – an organism that causes disease

period (of a wave) – the time for one cycle

peripheral nervous system (PNS) – the nervous system outside of the brain and spinal cord

permeable – allowing liquids and gases to flow through

pH meter – a device for measuring how acidic or alkaline a solution is

photodiode – a diode that converts light into electricity; used in light detectors

photosynthesis – process used by plants to convert carbon dioxide and water into sugars using sunlight

photovoltaic cell – a solar cell that is used to convert solar energy to electricity

physical dependence – the body's need to take a drug which if absent causes withdrawal symptoms

pitfall trap – a container buried in the ground to trap moving animals

polarised capacitor – a capacitor that needs to be connected the correct way round in a circuit

polymerisation – a chemical reaction in which individual monomer molecules join together to form a polymer

polynucleotide – a chain of nucleotide bases bonded together

pooter – a device used to collect insects and other small creatures

population – all of the organisms of a certain species within a community

potential difference – the voltage between two points in an electric circuit

potential divider circuit – circuit that is used to share voltages between components

potential energy – energy that an object stores

predator – an animal that survives by hunting and eating other organisms

prey – animals caught or hunted for food

producer – an organism, usually a plant, that produces its own food; found at the bottom of a food chain

product – a substance which is formed at the end of a chemical reaction

progesterone – female hormone involved in building and maintaining the lining of the uterus

protease – an enzyme that catalyses the breakdown of proteins

protozoa – one of the kingdoms of organisms; single-celled or small multi-cellular creatures

psychological dependence – the feeling by a user that they need a drug

psychotic illness – mental illness resulting in the inability to control actions and behaviour

quadrat – a square or rectangle used to mark an area to be sampled for plants

radiation – the transfer of thermal energy in the form of light

reabsorption – the re-uptake of useful molecules back into blood

reactant – a substance which reacts in a chemical reaction to form new products

recessive – (a gene) expressed only when homozygous (both alleles are the same); represented by a lower-case letter

red blood cell – specialised cell containing haemoglobin that carries oxygen around the body

red giant – a star which has used up all the hydrogen in its core and is red in colour

reed switch – device that uses the property of a magnetic field to switch on and off

reflex action – an automatic response, not requiring thought, to a stimulus

reflex arc – a very quick nervous response that may not require input from the brain

refraction – when waves (e.g. light, sound etc.) change direction as they pass from one material to another

relay neuron – a neuron that connects the motor and sensory neurons together

renewable energy – energy from a source that can be replaced

resistant – bacteria that become resistant to antibiotics have the ability to survive treatment with antibiotics

resistor – an electrical component that limits the flow of current

respiration – the process of cells making energy from oxygen and glucose

ribosome – organelle where proteins are made; found in the cell cytoplasm

rickets – a childhood disease in which bones are weak; caused by a deficiency in vitamin D

right ascension – the angular distance 'east' and 'west' of the vernal equinox; similar to longitude on Earth

right to withdraw – the right of a participant to stop taking part in a study at any time

sap – watery solution made up of sugars, salts and minerals providing nutrients to all parts of a plant

scurvy – a condition affecting skin and gums; caused by a deficiency in vitamin C

secondary cell – one or more of these make up a rechargeable battery

sensory neuron – a neuron that carries a nerve impulse from a receptor to the CNS

sensory receptor – a receptor that detects stimulus or change inside or outside the body

signal diode – a diode that can be used to protect transistors

Solar System – the Sun and the planets and objects that orbit around it

solution – a liquid in which another substance has been dissolved

somatic nervous system – consists of nerve fibers that send information between the central nervous system and skeletal muscle

specific (enzyme–substrate) – the fact that the shape of an enzyme active site is tailored to the substrate

specific immune response – involves identification of specific antigens on a pathogen

speed – change in distance with respect to time

stem cells – early embryonic cells that have not yet become specialised

step-down transformer – a device that reduces voltage

step-up transformer – a device that increases voltage

supernova – the end of life of a very dense star by a catastrophic explosion

T lymphocyte – type of white blood cell that kills viruses or cancerous cells

testosterone – a male hormone responsible for sexual maturation and characteristics; produced in the testes

thermal conductivity – a measurement of how well a substance conducts heat

thymine – a base that pairs with adenine to make a base pair in DNA

thyroxine – hormone produced by the thyroid gland that regulates metabolism

tilt switch – switch that turns on or off depending on its angular position

toggle switch – simple on/off switch that is used in homes

transcription – production of mRNA from DNA (the mRNA contains a copy of the genetic information from the DNA)

transfer RNA (tRNA) – puts amino acids in the right order to make a protein

translation – production of a protein using the genetic information in mRNA

transpose – to rearrange an equation

trigonometric parallax method – a mathematical method used to measure the distance to a nearby star

triplet code – three nucleotide bases that are required to code for a single amino acid

truth table – table showing the output for all possible inputs of a logic gate or logic circuit

universal indicator – a chemical solution which changes colour according to the pH

vaccination – an antigen introduced into the body to provide protection against a disease

valid results – results that measure what they say they measure

variable resistor – device whose resistance can be changed

vector (biological) – a living organism that can spread disease from one organism to another

vein – blood vessel that carries (usually deoxygenated) blood to the heart

virus – non-living microorganism

visible prints – fingerprints that contrast well with the surface and can be seen with the naked eye

voltage – the amount of electrical energy that is transferred to an electrical component per unit of charge

voltmeter – instrument that measures voltage

wavelength – the length of one single oscillation

white blood cell – class of specialised cell making up the immune system; found in blood

white dwarf – a small star approximately the size of the Earth but with the mass of our Sun

word equation – an equation giving the names of the reactants and products in a chemical reaction

The periodic table of the elements

Key

| relative atomic mass |
| **atomic symbol** |
| name |
| atomic (proton) number |

1	2											3	4	5	6	7	8
						1 **H** hydrogen 1											4 **He** helium 2
7 **Li** lithium 3	9 **Be** beryllium 4											11 **B** boron 5	12 **C** carbon 6	14 **N** nitrogen 7	16 **O** oxygen 8	19 **F** fluorine 9	20 **Ne** neon 10
23 **Na** sodium 11	24 **Mg** magnesium 12											27 **Al** aluminium 13	28 **Si** silicon 14	31 **P** phosphorous 15	32 **S** sulfur 16	35.5 **Cl** chlorine 17	40 **Ar** argon 18
39 **K** potassium 19	40 **Ca** calcium 20	45 **Sc** scandium 21	48 **Ti** titanium 22	51 **V** vanadium 23	52 **Cr** chromium 24	55 **Mn** manganese 25	56 **Fe** iron 26	59 **Co** cobalt 27	59 **Ni** nickel 28	64 **Cu** copper 29	65 **Zn** zinc 30	70 **Ga** gallium 31	73 **Ge** germanium 32	75 **As** arsenic 33	79 **Se** selenium 34	80 **Br** bromine 35	84 **Kr** krypton 36
85 **Rb** rubidium 37	88 **Sr** strontium 38	89 **Y** yttrium 39	91 **Zr** zirconium 40	93 **Nb** niobium 41	96 **Mo** molybdenum 42	[98] **Tc** technetium 43	101 **Ru** ruthenium 44	103 **Rh** rhodium 45	106 **Pd** palladium 46	108 **Ag** silver 47	112 **Cd** cadmium 48	115 **In** indium 49	119 **Sn** tin 50	122 **Sb** antimony 51	128 **Te** tellurium 52	127 **I** iodine 53	131 **Xe** xenon 54
133 **Cs** caesium 55	137 **Ba** barium 56	139 **La*** lanthanum 57	178 **Hf** hafnium 72	181 **Ta** tantalum 73	184 **W** tungsten 74	186 **Re** rhenium 75	190 **Os** osmium 76	192 **Ir** iridium 77	195 **Pt** platinum 78	197 **Au** gold 79	201 **Hg** mercury 80	204 **Tl** thallium 81	207 **Pb** lead 82	209 **Bi** bismuth 83	[209] **Po** polonium 84	[210] **At** astatine 85	[222] **Rn** radon 86
[223] **Fr** francium 87	[226] **Ra** radium 88	[227] **Ac*** actinium 89															

* The lanthanoids (atomic numbers 58–71) and the actinoids (atomic numbers 90–103) have been omitted.

Index

Page numbers in italics refer to items in the margins.